Algebraic and Geometric Methods in Discrete Mathematics

CONTEMPORARY MATHEMATICS

685

Algebraic and Geometric Methods in Discrete Mathematics

AMS Special Session on
Algebraic and Geometric Methods in
Applied Discrete Mathematics
January 11, 2015
San Antonio, TX

Heather A. Harrington
Mohamed Omar
Matthew Wright
Editors

American Mathematical Society
Providence, Rhode Island

2010 *Mathematics Subject Classification.* Primary 00B20, 13P25, 20C30, 46N30, 51D20, 52B05, 62-07, 62P10, 65C60, 91B12.

Library of Congress Cataloging-in-Publication Data

Names: Harrington, Heather A., 1984– editor. | Omar, Mohamed, 1982– editor. | Wright, Matthew, 1984– editor.

Title: Algebraic and geometric methods in applied discrete mathematics: AMS special session on algebraic and geometric methods in applied discrete mathematics, January 11, 2015, San Antonio, Texas / Heather A. Harrington, Mohamed Omar, Matthew Wright, editors.

Description: Providence, Rhode Island: American Mathematical Society, [2017] | Series: Contemporary mathematics; volume 685 | Includes bibliographical references.

Identifiers: LCCN 2016042006 | ISBN 9781470423216 (alk. paper)

Subjects: LCSH: Mathematics–Congresses. | Geometry, Algebraic – Congresses. | AMS: General – Conference proceedings and collections of papers – Proceedings of conferences of general interest. msc | Commutative algebra – Computational aspects and applications – Applications of commutative algebra (e.g., to statistics, control theory, optimization, etc.). msc | Group theory and generalizations – Representation theory of groups – Representations of finite symmetric groups. msc | Functional analysis – Miscellaneous applications of functional analysis – Applications in probability theory and statistics. msc | Geometry – Geometric closure systems – Combinatorial geometries. msc | Convex and discrete geometry – Polytopes and polyhedra – Combinatorial properties (number of faces, shortest paths, etc.). msc | Statistics – Data analysis. msc | Statistics – Applications – Applications to biology and medical sciences. msc | Numerical analysis – Probabilistic methods, simulation and stochastic differential equations – Computational problems in statistics. msc | Game theory, economics, social and behavioral sciences – Mathematical economics – Voting theory. msc

Classification: LCC QA39.3 .A5255 2017 | DDC 511/.1–dc23 LC record available at https://lccn.loc.gov/2016042006

Contemporary Mathematics ISSN: 0271-4132 (print); ISSN: 1098-3627 (online)

DOI: `http://dx.doi.org/10.1090/conm/685`

---

10 9 8 7 6 5 4 3 2 1   22 21 20 19 18 17

Contents

Preface

This Contemporary Mathematics volume was initiated by the same named special session at the 2015 AMS/MAA Joint Mathematics Meetings, and also celebrates the themes of an AMS Mathematics Research Community (MRC) in 2014. The MRC on Algebraic and Geometric Methods in Applied Discrete Mathematics was held June 15–21, 2014, at Snowbird Resort in Utah. The goal of the MRC was to introduce young mathematicians to research bridging pure and applied mathematics—specifically, the use of algebraic and geometric methods to solve applied discrete problems. Working groups at the MRC included combinatorial topology in the social sciences, representation theory in data analysis, combinatorics in molecular biology, algebraic and geometric approaches in neuroscience, and algebraic and geometric methods in optimization.

The 2015 AMS/MAA Joint Mathematics Meetings, held January 10–13 in San Antonio, featured a special session dedicated to the themes of the MRC. Talks in this session were presented by experts in fields connected to the MRC topics. Given the intense interest in this special session and the enthusiasm of the session speakers and MRC participants, it was suggested that proceedings of the session be collected in a book. The present volume is that book.

We are deeply grateful to Carina Curto, Jesus A. De Loera, Christine Heitsch, Michael Orrison, and Francis Edward Su for organizing the MRC program and supporting this volume. We have benefited immensely from your mathematical expertise, mentoring, and willingness to introduce young mathematicians to your research areas. We also express great thanks to the AMS staff, especially Ellen Maycock, Melissa Colton, and Robin Hagan Aguiar for your work in arranging the logistics of the MRC, and to Christine Thivierge for your assistance in preparing this volume. The MRC would not have been possible without funding from the National Science Foundation, which we also acknowledge with thanks.

Heather A. Harrington
Mohamed Omar
Matthew Wright

Contemporary Mathematics
Volume **685**, 2017
http://dx.doi.org/10.1090/conm/685/13717

Eigenconfigurations of tensors

Hirotachi Abo, Anna Seigal, and Bernd Sturmfels

ABSTRACT. Square matrices represent linear self-maps of vector spaces, and their eigenpoints are the fixed points of the induced map on projective space. Likewise, polynomial self-maps of a projective space are represented by tensors. We study the configuration of fixed points of a tensor or symmetric tensor.

1. Introduction

Square matrices A with entries in a field K represent linear maps of vector spaces, say $K^n \to K^n$, and hence linear maps $\psi : \mathbb{P}^{n-1} \dashrightarrow \mathbb{P}^{n-1}$ of projective spaces over K. If A is nonsingular then ψ is well-defined everywhere, and the eigenvectors of A correspond to the fixed points of ψ. The *eigenconfiguration* of A consists of n points in $\mathbb{P}^{n-1}$, provided A is generic and K is algebraically closed. Conversely, every spanning configuration of n points in $\mathbb{P}^{n-1}$ arises as the eigenconfiguration of an $n \times n$-matrix A. However, for special matrices A, we obtain multiplicities and eigenspaces of higher dimensions [**AE**]. Moreover, if $K = \mathbb{R}$ and A is symmetric then its complex eigenconfiguration consists of real points only.

This paper concerns the extension from linear to non-linear maps. Their fixed points are the eigenvectors of tensors. The spectral theory of tensors was pioneered by Lim [**Lim**] and Qi [**Qi**]. It is now a much-studied topic in applied mathematics.

For instance, consider a quadratic map $\psi : \mathbb{P}^{n-1} \dashrightarrow \mathbb{P}^{n-1}$, with coordinates

$$\psi_i(x_1, \ldots, x_n) = \sum_{j=1}^{n} \sum_{k=1}^{n} a_{ijk} x_j x_k \qquad \text{for } i = 1, \ldots, n. \tag{1.1}$$

One organizes the coefficients of ψ into a tensor $A = (a_{ijk})$ of format $n \times n \times n$.

In what follows, we assume that $A = (a_{i_1 i_2 \cdots i_d})$ is a d-dimensional tensor of format $n \times n \times \cdots \times n$. The entries $a_{i_1 i_2 \ldots i_d}$ lie in an algebraically closed field K of characteristic zero, usually the complex numbers $K = \mathbb{C}$. Such a tensor $A \in (K^n)^{\otimes d}$ defines polynomial maps $K^n \to K^n$ and $\mathbb{P}^{n-1} \dashrightarrow \mathbb{P}^{n-1}$ just as in the formula (1.1):

$$\psi_i(x_1, \ldots, x_n) = \sum_{j_2=1}^{n} \sum_{j_3=1}^{n} \cdots \sum_{j_d=1}^{n} a_{i j_2 j_3 \cdots j_d} x_{j_2} x_{j_3} \cdots x_{j_d} \quad \text{for } i = 1, \ldots, n.$$

2010 *Mathematics Subject Classification.* Primary 15A18; Secondary 13P25, 14M12, 15A69.
Key words and phrases. Tensors, eigenvectors, computational algebraic geometry.
We acknowledge support by the National Science Foundation (DMS-1419018) and the US-UK Fulbright Commission. This project started at the Simons Institute for the Theory of Computing. The authors are grateful to Manuel Kauers, Giorgio Ottaviani and Cynthia Vinzant for their help.

Thus each of the n coordinates of ψ is a homogeneous polynomial ψ_i of degree $d-1$ in $\mathbf{x} = (x_1, x_2, \ldots, x_n)$. The eigenvectors of A are the solutions of the constraint

$$\text{rank}\begin{pmatrix} x_1 & x_2 & \cdots & x_n \\ \psi_1(\mathbf{x}) & \psi_2(\mathbf{x}) & \cdots & \psi_n(\mathbf{x}) \end{pmatrix} \leq 1. \tag{1.2}$$

The *eigenconfiguration* is the variety defined by the 2×2-minors of this matrix. For a special tensor A, the ideal defined by (1.2) may not be radical, and in that case we can study its *eigenscheme*. Recent work in [**AE**] develops this for $d = 2$.

We note that every n-tuple $(\psi_1, \ldots, \psi_n)$ of homogeneous polynomials of degree $d-1$ in n variables can be represented by some tensor A as above. This representation is not unique unless we require that A is symmetric in the last $d-1$ indices. Our maps $\psi : \mathbb{P}^{n-1} \dashrightarrow \mathbb{P}^{n-1}$ are arbitrary polynomial dynamical system on projective space, in the sense of [**FS**]. Thus the study of eigenconfigurations of tensors is equivalent to the study of fixed-point configurations of polynomial maps.

Of most interest to us are *symmetric tensors* A, i.e. tensors whose entries $a_{i_1 i_2 \cdots i_d}$ are invariant under permuting the d indices. These are in bijection with homogeneous polynomials $\phi = \sum a_{i_1 i_2 \cdots i_d} x_{i_1} x_{i_2} \cdots x_{i_d}$, and we take $\psi_j = \partial\phi/\partial x_j$. The eigenvectors of a symmetric tensor correspond to fixed points of the *gradient map* $\nabla\phi : \mathbb{P}^{n-1} \dashrightarrow \mathbb{P}^{n-1}$, and our object of study is the variety in $\mathbb{P}^{n-1}$ defined by

$$\text{rank}\begin{pmatrix} x_1 & x_2 & \cdots & x_n \\ \partial\phi/\partial x_1 & \partial\phi/\partial x_2 & \cdots & \partial\phi/\partial x_n \end{pmatrix} \leq 1. \tag{1.3}$$

This paper uses the term *eigenpoint* instead of eigenvector to stress that we work in $\mathbb{P}^{n-1}$. In our definition of eigenpoints we include the common zeros of $\psi_1, \ldots, \psi_n$. These are the points where the map $\mathbb{P}^{n-1} \dashrightarrow \mathbb{P}^{n-1}$ is undefined. For a symmetric tensor ϕ, they are the singular points of the hypersurface $\{\phi = 0\}$ in $\mathbb{P}^{n-1}$. At those points the gradient $\nabla\phi$ vanishes, so condition (1.3) holds.

EXAMPLE 1.1. Let $n = d = 3$ and $\phi = xyz$. The corresponding symmetric $3 \times 3 \times 3$ tensor A has six nonzero entries $1/6$ and the other 21 entries are 0. Here $\nabla\phi : \mathbb{P}^2 \dashrightarrow \mathbb{P}^2, (x : y : z) \to (yz : xz : xy)$ is the classical *Cremona transformation.* This map has four fixed points, namely $(1 : 1 : 1)$, $(1 : 1 : -1)$, $(1 : -1 : 1)$ and $(-1 : 1 : 1)$. Also, the cubic curve $\{\phi = 0\}$ has the three singular points $(1 : 0 : 0), (0 : 1 : 0), (0 : 0 : 1)$. In total, the tensor A has seven eigenpoints in $\mathbb{P}^2$.

This paper is organized as follows. In Section 2 we count the number of eigenpoints, and we explore eigenconfigurations of Fermat polynomials, plane arrangements, and binary forms. Section 3 generalizes the fact that the left eigenvectors and right eigenvectors of a square matrix are distinct but compatible. We explore this compatibility for the d eigenconfigurations of a d-dimensional tensor with $n = 2$.

Section 4 concerns the *eigendiscriminant* of the polynomial system (1.2) and its variant in (2.2). This is the irreducible polynomial in the n^d unknowns $a_{i_1 i_2 \cdots i_d}$ which vanishes when two eigenpoints come together. We give a formula for its degree in terms of n, d and ℓ. Section 5 takes first steps towards characterizing eigenconfigurations among finite subsets of $\mathbb{P}^{n-1}$, starting with the case $n = d = 3$.

In Section 6 we focus on real tensors and their dynamics on real projective space $\mathbb{P}^{n-1}_{\mathbb{R}}$. We examine whether all complex eigenpoints can be real, and we use line arrangements to give an affirmative answer for $n = 3$. The paper concludes with a brief discussion of attractors for the dynamical systems $\psi : \mathbb{P}^{n-1}_{\mathbb{R}} \dashrightarrow \mathbb{P}^{n-1}_{\mathbb{R}}$. These are also known as the *robust eigenvectors* of the *tensor power method* [**AG**, **Rob**].

2. The count and first examples

In this section we assume that the given tensor A is generic, meaning that it lies in a certain dense open subset in the space $(K^n)^{\otimes d}$ of all $n\times \cdots \times n$-tensors. This set will be characterized in Section 4 as the nonvanishing locus of the eigendiscriminant.

THEOREM 2.1. *The number of solutions in* $\mathbb{P}^{n-1}$ *of the system* (1.2) *equals*

$$\frac{(d-1)^n-1}{d-2} \;=\; \sum_{i=0}^{n-1}(d-1)^i. \tag{2.1}$$

The same count holds for eigenconfigurations of symmetric tensors, given by (1.3). *In the matrix case* ($d=2$) *we use the formula on the right, which evaluates to* n.

This result appeared in the tensor literature in [**CS, OO**], but it had already been known in complex dynamics due to Fornaess and Sibony [**FS**, Corollary 3.2]. We shall present two proofs of Theorem 2.1, cast in a slightly more general context.

For certain applications (e.g. in spectral hypergraph theory [**LQY**]), it makes sense to focus on positive real numbers and to take the ℓth root after each iteration of the dynamical system ψ. This leads to the following generalization of our equations:

$$\operatorname{rank}\begin{pmatrix} x_1^\ell & x_2^\ell & \cdots & x_n^\ell \\ \psi_1(\mathbf{x}) & \psi_2(\mathbf{x}) & \cdots & \psi_n(\mathbf{x})\end{pmatrix} \;\leq\; 1. \tag{2.2}$$

We refer to the solutions as the ℓth *eigenpoints* of the given tensor A. For $\ell=1$, this is the definition in the Introduction. In the nomenclature devised by Qi [**CQZ, Qi**], one obtains *E-eigenvectors* for $\ell=1$ and *Z-eigenvectors* for $\ell=d-1$. The subvariety of $\mathbb{P}^{n-1}$ defined by (2.2) is called the ℓth *eigenconfiguration* of the tensor A.

THEOREM 2.2. *The* ℓ^{th} *eigenconfiguration of a generic tensor* A *consists of*

$$\frac{(d-1)^n \;-\; \ell^n}{d-1\;-\;\ell} \;=\; \sum_{i=0}^{n-1}(d-1)^i\ell^{n-1-i} \tag{2.3}$$

distinct points in $\mathbb{P}^{n-1}$. *If* $\ell=d-1$ *then the formula on the right is to be used.*

PROOF. Consider the $2\times n$-matrix in (2.2). Its rows are filled with homogeneous polynomials in $S=K[x_1,\ldots,x_n]$ of degrees ℓ and m respectively, where the ψ_i are generic. Requiring this matrix to have rank ≤ 1 defines a subscheme of $\mathbb{P}^{n-1}$. By the Thom-Porteous-Giambelli formula [**Fu1**, §14.4], this scheme is zero-dimensional, and its length is given by the complete homogeneous symmetric polynomial of degree $n-1$ in the row degrees, ℓ and m. This is precisely (2.3) if we set $m=d-1$.

Another approach, which also shows that the scheme is reduced, is to use vector bundle techniques. Consider the $2\times n$-matrix as a graded S-module homomorphism from $S(-\ell)\oplus S(-m)$ to $S^{\oplus n}$. The quotient module Q of $S^{\oplus n}$ by the submodule generated by the first row $(x_1^\ell,\ldots,x_n^\ell)$ is projective. In other words, the sheafification $\widetilde{Q}$ of Q is locally free. The scheme associated with the $2\times n$-matrix can therefore be thought of as the zero scheme of a generic global section of $\widetilde{Q}(m)$. Since $\widetilde{Q}(m)$ is globally generated, the scheme is reduced [**Ein**, Lemma 2.5]. □

Here is a brief remark about *eigenvalues*. If $\mathbf{x}\in K^n$ is an ℓth eigenvector of A then there exists a scalar $\lambda\in K$ such that $\psi_i(\mathbf{x})=\lambda x_i^\ell$ for all i. We call $(\mathbf{x},\lambda)$ an *eigenpair*. If this holds then $(\nu\mathbf{x},\nu^{d-1-\ell}\lambda)$ is also an eigenpair for all $\nu\in K\backslash\{0\}$. Such equivalence classes of eigenpairs correspond to the ℓth eigenpoints in $\mathbb{P}^{n-1}$.

The case $\ell = d-1$ is special because every eigenpoint has an associated eigenvalue. If $\ell \neq d-1$ then eigenpoints make sense but eigenvalues are less meaningful.

PROOF OF THEOREM 2.1. The first statement is the $\ell = 1$ case of Theorem 2.2. For the second assertion, it suffices to exhibit one symmetric tensor ϕ that has the correct number of eigenpoints. We do this for the *Fermat polynomial*

$$\phi(\mathbf{x}) \;=\; x_1^d + x_2^d + \cdots + x_n^d. \tag{2.4}$$

According to (1.2), the eigenconfiguration of ϕ is the variety in $\mathbb{P}^{n-1}$ defined by

$$\operatorname{rank}\begin{pmatrix} x_1 & x_2 & \cdots & x_n \\ x_1^{d-1} & x_2^{d-1} & \cdots & x_n^{d-1} \end{pmatrix} \;\leq\; 1. \tag{2.5}$$

We follow [**Rob**] in characterizing all solutions $\mathbf{x}$ in $\mathbb{P}^{n-1}$ to the binomial equations

$$x_i x_j (x_i^{d-2} - x_j^{d-2}) \;=\; 0 \qquad \text{for} \quad 1 \leq i < j \leq n.$$

For any non-empty subset $I \subseteq \{1, 2, \ldots, n\}$, there are $(d-2)^{|I|-1}$ solutions $\mathbf{x}$ with $\operatorname{supp}(\mathbf{x}) = \{i \mid x_i \neq 0\}$ equal to I. Indeed, we may assume $x_i = 1$ for the smallest index i in I, and the other values are arbitrary $(d-2)^{\text{nd}}$ roots of unity. In total,

$$\sum_I (d-2)^{|I|-1} \;=\; \sum_{i=1}^{n} \binom{n}{i} (d-2)^{i-1} \;=\; \frac{1}{d-2} \sum_{i=1}^{n} \binom{n}{i} (d-2)^i 1^{n-i} \;=\; \frac{(d-2+1)^n - 1}{d-2}.$$

This equals (2.1). Here we assume $d \geq 3$. The familiar matrix case is $d = 2$. □

EXAMPLE 2.3. Let $d = 4$. For each I, there are $2^{|I|-1}$ eigenpoints, with $x_i = \pm 1$ for $i \in I$ and $x_j = 0$ for $j \notin I$. The total number of eigenpoints in $\mathbb{P}^{n-1}$ is $(3^n - 1)/2$.

We note that the argument in the proof of Theorem 2.1 does not work for $\ell \geq 2$. For instance, if $\ell = d-1$ then every point in $\mathbb{P}^{n-1}$ is an eigenpoint of the Fermat polynomial. At present we do not know an analogue to that polynomial for $\ell \geq 2$.

PROBLEM 2.4. *Given any ℓ, d and n, exhibit explicit polynomials $\phi(\mathbf{x})$ of degree d in n variables such that (2.2) has (2.3) distinct isolated solutions in $\mathbb{P}^{n-1}$.*

We are looking for solutions with interesting combinatorial structure. In Section 6 we shall examine the case when $\phi(\mathbf{x})$ factors into linear factors, and we shall see how the geometry of hyperplane arrangements can be used to derive an answer. A first instance was the Cremona map in Example 1.1. Here is a second example.

EXAMPLE 2.5. For $n = 4$ the count of the eigenpoints in (2.1) gives $d^3 - 2d^2 + 2d$. We now fix $d = 5$, so this number equals 85. Consider the special symmetric tensor

$$\phi(\mathbf{x}) \;=\; x_1 x_2 x_3 x_4 (x_1 + x_2 + x_3 + x_4).$$

The surface defined by ϕ consists of five planes in $\mathbb{P}^3$. These intersect pairwise in ten lines. Each point on such a line is an eigenpoint because it is singular on the surface. Furthermore, there are 15 isolated eigenpoints; these have real coordinates:

$$\begin{gathered} (2:2:-1:-1),\ (2:-1:2:-1),\ (2:-1:-1:2),\ (-1:2:2:-1), \\ (-1:2:-1:2),\ (-1:-1:2:2),\ (1:1:1:1),\ \left(\tfrac{1}{2}(5 \pm \sqrt{13}):1:1:1\right), \\ \left(1:\tfrac{1}{2}(5 \pm \sqrt{13}):1:1\right),\ \left(1:1:\tfrac{1}{2}(5 \pm \sqrt{13}):1\right),\ \left(1:1:1:\tfrac{1}{2}(5 \pm \sqrt{13})\right). \end{gathered} \tag{2.6}$$

The five planes divide $\mathbb{P}^3_{\mathbb{R}}$ into 15 regions. Each region contains one point in (2.6).

Now, take a generic quintic $\phi'(\mathbf{x})$ in $\mathbb{R}[x_1, x_2, x_3, x_4]$, and consider the eigenconfiguration of $\phi(\mathbf{x}) + \epsilon\phi'(\mathbf{x})$. This consists of 85 points in $\mathbb{P}^3$. These are algebraic functions of ϵ. For $\epsilon > 0$ small, we find 15 real eigenpoints near (2.6). The other 70 eigenpoints arise from the 10 lines. How many are real depends on the choice of ϕ'.

The situation is easier for $n = 2$, when the tensor A has format $2\times 2\times \cdots \times 2$. It determines two binary forms ψ_1 and ψ_2. The eigenpoints of A are defined by

$$y \cdot \psi_1(x,y) - x \cdot \psi_2(x,y) \,=\, 0. \tag{2.7}$$

This is a binary form of degree d, so it has d zeros in $\mathbb{P}^1$, as predicted by (2.1). Conversely, every binary form of degree d can be written as $y\psi_1 - x\psi_2$. This implies:

REMARK 2.6. Every set of d points in $\mathbb{P}^1$ is the eigenconfiguration of a tensor.

The discussion is more interesting when we restrict ourselves to symmetric tensors. These correspond to binary forms $\phi(x,y)$ and their eigenpoints are defined by

$$y \cdot \frac{\partial \phi}{\partial x} - x \cdot \frac{\partial \phi}{\partial y} \,=\, 0.$$

The matrix case ($d = 2$) shows that Remark 2.6 cannot hold as stated for symmetric tensors. Indeed, if $A = \begin{pmatrix} a & b \\ b & c \end{pmatrix}$ and $\phi = ax^2 + 2bxy + cy^2$ then $\frac{1}{2}(y\frac{\partial \phi}{\partial x} - x\frac{\partial \phi}{\partial y}) = -bx^2 + (a-c)xy + by^2$. This confirms the familiar facts that the two eigenpoints $(u_1 : v_1)$ and $(u_2 : v_2)$ are real when $a, b, c \in \mathbb{R}$ and they satisfy $u_1u_2 + v_1v_2 = 0$. The following result generalizes the second fact from symmetric matrices to tensors.

THEOREM 2.7. *A set of d points $(u_i : v_i)$ in $\mathbb{P}^1$ is the eigenconfiguration of a symmetric tensor if and only if either d is odd, or d is even and the operator*

$$\left(\frac{\partial^2}{\partial x^2} + \frac{\partial^2}{\partial y^2}\right)^{d/2} \quad \textit{annihilates the corresponding binary form} \quad \prod_{i=1}^{d}(v_i x - u_i y).$$

PROOF. The only-if direction follows from the observation that the Laplace operator $\frac{\partial^2}{\partial x^2} + \frac{\partial^2}{\partial y^2}$ commutes with the vector field $y\frac{\partial}{\partial x} - x\frac{\partial}{\partial y}$. Hence, for any ϕ of degree d, we obtain zero when $\frac{\partial^2}{\partial x^2} + \frac{\partial^2}{\partial y^2}$ gets applied $d/2$ times to $y\frac{\partial \phi}{\partial x} - x\frac{\partial \phi}{\partial y}$.

For the if direction, we examine the $(d+1) \times (d+1)$-matrix that represents the endomorphism $\phi \mapsto y\frac{\partial \phi}{\partial x} - x\frac{\partial \phi}{\partial y}$ on the space of binary forms of degree d. This matrix is invertible when d is odd, and its kernel is one-dimensional when d is even. Hence the map is surjective when d is odd, and it maps onto a hyperplane when d is even. The only-if part shows that this hyperplane equals $\left(\left(\frac{\partial^2}{\partial x^2} + \frac{\partial^2}{\partial y^2}\right)^{d/2} \right)^{\perp}$. □

After completion of our manuscript we learned that Theorem 2.7 was also found independently by Mauro Maccioni [**M**], as part of his PhD dissertation at Firenze, Italy.

EXAMPLE 2.8 ($d = 4$). Four points $(u_1{:}v_1), (u_2{:}v_2), (u_3{:}v_3), (u_4{:}v_4)$ on the line $\mathbb{P}^1$ arise as the eigenconfiguration of a symmetric $2\times 2\times 2\times 2$-tensor if and only if

$$3u_1u_2u_3u_4 + u_1u_2v_3v_4 + u_1u_3v_2v_4 + u_1u_4v_2v_3 + \cdots + u_3u_4v_1v_2 + 3v_1v_2v_3v_4 = 0.$$

This equation generalizes the orthogonality of the two eigenvectors of a symmetric 2×2-matrix. For instance, the columns of $U = \begin{pmatrix} 1 & 0 & 1 & 1 \\ 0 & 1 & 1 & -1 \end{pmatrix}$ represent the eigenconfiguration of a symmetric $2\times 2\times 2\times 2$-tensor, but this does not hold for $\begin{pmatrix} 1 & 1 \\ 0 & 1 \end{pmatrix} U$.

Example 2.8 underscores the fact that the constraints on eigenconfigurations of symmetric tensors A are not invariant under projective transformations. They are only invariant under the orthogonal group $O(n)$, like the Laplace operator in Theorem 2.7. By contrast, the constraints on eigenconfigurations of general (non-symmetric) tensors, such as Theorem 5.1, will be properties of projective geometry.

We are familiar with this issue from comparing the eigenconfigurations of real symmetric matrices with those of general square matrices. These are respectively the $O(n)$-orbit and the $GL(n)$-orbit of the standard coordinate basis.

3. Compatibility of eigenconfigurations

When defining the eigenvectors of a tensor A, the symmetry was broken by fixing the first index and summing over the last $d-1$ indices. There is nothing special about the first index. For any $k \in \{1, \ldots, d\}$ we can regard A as the self-map

$$\psi^{[k]} : \mathbb{P}^{n-1} \dashrightarrow \mathbb{P}^{n-1}$$

whose i^{th} coordinate is the following homogeneous polynomial of degree $d-1$ in $\mathbf{x}$:

$$\psi_i^{[k]}(\mathbf{x}) = \sum_{j_1=1}^{n} \cdots \sum_{j_{k-1}=1}^{n} \sum_{j_{k+1}=1}^{n} \cdots \sum_{j_d=1}^{n} a_{j_1 \cdots j_{k-1} i j_{k+1} \cdots j_d} x_{j_1} \cdots x_{j_{k-1}} x_{j_{k+1}} \cdots x_{j_d}.$$

Let $\mathrm{Eig}^{[k]}(A)$ denote the subvariety of $\mathbb{P}^{n-1}$ consisting of the fixed points of $\psi^{[k]}$. For a generic tensor A, this is a finite set of points in $\mathbb{P}^{n-1}$ of cardinality

$$D = \frac{(d-1)^n - 1}{d-2} = \#(\mathrm{Eig}^{[k]}(A)) \quad \text{for } d \geq 3.$$

This raises the following question: Suppose we are given d configurations, each consisting of D points in $\mathbb{P}^{n-1}$, and known to be the eigenconfiguration of some tensor. Under what condition do they come from the same tensor A?

We begin to address this question by considering the case of matrices ($d = 2$), where $D = n$. Our question is as follows: given an $n \times n$-matrix A, what is the relationship between the left eigenvectors and the right eigenvectors of A?

PROPOSITION 3.1. *Let $\{\mathbf{v}_1, \mathbf{v}_2, \ldots, \mathbf{v}_n\}$ and $\{\mathbf{w}_1, \mathbf{w}_2, \ldots, \mathbf{w}_n\}$ be two spanning subsets of $\mathbb{P}^{n-1}$. These arise as the left and right eigenconfigurations of some $n \times n$-matrix A if and only if, up to relabeling, the dot products of vectors corresponding to $\mathbf{w}_i$ and $\mathbf{v}_j$ are zero whenever $i \neq j$.*

PROOF. Let V be a square matrix whose columns are the eigenvectors of A. Then the columns of $(V^{-1})^T$ form a basis of eigenvectors for A^T. □

The condition in Proposition 3.1 defines an irreducible variety, denoted $\mathrm{EC}_{n,2}$ and called the *eigencompatibility variety* for $n \times n$-matrices. It lives naturally in the space of pairs of unordered configurations of n points in $\mathbb{P}^{n-1}$. In symbols,

$$\mathrm{EC}_{n,2} \subset \mathrm{Sym}_n(\mathbb{P}^{n-1}) \times \mathrm{Sym}_n(\mathbb{P}^{n-1}). \tag{3.1}$$

It has middle dimension $n(n-1)$, and it maps birationally onto either factor. We may identify $\mathrm{Sym}_n(\mathbb{P}^{n-1})$ with the Chow variety of products of n linear forms in n variables. Here, each configuration $\{\mathbf{v}_1, \mathbf{v}_2, \ldots, \mathbf{v}_n\}$ is represented by $\prod_{i=1}^n (\mathbf{v}_i \cdot \mathbf{x})$. The coefficients of this homogeneous polynomial serve as coordinates on $\mathrm{Sym}_n(\mathbb{P}^{n-1})$. It would be worthwhile to express Proposition 3.1 in these coordinates.

EXAMPLE 3.2 ($n = 2$). The eigencompatibility variety $\mathrm{EC}_{2,2}$ for 2×2-matrices is a surface in $\big(\mathrm{Sym}_2(\mathbb{P}^1)\big)^2$. This ambient space equals $(\mathbb{P}^2)^2$, by representing a pair of unlabeled points on the line $\mathbb{P}^1$ with the binary quadric that defines it. To be precise, a point $\big((u_0{:}u_1{:}u_2),(v_0{:}v_1{:}v_2)\big)$ in $(\mathbb{P}^2)^2$ is identified with the binary forms

$$f(s,t) \;=\; u_0 s^2 + u_1 st + u_2 t^2 \quad \text{and} \quad g(s,t) \;=\; v_0 s^2 + v_1 st + v_2 t^2.$$

We want the zeros of $f(s,t)$ and $g(s,t)$ to be the right and left eigenconfigurations of the same 2×2-matrix. Proposition 3.1 tells us that this is equivalent to

$$f(s,t) \;=\; \lambda(as+bt)(cs+dt) \quad \text{and} \quad g(s,t) \;=\; \mu(bs-at)(ds-ct).$$

By eliminating the parameters a, b, c, d, λ, and μ, we find that the surface $\mathrm{EC}_{2,2}$ is essentially the diagonal in $(\mathbb{P}^2)^2$. It is defined by the determinantal condition

$$\text{(3.2)} \qquad \mathrm{rank}\begin{pmatrix} u_0 & u_1 & u_2 \\ v_2 & -v_1 & v_0 \end{pmatrix} \;\leq\; 1.$$

Our aim in this section is to generalize this implicit representation of $\mathrm{EC}_{2,2}$.

Let $\mathrm{EC}_{n,d}$ denote the eigencompatibility variety of d-dimensional tensors of format $n{\times}n{\times}\cdots{\times}n$. This is defined as follows. Every generic tensor A has d eigenconfigurations. The *eigenconfiguration with index* k of the tensor A is the fixed locus of the map $\psi^{[k]}$. Each configuration is a set of unlabeled D points in $\mathbb{P}^{n-1}$, which we regard as a point in $\mathrm{Sym}_D(\mathbb{P}^{n-1})$. The d-tuples of eigenconfigurations, one for each index k, parametrize

$$\text{(3.3)} \qquad \mathrm{EC}_{n,d} \,\subset\, \big(\mathrm{Sym}_D(\mathbb{P}^{n-1})\big)^d.$$

Thus $\mathrm{EC}_{n,d}$ is the closure of the locus of d-tuples of eigenconfigurations of tensors.

Already the case of binary tensors ($n = 2$) is quite interesting. We shall summarize what we know about this. Let A be a tensor of format $2 \times 2 \times \cdots \times 2$, with d factors. Each of its d eigenconfigurations consists of $D = d$ points on the line $\mathbb{P}^1$. The symmetric power $\mathrm{Sym}_d(\mathbb{P}^1)$ is identified with the $\mathbb{P}^d$ of binary forms of degree d. The zeros of such a binary form is an unlabeled configuration of d points in $\mathbb{P}^1$. Thus, the eigencompatibility variety for binary tensors is a subvariety

$$\mathrm{EC}_{2,d} \;\subset\; (\mathbb{P}^d)^d.$$

The case $d = 2$ was described in Example 3.2. Here are the next few cases.

EXAMPLE 3.3 ($d = 3$). Points in $(\mathbb{P}^3)^3$ are triples of binary cubics

$$\begin{aligned} f(s,t) &= u_0 s^3 + u_1 s^2 t + u_2 s t^2 + u_3 t^3, \\ g(s,t) &= v_0 s^3 + v_1 s^2 t + v_2 s t^2 + v_3 t^3, \\ h(s,t) &= w_0 s^3 + w_1 s^2 t + w_2 s t^2 + w_3 t^3, \end{aligned}$$

where two binary cubics are identified if they differ by a scalar multiple. The three eigenconfigurations of a $2{\times}2{\times}2$-tensor $A = (a_{ijk})$ are defined by the binary cubics

$$\begin{aligned} f(s,t) &= \lambda \cdot \big(a_{211}s^3 - (a_{111} - a_{212} - a_{221})s^2t + (a_{222} - a_{112} - a_{121})st^2 - a_{122}t^3\big) \\ g(s,t) &= \mu \cdot \big(a_{121}s^3 - (a_{111} - a_{122} - a_{221})s^2t + (a_{222} - a_{112} - a_{211})st^2 - a_{212}t^3\big) \\ h(s,t) &= \nu \cdot \big(a_{112}s^3 - (a_{111} - a_{122} - a_{212})s^2t + (a_{222} - a_{121} - a_{211})st^2 - a_{221}t^3\big) \end{aligned}$$

Our task is to eliminate the 11 parameters a_{ijk} and λ, μ, ν from these formulas. Geometrically, our variety $\mathrm{EC}_{2,3}$ is represented as the image of a rational map

$$\text{(3.4)} \qquad \mathbb{P}^7 \dashrightarrow (\mathbb{P}^3)^3, \;\; A \mapsto (f,g,h).$$

This is linear in the coefficients a_{ijk} of A and maps the tensor to a triple of binary forms. To characterize the image of (3.4), in Theorem 3.6 we introduce the matrix

$$\mathbf{E}_3 = \begin{pmatrix} u_1-u_3 & u_1-u_3 & u_0-u_2 & u_0-u_2 \\ v_1-v_3 & 0 & v_0-v_2 & 0 \\ 0 & w_1-w_3 & 0 & w_0-w_2 \end{pmatrix}. \tag{3.5}$$

Let I be the ideal generated by the 3×3-minors of $\mathbf{E}_3$. Its zero set has the eigencompatibility variety $\mathrm{EC}_{2,3}$ as an irreducible component. There are also three extraneous irreducible components, given by the rows of the matrix:

$$I_1 = \langle u_0-u_2, u_1-u_3\rangle,\ I_2 = \langle v_0-v_2, v_1-v_3\rangle, \text{ and } I_3 = \langle w_0-w_2, w_1-w_3\rangle.$$

The homogeneous prime ideal of $\mathrm{EC}_{2,3}$ is found to be the ideal quotient

$$(I : I_1I_2I_3) \quad = \quad \left\langle 2\times 2\text{-minors of } \begin{pmatrix} u_0-u_2 & v_0-v_2 & w_0-w_2 \\ u_1-u_3 & v_1-v_3 & w_1-w_3 \end{pmatrix} \right\rangle. \tag{3.6}$$

We conclude that the eigencompatibility variety $\mathrm{EC}_{2,3}$ has codimension 2 in $(\mathbb{P}^3)^3$.

EXAMPLE 3.4 ($d=4$). Points in $(\mathbb{P}^4)^4$ are quadruples of binary quartics

$$\begin{gathered} u_0s^4+u_1s^3t+u_2s^2t^2+u_3st^3+u_4t^4, \\ v_0s^4+v_1s^3t+v_2s^2t^2+v_3st^3+v_4t^4, \\ w_0s^4+w_1s^3t+w_2s^2t^2+w_3st^3+w_4t^4, \\ x_0s^4+x_1s^3t+x_2s^2t^2+x_3st^3+x_4t^4. \end{gathered}$$

One can represent the homogeneous ideal of the eigencompatibility variety $\mathrm{EC}_{2,4}$ in a similar way to Example 3.3. Let I be the ideal generated by the 4×4-minors of

$$\begin{bmatrix} u_1{-}u_3 & u_1{-}u_3 & u_1{-}u_3 & u_2{-}u_0{+}u_4 & u_2{-}u_0{-}u_4 & 2u_0{-}u_2{+}2u_4 & 3u_0{-}2u_2{+}3u_4 \\ v_3{-}v_1 & 0 & 0 & v_0{-}v_2{+}v_4 & 0 & v_2 & v_2 \\ 0 & w_3{-}w_1 & 0 & 0 & w_0{-}w_2{+}w_4 & w_2 & w_2 \\ 0 & 0 & x_3{-}x_1 & 0 & 0 & x_0+x_4 & x_2 \end{bmatrix} \tag{3.7}$$

Let I_{ij} be the ideal generated by the 2×2-minors of the 2×7-submatrix consisting of the ith and j^{th} rows in (3.7). The homogeneous prime ideal of $\mathrm{EC}_{2,4}\subset(\mathbb{P}^4)^4$ is obtained as the ideal quotient $\big(I : I_{12}I_{13}I_{14}I_{23}I_{24}I_{34}\big)$. We obtain $\dim(\mathrm{EC}_{2,4}) = 12$.

EXAMPLE 3.5 ($d=5$). The eigencompatibility variety $\mathrm{EC}_{2,5}$ has codimension 4 in $(\mathbb{P}^5)^5$, so $\dim(\mathrm{EC}_{2,5}) = 21$. We represent this variety by the 5×8-matrix

$$\begin{bmatrix} -u_1+u_3-u_5 & v_1-v_3+v_5 & 0 & 0 & 0 \\ u_1-u_3+u_5 & 0 & w_1-w_3+w_5 & 0 & 0 \\ -u_1+u_3-u_5 & 0 & 0 & x_1-x_3+x_5 & 0 \\ -u_1+u_3-u_5 & 0 & 0 & 0 & y_1-y_3+y_5 \\ u_0-u_2+u_4 & 0 & w_0-w_2+w_4 & 0 & 0 \\ 0 & v_0-v_2+v_4 & w_0-w_2+w_4 & 0 & 0 \\ 0 & 0 & w_0-w_2+w_4 & x_0-x_2+x_4 & 0 \\ 0 & 0 & w_0-w_2+w_4 & 0 & y_0-y_2+y_4 \end{bmatrix}^T \tag{3.8}$$

As before, the variety of maximal minors of this 5×8-matrix has multiple components. Our variety $\mathrm{EC}_{2,5}$ is the main component, obtained by taking the ideal quotient by determinantal ideals that are given by proper subsets of the rows.

In what follows we derive a general result for binary tensors. This will explain the origin of the matrices (3.5), (3.7) and (3.8) that were used to represent $\mathrm{EC}_{2,d}$.

Fix $V = K^n$. Tensors A live in the space $V^{\otimes d}$. For each k, the map $A \mapsto \psi^{[k]}$ factors through the linear map that symmetrizes the factors indexed by $[d]\backslash\{k\}$:

$$V^{\otimes d} \longrightarrow \mathrm{Sym}_{d-1}(V) \otimes V. \tag{3.9}$$

Taking the wedge product with $(x_1, x_2, \ldots, x_n)$ defines a further linear map

$$\begin{array}{ccc} \mathrm{Sym}_{d-1}(V) \otimes V & \longrightarrow & \mathrm{Sym}_d(V) \otimes \bigwedge^2 V \\ \Psi & & \Psi \\ \sum_{i=1}^{n} \psi_i \otimes e_i & \longmapsto & \sum_{1 \leq i < j \leq n} (\psi_i x_j - \psi_j x_i) \otimes (e_i \wedge e_j), \end{array} \tag{3.10}$$

where $\{e_1, \ldots, e_n\}$ is a basis for V. Write $\ell^{[k]}$ for the composition of (3.10) after (3.9). Thus $\ell^{[k]}(A)$ is a vector of length $\binom{n}{2}$ whose entries are polynomials of degree d that define the eigenconfiguration with index k. For instance, in Example 3.3, $f = \ell^{[1]}(A)$, $g = \ell^{[2]}(A)$, $h = \ell^{[3]}(A)$.

The kernel of $\ell^{[k]}$ consists of all tensors whose eigenconfiguration with index k is all of $\mathbb{P}^{n-1}$. We are interested in the space of tensors where this happens simultaneously for all indices k:

$$K_{n,d} \;=\; \bigcap_{k=1}^{d} \mathrm{kernel}(\ell^{[k]}). \tag{3.11}$$

The tensors in $K_{n,d}$ can be regarded as being trivial as far as eigenvectors are concerned. For instance, in the classical matrix case ($d = 2$), we have

$$K_{n,2} \;=\; \mathrm{kernel}(\ell^{[1]}) \;=\; \mathrm{kernel}(\ell^{[2]}),$$

and this is the 1-dimensional space spanned by the identity matrix.

In what follows we restrict our attention to binary tensors ($n = 2$). We regard $\ell^{[k]}$ as a linear map $\mathbb{P}^{2^d-1} \dashrightarrow \mathbb{P}^d$. The eigencompatibility variety $\mathrm{EC}_{2,d}$ is the closure of the image of the map $\mathbb{P}^{2^d-1} \dashrightarrow (\mathbb{P}^d)^d$ given by the tuple $(\ell^{[1]}, \ldots, \ell^{[d]})$. Let $\mathbf{u}^{[k]}$ be a column vector of unknowns representing points in the kth factor $\mathbb{P}^d$.

THEOREM 3.6. *There exists a $d \times e$-matrix $\mathbf{E}_d$ with $e = \dim(K_{2,d}) + d(d+1) - 2^d$, whose entries in the kth row are $\mathbb{Z}$-linear forms in $\mathbf{u}^{[k]}$, such that $\mathrm{EC}_{2,d}$ is an irreducible component in the variety defined by the $d \times d$-minors of $\mathbf{E}_d$. Its ideal is obtained from those $d \times d$-minors by taking the ideal quotient (or saturation) with respect to the maximal minor ideals of proper subsets of the rows of $\mathbf{E}_d$.*

PROOF. We shall derive this using the linear algebra method in [**AST**, §2]. We express $\ell^{[k]}$ as a $(d{+}1) \times 2^d$-matrix, and we form the $d(d+1) \times (2^d + d)$-matrix

$$\begin{pmatrix} \ell^{[1]} & \mathbf{u}^{[1]} & 0 & \cdots & 0 \\ \ell^{[2]} & 0 & \mathbf{u}^{[2]} & \cdots & 0 \\ \vdots & \vdots & \vdots & \ddots & \vdots \\ \ell^{[d]} & 0 & 0 & \cdots & \mathbf{u}^{[d]} \end{pmatrix}. \tag{3.12}$$

The left $d(d+1) \times 2^d$-submatrix has entries in $\{-1, 0, +1\}$ and its kernel is $K_{2,d}$. The rank of that submatrix is $r = 2^d - \dim(K_{2,d})$. Using row operations, we can transform (3.12) into a matrix $\begin{pmatrix} A & B \\ 0 & C \end{pmatrix}$ where A is an $r \times 2^d$ matrix of rank r, and C is an $e \times d$-matrix whose k^{th} column has linear entries in the coordinates of $\mathbf{u}^{[k]}$.

The variety $\mathrm{EC}_{2,d}$ is the set of all points $(\mathbf{u}^{[1]}, \mathbf{u}^{[2]}, \ldots, \mathbf{u}^{[d]})$ in $(\mathbb{P}^d)^d$ such that the kernel of (3.12) contains a vector whose last d coordinates are non-zero. Equivalently, the kernel of C contains a vector whose d coordinates are all non-zero.

Let $\mathbf{E}_d$ be the transpose of C. This is a $d \times e$-matrix whose k^{th} row has entries that are $\mathbb{Z}$-linear in $\mathbf{u}^{[k]}$. By construction, $\mathrm{EC}_{2,d}$ is the set of points $(\mathbf{u}^{[1]}, \ldots, \mathbf{u}^{[d]})$ in $(\mathbb{P}^d)^d$ such that $\mathbf{v} \cdot \mathbf{E}_d = 0$ for some $\mathbf{v} \in (K \backslash \{0\})^d$. This completes the proof. □

By our matrix representation, the codimension of $\mathrm{EC}_{2,d}$ is at most $e-d+1$, so

$$\dim(\mathrm{EC}_{2,d}) \;\geq\; d^2 - (e-d+1). \tag{3.13}$$

Examples 3.2, 3.3, and 3.4 suggest that (3.13) is an equality.

CONJECTURE 3.7. *The dimension of* $\mathrm{EC}_{2,d}$ *equals* $d^2 - (e-d+1)$.

We do not know the dimension of $K_{2,d} = \bigcap_k \mathrm{kernel}(\ell^{[k]})$. In our examples, we saw that $\dim(K_{2,d}) = 1, 0, 3, 10$ for $d = 2, 3, 4, 5$ respectively. It would be desirable to better understand the common kernel $K_{n,d}$ for arbitrary n and d:

PROBLEM 3.8. *Find the dimension of the space* $K_{n,d}$ *in (*3.11*).*

Another problem is to understand the diagonal of $\mathrm{EC}_{n,d}$ in the embedding (3.3). This diagonal parametrizes *simultaneous eigenconfigurations*, arising from special tensors A whose d maps $\psi^{[1]}, \ldots, \psi^{[d]}$ all have the same fixed point locus in $\mathbb{P}^{n-1}$. Symmetric tensors A have this property, and the issue is to characterize all others.

EXAMPLE 3.9 ($n=2$). The diagonal of $\mathrm{EC}_{2,d}$ is computed by setting $\mathbf{u}^{[1]} = \mathbf{u}^{[2]} = \cdots = \mathbf{u}^{[d]}$ in the prime ideal described in Theorem 3.6. If d is odd, then there is no constraint, by Theorem 2.7. However, for d even, the diagonal of $\mathrm{EC}_{2,d}$ is interesting. For instance, for $d=2$, equating the rows in (3.2) gives two components

$$\left\langle 2 \times 2\text{-minors of } \begin{pmatrix} a_0 & a_1 & a_2 \\ a_2 & -a_1 & a_0 \end{pmatrix} \right\rangle \;=\; \langle a_0 + a_2 \rangle \cap \langle a_0 - a_2, a_1 \rangle.$$

The first component is the known case of symmetric 2×2-matrices. The second component is a point in $\mathrm{Sym}_2(\mathbb{P}^1)$, namely the binary form $s^2 + t^2 = (s-it)(s+it)$. This is the simultaneous eigenconfiguration of any matrix $A = \begin{pmatrix} a & b \\ -b & a \end{pmatrix}$ with $b \neq 0$.

4. The eigendiscriminant

The d-dimensional tensors of format $n \times n \times \cdots \times n$ represent points in a projective space $\mathbb{P}^N$ where $N = n^d - 1$. For a generic tensor $A \in \mathbb{P}^N$, the ℓth eigenconfiguration, in the sense of (2.2), consists of a finite set of reduced points in $\mathbb{P}^{n-1}$. We know from Theorem 2.2 that the number of these points equals

$$\rho(n,d,\ell) \;=\; \sum_{i=0}^{n-1} (d-1)^i \ell^{n-1-i}.$$

In this section we study the set $\Delta_{n,d,\ell}$ of all tensors A for which the eigenconfiguration consists of fewer than $\rho(n,d,\ell)$ points or is not zero-dimensional. This set is a subvariety of $\mathbb{P}^N$, called the ℓth *eigendiscriminant*. We also abbreviate

$$\gamma(n,d,\ell) \;=\; \sum_{j=2}^{n-1} (-1)^{n-1+j}(j-1) \left[\sum_{k=0}^{j} (-1)^k \binom{n}{j-k} \binom{j(d-1) - k\ell - 1}{n-1} \right]. \tag{4.1}$$

The following is our main result in this section:

THEOREM 4.1. *The ℓth eigendiscriminant is an irreducible hypersurface with*

$$\text{degree}(\Delta_{n,d,\ell}) \quad = \quad 2\gamma(n,d,\ell) + 2\rho(n,d,\ell) - 2. \tag{4.2}$$

We identify $\Delta_{n,d,\ell}$ with the unique (up to sign) irreducible polynomial with integer coefficients in the n^d unknowns $a_{i_1 i_2 \cdots i_d}$ that vanishes on this hypersurface. From now on we use the term eigendisciminant to refer to the polynomial $\Delta_{n,d,\ell}$.

The case of most interest is $\ell = 1$, which pertains to the eigenconfiguration of a tensor in the usual sense of (1.2). For that case, we write $\Delta_{n,d} = \Delta_{n,d,1}$ for the eigendiscriminant, and the formula for its degree can be simplified as follows:

COROLLARY 4.2. *The eigendiscriminant is a homogeneous polynomial of degree*

$$\text{degree}(\Delta_{n,d}) \;=\; n(n-1)(d-1)^{n-1}.$$

The following proof is due to Manuel Kauers. We are grateful for his help.

PROOF. We set $\ell = 1$ in the expression (4.2). Our claim is equivalent to

$$\gamma(n,d,1) \;=\; \binom{n}{2}(d-1)^{n-1} - \frac{(d-1)^n - 1}{d-2} + 1. \tag{4.3}$$

We abbreviate the innermost summand in (4.1) as

$$s_{n,d,j}(k) \;=\; (-1)^k \binom{n}{j-k}\binom{j(d-1)-k-1}{n-1}.$$

Using Gosper's algorithm [**PWZ**, Chapter 5], we find the multiple

$$S_{n,d,j}(k) \;:=\; \frac{(j(d-1)-k)(j-k-n)}{nj(d-2)} s_{n,d,j}(k).$$

It can now be checked by hand that this satisfies

$$S_{n,d,j}(k+1) - S_{n,d,j}(k) \;=\; s_{n,d,j}(k).$$

Summing over the range $k = 0, \ldots, j-1$ and simplifying expressions lead to

$$\sum_{k=0}^{j} s_{n,d,j}(k) \;=\; S_{n,d,j}(j+1) - S_{n,d,j}(0) \;=\; \frac{d-1}{d-2}\binom{n-1}{j}\binom{j(d-1)-1}{n-1}. \tag{4.4}$$

This is valid for all $j \geq 1$.

Next we introduce the expression

$$A(n,d) \;=\; \sum_{j=0}^{n} (-1)^j \binom{n}{j}\binom{dj-1}{n}(j-1). \tag{4.5}$$

Consider $\binom{dj-1}{n}(j-1)$ as a polynomial in j of degree $n+1$. In the binomial basis,

$$\binom{dj-1}{n}(j-1) \;=\; (n+1)d^n\binom{j}{n+1} + \left((d-1)d^{n-1}\binom{n+1}{2} - d^n\right)\binom{j}{n}$$
$$+ \text{ lower degree terms,}$$

Recall from [**GKP**, page 190] that

$$\sum_{j=0}^{n} (-1)^j \binom{n}{j}\binom{j}{k} \;=\; \begin{cases} (-1)^n & \text{if } k = n, \\ 0 & \text{otherwise.} \end{cases}$$

This implies

$$A(n,d) \;=\; (-1)^n\Big((d-1)d^{n-1}\binom{n+1}{2} - d^n\Big).$$

Combining this identity with (4.1), (4.4) and (4.5), we now derive

$$\begin{aligned}\gamma(n,d,1) &= (-1)^{n-1}\frac{d-1}{d-2}\left(A(n-1,d-1)+\binom{-1}{n-1}\right)\\ &= (-1)^{n-1}\frac{d{-}1}{d{-}2}\left((-1)^{n-1}\left((d{-}2)(d{-}1)^{n-2}\binom{n}{2} - (d{-}1)^{n-1}\right) + (-1)^{n-1}\right)\\ &= \frac{d-1}{d-2}\left((d-2)(d-1)^{n-2}\binom{n}{2} - (d-1)^{n-1}\right) + \frac{d-1}{d-2}\\ &= \binom{n}{2}(d-1)^n - \frac{(d-1)^n-(d-1)}{d-2}.\end{aligned}$$

This equals the desired expression for $\gamma(n,d,1)$ on the right hand side of (4.3). □

The proof of Theorem 4.1 involves some algebraic geometry and will be presented later in this section. We first discuss a few examples to illustrate $\Delta_{n,d}$.

EXAMPLE 4.3 ($d=2$). The eigendiscriminant of an $n\times n$-matrix $A=(a_{ij})$ is the discriminant of its characteristic polynomial. In symbols,

$$\Delta_{n,2} \;=\; \mathrm{discr}_\lambda\big(\det(\,A-\lambda\cdot \mathrm{Id}_n)\,\big).$$

This is a homogeneous polynomial of degree $n(n-1)$ in the matrix entries a_{ij}. For instance, for a 3×3-matrix, the eigendiscriminant is a polynomial with 144 terms:

$$\Delta_{3,2} \;=\; a_{11}^4a_{22}^2 - 2a_{11}^4a_{22}a_{33} + 4a_{11}^4a_{23}a_{32} + a_{11}^4a_{33}^2 - 2a_{11}^3a_{12}a_{21}a_{22} + \cdots + a_{23}^2a_{32}^2a_{33}^2.$$

This polynomial vanishes whenever two of the eigenvalues of A coincide.

There is a beautiful theory behind $\Delta_{n,2}$ in the case when A is real symmetric, so the eigenconfiguration is defined over $\mathbb{R}$. The resulting *symmetric eigendiscriminant* is a nonnegative polynomial of degree $n(n-1)$ in the $\binom{n+1}{2}$ matrix entries. Its real variety has codimension 2 and degree $\binom{n+1}{3}$, and its determinantal representation governs expressions of $\Delta_{n,2}$ as a sum of squares of polynomials of degree $\binom{n}{2}$. For further reading on this topic see [**Stu**, Section 7.5] and the references given there.

EXAMPLE 4.4 ($n=2$). The eigendiscriminant of a d-dimensional tensor of format $2\times 2\times\cdots\times 2$ is the discriminant of the associated binary form in (2.7), i.e.

$$\Delta_{2,d} \;=\; \mathrm{disc}_{(x,y)}\big(\,y\cdot\psi_1(x,y) - x\cdot\psi_2(x,y)\,\big).$$

This is a homogeneous polynomial of degree $2d-2$ in the 2^d tensor entries $a_{i_1i_2\cdots i_d}$.

EXAMPLE 4.5 ($n=d=3$). The eigendiscriminant $\Delta_{3,3}$ of a $3{\times}3{\times}3$-tensor $A=(a_{ijk})$ is a homogeneous polynomial of degree 24 in the 27 entries a_{ijk}. If we specialize A to a symmetric tensor, corresponding to a ternary cubic

$$\phi(x,y,z) \;=\; c_{300}x^3 + c_{210}x^2y + c_{201}x^2z + \cdots + c_{003}z^3,$$

then $\Delta_{3,3}$ remains irreducible. The resulting irreducible polynomial of degree 24 in the ten coefficients c_{ijk} is the *eigendiscriminant of a tenary cubic.* At present we do not know an explicit formula for $\Delta_{3,3}$, but it is fun to explore specializations of the eigendiscriminant. For instance, if $\phi = (2x+y)(2x+z)(2y+z) + u \cdot xyz$ then

$$\begin{aligned}\Delta_{3,3} \;=\; & 16u^{24} + 2304u^{23} + 152784u^{22} + 6097536u^{21} + 159761808u^{20} + 2779161840u^{19} + 29727588168u^{18} \\ & +124641852624u^{17} - 1234078589016u^{16} - 18314627517360u^{15} - 8929524942432u^{14} + 1200933047925648u^{13} \\ & +3722203539791685u^{12} - 63418425922462464u^{11} - 257381788882972176u^{10} + 2676970903961440800u^{9} \\ & +7927655114836286496u^{8} - 8901348223990895539 2u^{7} - 139343550261710123 52u^{6} \\ & +17292503563715567923 20u^{5} - 51592223249011929300 48u^{4} - 11838757458480721920u^{3} \\ & +28255456641734116982784u^{2} - 56809371779894977339392u + 37304830510913780269056,\end{aligned}$$

and if $\phi = u \cdot x^3 + v \cdot y^3 + w \cdot z^3 + xyz$ then $\Delta_{3,3}$ is the square of polynomial

$$\begin{aligned}& 531441u^4v^4w^4 - 708588u^5v^3w^3 - 708588u^3v^5w^3 - 708588u^3v^3w^5 + 1062882u^4v^4w^2 + 1062882u^4v^2w^4 \\ & +1062882u^2v^4w^4 - 1810836u^3v^3w^3 - 177147u^4v^4 + 39366u^4v^2w^2 + 39366u^2v^4w^2 - 177147u^4w^4 \\ & +39366u^2v^2w^4 - 177147v^4w^4 + 314928u^3v^3w + 314928u^3vw^3 + 314928uv^3w^3 - 244944u^2v^2w^2 - 46656u^3vw \\ & -46656uv^3w - 46656uvw^3 + 23328u^2v^2 + 23328u^2w^2 + 23328v^2w^2 + 6912uvw - 2304u^2 - 2304v^2 - 2304w^2 + 256.\end{aligned}$$

We now embark towards the proof of Theorem 4.1. Let A and B be generic tensors of the same format, and write $(\psi_1, \ldots, \psi_n)$ and $(\omega_1, \ldots, \omega_n)$ for the vectors of degree $d-1$ polynomials that represent the corresponding maps $\mathbb{P}^{n-1} \dashrightarrow \mathbb{P}^{n-1}$. Let $\mathcal{C}$ denote the subvariety of $\mathbb{P}^{n-1}$ defined by the determinantal constraints

$$\text{rank}\begin{pmatrix} \psi_1(\mathbf{x}) & \psi_2(\mathbf{x}) & \cdots & \psi_n(\mathbf{x}) \\ \omega_1(\mathbf{x}) & \omega_2(\mathbf{x}) & \cdots & \omega_n(\mathbf{x}) \\ x_1^\ell & x_2^\ell & \cdots & x_n^\ell \end{pmatrix} \;\leq\; 2. \tag{4.6}$$

Since the ψ_i and ω_j are generic, this defines a variety of codimension $n-2$. We find that $\mathcal{C}$ is a curve that is smooth and irreducible, by an argument similar to that in the proof of Theorem 2.2. The following lemma is the key to Theorem 4.1.

LEMMA 4.6. *The expression in (4.1) is the genus of the curve $\mathcal{C}$. In symbols,*

$$\text{genus}(\mathcal{C}) \;=\; \gamma(n, d, \ell).$$

Using this lemma, we now derive the degree of the eigendiscriminant.

PROOF OF THEOREM 4.1. We define a map $\mu : \mathcal{C} \to \mathbb{P}^1$ as follows. For any point $\mathbf{x}$ on the curve $\mathcal{C}$, the matrix in (4.6) has rank 2, so, up to scaling, there exists a unique row vector $(a, b, c) \in K^3$ that spans the left kernel of that $3 \times n$-matrix. We define the image of $\mathbf{x} \in \mathcal{C}$ to be the point $\mu(\mathbf{x}) = (a : b)$ on the projective line $\mathbb{P}^1$. This condition means that $\mathbf{x}$ is an eigenpoint of the tensor $aA + bB$. Conversely, for any $(a : b) \in \mathbb{P}^1$, the fiber $\mu^{-1}(a : b)$ consists precisely of the eigenpoints of $aA + bB$. Hence, since A and B are generic, the generic fiber is finite and reduced of cardinality $\rho(n, d, \ell)$. In other words, $\mu : \mathcal{C} \to \mathbb{P}^1$ is a map of degree $\rho(n, d, \ell)$.

We restrict the eigendiscriminant to our $\mathbb{P}^1$ of tensors. The resulting binary form $\Delta_{n,d,\ell}(aA+bB)$ is squarefree, and its degree is the left hand side in (4.2). The points $(a : b) \in \mathbb{P}^1$ where $\Delta_{n,d,\ell}(aA + bB) = 0$ are the *branch points* of the map μ. The corresponding multiplicity-two eigenpoints $\mathbf{x}$ form the *ramification divisor* on $\mathcal{C}$. The number of branch points of μ is the degree of the eigendiscriminants $\Delta_{n,d,\ell}$.

The Riemann-Hurwitz Formula [**Fu2**, Exercise 8.36] states that the number of branch points of the map $\mu : \mathcal{C} \to \mathbb{P}^1$ is $2 \cdot \text{genus}(\mathcal{C}) + 2 \cdot \text{degree}(\mu) - 2$. By the first paragraph, and by Lemma 4.6, this expression is the right hand side of (4.2). $\square$

Our proofs of Lemma 4.6 and the irreducibility of the eigendiscriminant are fairly complicated; therefore we decided not to include them here. They are based on the properties of vector bundles, like those seen in the proof of Theorem 2.2. We plan to develop this further and publish it in a later paper on discriminants arising from maximal minors of matrices with more than two rows.

What we shall do instead is to prove an alternative combinatorial formula for the genus of $\mathcal{C}$ that is equivalent to (4.1). This does not prove Lemma 4.6 because we presently do not know a direct argument to show that they are equal. Nevertheless, the following discussion is an illustration of useful commutative algebra techniques.

INSTEAD OF LEMMA 4.6. The Hilbert polynomial $H_{\mathcal{C}}(t)$ of the curve $\mathcal{C}$ equals

$$H_{\mathcal{C}}(t) \quad = \quad \text{degree}(\mathcal{C}) \cdot t \, + \, (1 - \text{genus}(\mathcal{C})).$$

Recall that $\mathcal{C}$ is a linear section of the variety defined by the maximal minors of a $3 \times n$-matrix whose rows are homogeneous of degrees $d-1, d-1$ and ℓ. That variety is Cohen-Macaulay. We shall compute the Hilbert polynomial of the coordinate ring of $\mathcal{C}$ from its graded minimal free resolution over $S = K[x_1, \ldots, x_n]$.

Consider the S-linear map from $F = S^{\oplus n}$ to $G = S(d-1)^{\oplus 2} \oplus S(\ell)$ given by

$$\alpha = \begin{pmatrix} \psi_1(\mathbf{x}) & \psi_2(\mathbf{x}) & \cdots & \psi_n(\mathbf{x}) \\ \omega_1(\mathbf{x}) & \omega_2(\mathbf{x}) & \cdots & \omega_n(\mathbf{x}) \\ x_1^\ell & x_2^\ell & \cdots & x_n^\ell \end{pmatrix}. \tag{4.7}$$

By [**Eis**, Section A2H], the corresponding *Eagon-Northcott complex* $\text{EN}(\alpha)$ equals

$$0 \to \text{Sym}_{n-3}(G^\vee) \otimes \bigwedge^{n} F \to \text{Sym}_{n-4}(G^\vee) \otimes \bigwedge^{n-1} F \to \cdots \to G^\vee \otimes \bigwedge^{4} F \to \bigwedge^{3} F \to \bigwedge^{3} G,$$

where $\text{Sym}_i(G^\vee)$ is the ith symmetric power of $G^\vee$ and $\bigwedge^i F$ is the ith exterior power of F. We compute the Hilbert polynomial $H_M(t)$ of each module M in $\text{EN}(\alpha)$.

Since $\mathcal{C}$ has codimension $n - 2 = \text{rank}\, F - \text{rank}\, G + 1$, the complex $\text{EN}(\alpha) \otimes S(-2d - \ell + 2)$ is a free resolution of the coordinate ring of $\mathcal{C}$. In particular,

$$H_{\mathcal{C}}(t) \; = \; H_{\bigwedge^3 G}(t - 2d - \ell + 2) + \sum_{j=3}^{n} (-1)^j H_{E_j}(t - 2d - \ell + 2), \tag{4.8}$$

where $E_j = \text{Sym}_{j-3}(G^\vee) \otimes \bigwedge^j F$. Since $G^\vee = S(-d+1)^{\oplus 2} \oplus S(-\ell)$ and $F = S^{\oplus n}$,

$$\text{Sym}_{j-3}(G^\vee) = \bigoplus_{k=0}^{j-3} S\left((j-k-3)(-d+1) - k\ell\right)^{\oplus j-k+2} \text{ and } \bigwedge^j F = S^{\oplus \binom{n}{j}}.$$

Their tensor product is the jth term in $\text{EN}(\alpha)$. As a graded S-module, it equals

$$E_j \; = \; \bigoplus_{k=0}^{j-3} S\left((j-k-3)(-d+1) - k\ell\right)^{\oplus (j-k-2)\binom{n}{j}}.$$

The shifted Hilbert series of this module is the summand on the right of (4.8):

$$H_{E_j}(t - 2d - \ell + 2) \; = \; \sum_{k=0}^{j-3} (j-k-2) \binom{n}{j} \binom{t + (n-1) + (j-k-1)(1-d) - (k+1)\ell}{n-1}.$$

We conclude that the Hilbert polynomial $H_{\mathcal{C}}(t)$ of the curve $\mathcal{C}$ equals

$$\binom{t+n-1}{n-1} + \sum_{j=3}^{n} (-1)^j \sum_{k=0}^{j-3} (j-k-2) \binom{n}{j} \binom{t + (n-1) + (j-k-1)(1-d) - (k+1)\ell}{n-1}.$$

The genus of $\mathcal{C}$ is obtained by substituting $t = 0$ and subtracting the result from 1:

$$\gamma(n, d, \ell) = \sum_{j=3}^{n} (-1)^{j-1} \sum_{k=0}^{j-3} (j-k-2) \binom{n}{j} \binom{(n-1) + (j-k-1)(1-d) - (k+1)\ell}{n-1}.$$

This formula is equivalent to (4.1). □

EXAMPLE 4.7 ($n = 4$). The last formula seen above specializes to

$$\gamma(4,d,\ell) \;=\; 4\binom{3+2(1-d)-\ell}{3} - \left[2\binom{3+3(1-d)-\ell}{3} + \binom{3+2(1-d)-2\ell}{3}\right],$$

while the genus formula in (4.1) states

$$\begin{aligned}\gamma(4,d,\ell) \;=\; & -\left[6\tbinom{2(d-1)-1}{3} - 4\tbinom{2(d-1)-\ell-1}{3} + \tbinom{2(d-1)-2\ell-1}{3}\right] \\ & +2\left[4\tbinom{3(d-1)-1}{3} - 6\tbinom{3(d-1)-\ell-1}{3} + 4\tbinom{3(d-1)-2\ell-1}{3} - \tbinom{3(d-1)-3\ell-1}{3}\right].\end{aligned}$$

Both of these evaluate to the cubic polynomial

$$\gamma(4,d,\ell) \;=\; 5d^3+5d^2\ell+3d\ell^2+\ell^3-21d^2-14d\ell-5\ell^2+27d+9\ell-10.$$

Therefore, by Theorem 4.1, the degree of the ℓth eigendiscriminant for $n = 4$ equals

$$\text{degree}(\Delta_{4,d,\ell}) \;=\; 3d^3+3d^2\ell+2d\ell^2+\ell^3-12d^2-8d\ell-3\ell^2+15d+5\ell-6.$$

For $\ell = 1$, this factorizes as promised in Corollary 4.2: $\text{degree}(\Delta_{4,d,1}) = 12(d-1)^3$.

5. Seven points in the plane

Our study had been motivated by the desire to find a geometric characterization of eigenconfigurations among all finite subsets of $\mathbb{P}^{n-1}$. The solution for $n = 2$ was presented in Theorem 2.7. However, the relevant geometry is more difficult in higher dimensions. In this section we take some steps towards a characterization for $n = 3$. The eigenconfiguration of a general tensor A in $(K^3)^{\otimes d}$ consists of d^2-d+1 points in $\mathbb{P}^2$. So, our question can be phrased like this: given a configuration $Z \in (\mathbb{P}^2)^{d^2-d+1}$, decide whether it is an eigenconfiguration. If yes, construct a corresponding tensor $A \in (K^3)^{\otimes d}$, and decide whether A can be chosen to be symmetric.

The first interesting case is $d = n = 3$. Here the following result holds.

THEOREM 5.1. *A configuration of seven points in $\mathbb{P}^2$ is the eigenconfiguration of a $3 \times 3 \times 3$-tensor if and only if no six of the seven points lie on a conic.*

The only-if part of this theorem appears also in [**OS**, Proposition 2.1], where Ottaviani and Sernesi studied the degree 54 hypersurface of all *Lüroth quartics* in $\mathbb{P}^2$. We note that part (i) in [**OS**, Proposition 2.1] is not quite correct. A counterexample is the configuration Z consisting of four points on a line and three other general points. It is precisely this gap that makes our proof of Theorem 5.1 a bit lengthy.

This proof will be presented later in this section. Example 1.1 shows that some triples among the seven eigenpoints in $\mathbb{P}^2$ can be collinear. Another interesting point is that being an eigenconfiguration is not a closed condition. For a general d it makes sense to pass to the Zariski closure. We define Eig_d to be the closure in $(\mathbb{P}^2)^{d^2-d+1}$ of the set of all eigenconfigurations. Readers from algebraic geometry may prefer unlabeled configurations, and they would take the closure in the Chow variety $\text{Sym}_{d^2-d+1}(\mathbb{P}^2)$ or in the Hilbert scheme $\text{Hilb}_{d^2-d+1}(\mathbb{P}^2)$. For simplicity of exposition, we work in the space of labeled point configurations. We also consider the *variety of symmetric eigenconfigurations*, denoted $\text{Eig}_{d,\text{sym}}$. This is the Zariski closure in $(\mathbb{P}^2)^{d^2-d+1}$ of the set of eigenconfigurations of ternary forms ϕ of degree d. Towards the end of this section we examine the dimensions of Eig_d and $\text{Eig}_{d,\text{sym}}$.

We begin by approaching our problem with a pinch of commutative algebra. Let $Z \in (\mathbb{P}^2)^{d^2-d+1}$ and write I_Z for the ideal of all polynomials in $S = K[x,y,z]$ that vanish at all points in the configuration Z. This homogeneous radical ideal is Cohen-Macaulay because it has a free resolution of length 1 (see, for example, [**Eis**, Proposition 3.1]). By the Hilbert-Burch Theorem, the minimal free resolution of I_Z has the form

$$0 \to S^{\oplus(m-1)} \xrightarrow{\Phi} S^{\oplus m} \to I_Z \to 0.$$

The $m \times (m-1)$-matrix Φ is the *Hilbert-Burch matrix* of Z. The minimal free resolution of I_Z is unique up to change of bases in the graded S-modules. In that sense, we write $\Phi_Z := \Phi$. The ideal I_Z is generated by the maximal minors of Φ_Z. The following proposition is due to Ottaviani and Sernesi (see [**OS**, Proposition 2.1]).

PROPOSITION 5.2. *Let Z be a configuration in $(\mathbb{P}^2)^{d^2-d+1}$. Then Z is the eigenconfiguration of a tensor if and only if its Hilbert-Burch matrix has the form*

$$\Phi_Z = \begin{pmatrix} L_1 & F_1 \\ L_2 & F_2 \\ L_3 & F_3 \end{pmatrix}, \tag{5.1}$$

where L_1, L_2, L_3 are linear forms that are linearly independent over K.

This statement makes sense because the condition on Φ_Z is invariant under row operations over K. The ternary forms F_1, F_2, F_3 must all have the same degree, and the hypothesis on Z ensures that this common degree is $d-1$.

PROOF. We start with the only-if direction. Suppose that Z is an eigenconfiguration. Then there exist ternary forms ψ_1, ψ_2, ψ_3 of degree $d-1$ such that Z is defined set-theoretically by the 2×2-minors of

$$\begin{pmatrix} x & y & z \\ \psi_1(x,y,z) & \psi_2(x,y,z) & \psi_3(x,y,z) \end{pmatrix}. \tag{5.2}$$

The ideal generated by these minors is Cohen-Macaulay of codimension 2 and its degree equals the cardinality of Z. This implies that this ideal coincides with I_Z. The Hilbert-Burch Theorem ensures that the transpose of (5.2) equals Φ_Z. Since x, y, z are linearly independent, we see that Φ_Z has the form required in (5.1).

For the converse, suppose that the Hilbert-Burch matrix of Z has size 3×2 as in (5.1) with L_1, L_2, L_3 linearly independent. By performing row operations over K, we can replace L_1, L_2, L_3 by x, y, z. This means that the transpose of Φ_Z is (5.2) for some ψ_1, ψ_2, ψ_3. Any such triple of ternary forms of degree $d-1$ arises from some tensor $A \in (K^3)^{\otimes d}$. By construction, Z is the eigenconfiguration of A. ☐

Proposition 5.2 translates into an algorithm for testing whether a given $Z \in (\mathbb{P}^2)^{d^2-d+1}$ is an eigenconfiguration. The algorithm starts by computing the ideal

$$I_Z = \bigcap_{(\alpha:\beta:\gamma)\in Z} \langle x\beta - y\alpha,\, x\gamma - z\alpha,\, y\gamma - z\beta \rangle.$$

This ideal must have three minimal generators of degree d; otherwise Z is not an eigenconfiguration. If I_Z has three generators, then we compute the two syzygies. They must have degrees 1 and $d-1$, so the minimal free resolution of I_Z looks like

$$0 \to S(-d-1) \oplus S(-2d+1) \xrightarrow{\Phi} S(-d)^{\oplus 3} \to I_Z \to 0.$$

At this point we examine the matrix Φ. If the linear entries L_1, L_2, L_3 in the left column are linearly dependent, then Z is not an eigenconfiguration. Otherwise we perform row operations so that Φ^T looks like (5.2). The last step is to pick a tensor $A \in (K^3)^{\otimes d}$ that gives rise to the ternary forms ψ_1, ψ_2, ψ_3 in the second row of Φ^T.

The remaining task is to find a geometric interpretation of the criterion in Proposition 5.2. This was given for $d = 3$ in the result whose proof we now present.

PROOF OF THEOREM 5.1. Fix a configuration $Z \in (\mathbb{P}^2)^7$. Our claim states that the Hilbert-Burch matrix Φ_Z has format 3×2 as in (5.1), with L_1, L_2, L_3 linearly independent, if and only if no six of the points in Z lie on a conic.

We begin with the only-if direction. Take $p \in Z$ such that $Z\backslash\{p\}$ lies on a conic C in $\mathbb{P}^2$. Fix linear forms L_1 and L_2 that cut out p. The cubics CL_1 and CL_2 vanish on Z. By Proposition 5.2, we have $I_Z = \langle CL_1, CL_2, F\rangle$ where F is another cubic. Since F vanishes at p, there exist quadrics Q_1 and Q_2 such that $F = Q_2L_1 - Q_1L_2$. The generators of the ideal I_Z are the 2×2-minors of

$$\Psi = \begin{pmatrix} L_1 & Q_1 \\ L_2 & Q_2 \\ 0 & C \end{pmatrix}.$$

This means that Ψ is a Hilbert-Burch matrix Φ_Z for Z. However, by Proposition 5.2, the left column in any Φ_Z must consist of linearly independent linear forms. This is a contradiction, which completes the proof of the only-if direction.

We now establish the if direction. Fix any configuration $Z \in (\mathbb{P}^2)^7$ of seven points that do not lie on a conic. We first prove that the minimal free resolution of I_Z has the following form, where c is either 0 or 1:

$$0 \to S(-4)^{\oplus(c+1)} \oplus S(-5) \to S(-3)^{\oplus 3} \oplus S(-4)^{\oplus c} \to I_Z \to 0,. \tag{5.3}$$

By the Hilbert-Burch Theorem, the resolution of I_Z equals

$$0 \to \bigoplus_{i=1}^{t} S(-b_i) \xrightarrow{\Phi_Z} \bigoplus_{i=1}^{t+1} S(-a_i) \to I_Z \to 0,$$

where $t, a_1, \ldots, a_{t+1}, b_1, \ldots, b_t \in \mathbb{N}$ with $a_1 \geq \cdots \geq a_{t+1}$ and $b_1 \geq \cdots \geq b_t$. We abbreviate $e_i = b_i - a_i$ and $f_i = b_i - a_{i+1}$ for $i \in \{1, \ldots, t\}$. These invariants satisfy

(i) $f_i \geq e_i, e_{i+1}$,
(ii) $e_i, f_i \geq 1$.

Furthermore, Eisenbud shows in [**Eis**, Proposition 3.8] that

(iii) $a_i = \sum_{j=1}^{i-1} e_j + \sum_{j=i}^{t} f_j$.

There exist 3 linearly independent cubics that vanish on the seven points in Z. By [**Eis**, Corollary 3.9], the ideal I_Z has either 3 or 4 minimal generators, so $t \in \{2, 3\}$.

Suppose $t = 2$. Then $a_1 = a_2 = a_3 = 3$, and it follows from (iii) that

$$f_1 + f_2 = 3 \quad \text{and} \quad e_1 + f_2 = 3.$$

So, by (i) and (ii), we obtain $e_1 = f_1 = 1$ and $f_2 = 2$. This implies $b_1 = 5$ and $b_2 = 4$. Therefore, I_Z has a minimal free resolution of type (5.3) with $c = 0$.

Next, suppose $t = 3$. Then $a_1 \geq a_2 = a_3 = a_4 = 3$. From (iii) we now get

$$\begin{array}{ccccccc} f_1 & + & f_2 & + & f_3 & = & a_1 \\ e_1 & + & f_2 & + & f_3 & = & 3 \\ e_1 & + & e_2 & + & f_3 & = & 3. \end{array}$$

By (ii), $e_1 = e_2 = f_2 = f_3 = 1$. Therefore, $b_2 = a_2 + e_2 = 4$ and $b_3 = a_2 + f_2 = 4$. Corollary 3.10 in [**Eis**] says that

$$\sum_{i\leq j} e_i f_j \;=\; e_1(f_1+f_2+f_3) + e_2(f_2+f_3) + e_3 f_3 \;=\; a_1 + 3 \;=\; \deg Z \;=\; 7.$$

Hence $a_1 = 4$, $b_1 = 5$, and I_Z has a minimal free resolution of type (5.3) with $c = 1$.

To complete the proof, we now assume that no six points of Z lie on a conic. In particular, no conic contains Z, so the minimal free resolution of I_Z equals (5.3), with $c \in \{0,1\}$. Suppose that $c = 1$. The Hilbert-Burch matrix must be

$$\Phi_Z \;=\; \begin{pmatrix} L_{00} & L_{01} & Q_0 \\ L_{10} & L_{11} & Q_1 \\ L_{20} & L_{21} & Q_2 \\ 0 & 0 & L \end{pmatrix}$$

with L, L_{ij} are linear and Q_k are quadrics. Then $I_Z = \langle LQ'_0, LQ'_1, LQ'_2, Q\rangle$, where

$$Q'_0 = \begin{vmatrix} L_{00} & L_{01} \\ L_{10} & L_{11} \end{vmatrix}, Q'_1 = \begin{vmatrix} L_{00} & L_{01} \\ L_{20} & L_{21} \end{vmatrix},\; Q'_2 = \begin{vmatrix} L_{10} & L_{11} \\ L_{20} & L_{21} \end{vmatrix},\; Q = \begin{vmatrix} L_{00} & L_{01} & Q_0 \\ L_{10} & L_{11} & Q_1 \\ L_{20} & L_{21} & Q_2 \end{vmatrix}.$$

The ideal generated by L and Q contains I_Z. The intersection of the curves $\{L=0\}$ and $\{Q=0\}$ is contained in Z. Note that these curves share no positive-dimensional component, since Z is zero-dimensional. Thus $\{L = Q = 0\}$ consists of four points. Let L' be a linear form vanishing on two of the three other points. Then the conic $\{LL' = 0\}$ contains six points of Z, which contradicts our assumption.

Hence, $c = 0$. The resolution (5.3) tells us that the Hilbert-Burch matrix equals

$$\Phi_Z \;=\; \begin{pmatrix} L_1 & Q_1 \\ L_2 & Q_2 \\ L_3 & Q_3 \end{pmatrix},$$

with linear forms L_i and conics Q_j. If L_1, L_2, L_3 were linearly dependent then we can take $L_3 = 0$. So, the conic Q_3 contains the six points in $Z\backslash\{L_1 = L_2 = 0\}$. Consequently, the linear forms L_1, L_2, L_3 must be linearly independent. Proposition 5.2 now implies that Z is the eigenconfiguration of some $3\times 3\times 3$-tensor. □

After taking the Zariski closure, we have $\mathrm{Eig}_3 = (\mathbb{P}^2)^7$. We shall now discuss the subvariety $\mathrm{Eig}_{3,\mathrm{sym}}$ of those eigenconfigurations that come from symmetric tensors. Consider the three quadrics in the second row of (5.2). We write these as

$$\begin{aligned} \psi_1(\mathbf{x}) &= a_1x^2 + a_2xy + a_3xz + a_4y^2 + a_5yz + a_6z^2, \\ \psi_2(\mathbf{x}) &= b_1x^2 + b_2xy + b_3xz + b_4y^2 + b_5yz + b_6z^2, \\ \psi_3(\mathbf{x}) &= c_1x^2 + c_2xy + c_3xz + c_4y^2 + c_5yz + c_6z^2. \end{aligned}$$

We shall characterize the case of symmetric tensors in terms of these coefficients.

PROPOSITION 5.3. *The variety* $\mathrm{Eig}_{3,\mathrm{sym}}$ *is irreducible of dimension* 9 *in* $(\mathbb{P}^2)^7$. *An eigenconfiguration* Z *comes from a symmetric tensor* ϕ *as in* (1.3) *if and only if*

$$a_5 - b_3 = b_3 - c_2 = 2a_4 - 2a_6 - b_2 + c_3 = 2b_6 - 2b_1 - c_5 + a_2 = 2c_1 - 2c_4 - a_3 + b_5 = 0.$$

PROOF. There exists a symmetric tensor with eigenconfiguration Z if and only if there exist a cubic ϕ and a linear form $L = u_1x + u_2y + u_3z$ such that

$$\psi_1 + Lx = \frac{\partial\phi}{\partial x},\;\; \psi_2 + Ly = \frac{\partial\phi}{\partial y},\;\; \psi_3 + Lz = \frac{\partial\phi}{\partial z}. \tag{5.4}$$

We eliminate the cubic ϕ from this system by taking crosswise partial derivatives:

$$\frac{\partial\psi_1}{\partial y} + x\frac{\partial L}{\partial y} = \frac{\partial\psi_2}{\partial x} + y\frac{\partial L}{\partial x}, \ldots, \frac{\partial\psi_2}{\partial z} + y\frac{\partial L}{\partial z} = \frac{\partial\psi_3}{\partial y} + z\frac{\partial L}{\partial x}.$$

This is a system of linear equations in the 21 unknowns $a_1, \ldots, a_6, b_1, \ldots, b_6, c_1, \ldots, c_6, u_1, u_2, u_3$. By eliminating the last three unknowns u_1, u_2, u_3 from that system, we arrive at the five linearly independent equations in a_i, b_j, c_k stated above. □

Proposition 5.3 translates into an algorithm for testing whether a given configuration Z is the eigenconfiguration of a ternary cubic ϕ. Namely, we compute the syzygies of I_Z, we check that the Hilbert-Burch matrix has the form (5.2), and then we check the five linear equations. If these hold then ϕ is found by solving (5.4).

While the equations in Proposition 5.3 are linear, we did not succeed in computing the prime ideal of $\mathrm{Eig}_{3,\mathrm{sym}}$ in the homogeneous coordinate ring of $(\mathbb{P}^2)^7$. This is a challenging elimination problem. Some insight can be gained by intersecting $\mathrm{Eig}_{3,\mathrm{sym}}$ with natural subfamilies of $(\mathbb{P}^2)^7$. For instance, assume that Z contains the three coordinate points, so we restrict to the subspace $(\mathbb{P}^2)^4$ defined by

$$Z = \big\{(1{:}0{:}0), (0{:}1{:}0), (0{:}0{:}1), (\alpha_1 : \alpha_2 {:} \alpha_3), (\beta_1 {:} \beta_2 {:} \beta_3), (\gamma_1 {:} \gamma_2 {:} \gamma_3), (\delta_1 {:} \delta_2 {:} \delta_3)\big\}.$$

At this point it is important to recall that our problem is not projectively invariant.

THEOREM 5.4. *The variety* $\mathrm{Eig}_{3,\mathrm{sym}} \cap (\mathbb{P}^2)^4$ *is three-dimensional, and it represents the eigenconfigurations Z of the ternary cubics in the Hesse family*

$$\phi = ax^3 + by^3 + cz^3 + 3dxyz. \tag{5.5}$$

If a, b, c, d are real then the eigenconfiguration Z contains at least five real points.

PROOF. A ternary cubic $\phi = \sum_{i+j+k=3} c_{ijk} x^i y^j z^k$ has $(1 : 0 : 0)$ as an eigenpoint of ϕ if and only if $c_{210} = c_{201} = 0$. Likewise, $(0 : 1 : 0)$ is an eigenpoint if and only if $c_{120} = c_{021} = 0$, and $(0 : 0 : 1)$ is an eigenpoint if and only if $c_{102} = c_{012} = 0$. Hence the eigenconfiguration of ϕ contains all three coordinate points if and only if ϕ is in the Hesse family (5.5). Since $\mathrm{Eig}_{3,\mathrm{sym}}$ has codimension 5 in $(\mathbb{P}^2)^7$, the intersection $\mathrm{Eig}_{3,\mathrm{sym}} \cap (\mathbb{P}^2)^4$ has codimension ≤ 5, so its dimension is ≥ 3. The Hesse family is 3-dimensional, and so we conclude that $\dim\big(\mathrm{Eig}_{3,\mathrm{sym}} \cap (\mathbb{P}^2)^4\big) = 3$.

The four other eigenpoints of (5.5) are $\big(d\chi(\chi^2 - 1) : \chi(a\chi - c) : d(\chi^2 - 1)\big)$, where χ runs over the zeros of the polynomial

$$d(a^2 - d^2)\chi^4 - a(ab + cd)\chi^3 + 2(abc + d^3)\chi^2 - c(bc + ad)\chi + d(c^2 - d^2). \tag{5.6}$$

We claim that this quartic polynomial has at least two real roots for all $a, b, c, d \in \mathbb{R}$.

Inside the projective space $\mathbb{P}^4$ of quartics $f(\chi) = k_4\chi^4 + k_3\chi^3 + k_2\chi^2 + k_1\chi + k_0$, the family (5.6) is contained in the hypersurface defined by the quadric

$$\mathcal{S} = 2k_4k_2 + k_2^2 - 4k_1k_3 + 4k_0k_4 + 2k_0k_2.$$

The discriminant of $f(\chi)$ defines a hypersurface of degree 6 in $\mathbb{P}^4$. One of the open regions in the complement of the discriminant consists of quartics $f(\chi)$ with no real roots. In polynomial optimization (cf. [**BPT**, Lemma 3.3]) one represents this region by a formula of the following form, where κ is a new indeterminate:

$$f(\chi) = \begin{pmatrix}\chi^2 & \chi & 1\end{pmatrix} \cdot \begin{pmatrix} k_0 & k_1/2 & \kappa \\ k_1 & k_2 - 2\kappa & k_3/2 \\ \kappa & k_3/2 & k_4 \end{pmatrix} \cdot \begin{pmatrix}\chi^2 \\ \chi \\ 1\end{pmatrix},$$

The symmetric 3×3-matrix is required to be positive definite for some $\kappa \in \mathbb{R}$. The condition of being positive definite is expressed by the leading principal minors:

$$\mathcal{P} = k_0\,,\quad \mathcal{Q} = \det\begin{pmatrix} k_0 & k_1/2 \\ k_1/2 & k_2 - 2\kappa \end{pmatrix},\quad \mathcal{R} = \det\begin{pmatrix} k_0 & k_1/2 & \kappa \\ k_1/2 & k_2 - 2\kappa & k_3/2 \\ \kappa & k_3/2 & k_4 \end{pmatrix}.$$

It remains to be proved that there is no solution $(k_0, k_1, k_2, k_3, k_4, \kappa) \in \mathbb{R}^6$ to

$$\mathcal{P} > 0,\ \mathcal{Q} > 0,\ \mathcal{R} > 0 \text{ and } \mathcal{S} = 0.$$

We showed this by computing a *sum-of-squares proof*, in the sense of [**BPT**, Chapter 3]. More precisely, using the software `SOSTools` [**SOS**], we found explicit polynomials $p, q, r, s \in \mathbb{R}[k_0, k_1, k_2, k_3, k_4, \kappa]$ with floating point coefficients such that

$$p, q, r \text{ are sums of squares} \quad \text{and} \quad p\mathcal{P} + q\mathcal{Q} + r\mathcal{R} + s\mathcal{S} = -1.$$

We are grateful to Cynthia Vinzant for helping us with this computation. □

We close this section by returning to tensors in $(K^3)^{\otimes d}$ for general $d \geq 3$.

THEOREM 5.5. *Consider the spaces of eigenconfigurations of ternary tensors,*

$$\mathrm{Eig}_{d,\mathrm{sym}} \subset \mathrm{Eig}_d \subset (\mathbb{P}^2)^{d^2-d+1} \qquad \text{for } d \geq 3.$$

These projective varieties are irreducible, and their dimensions are

$$\dim(\mathrm{Eig}_{d,\mathrm{sym}}) = \frac{1}{2}(d^2 + 3d) \quad \text{and} \quad \dim(\mathrm{Eig}_d) = d^2 + 2d - 1.$$

PROOF. First we show $\dim(\mathrm{Eig}_d) = d^2 + 2d - 1$. Let W be the set of 2×3 matrices (5.2). This is a $3\binom{d+1}{2}$-dimensional vector space over K. The group

$$G = \left\{ \begin{pmatrix} 1 & 0 \\ f & a \end{pmatrix} \,\middle|\, a \in K\backslash\{0\} \text{ and } f \text{ is a ternary form of degree } d-2 \right\}$$

acts on W by left multiplication. Consider $\varphi, \omega \in W$. It is immediate to see that if $\omega = g \cdot \varphi$ for some $g \in G$, then the variety defined by the 2×2-minors of φ equals the variety defined by the 2×2-minors of ω. The converse also holds because of the uniqueness of the Hilbert-Burch matrix. The set W° of elements in W whose 2×3-minors define $d^2 - d + 1$ distinct points is an open subset of W. Therefore,

$$\begin{aligned} \dim(\mathrm{Eig}_d) \;&=\; \dim W^\circ/G \;=\; \dim W^\circ - \dim G \\ &=\; 3\binom{d+1}{2} - \left[\binom{d}{2} + 1\right] \;=\; d^2 + 2d - 1. \end{aligned}$$

Next we prove $\dim(\mathrm{Eig}_{d,\mathrm{sym}}) = \frac{1}{2}(d^2 + 3d)$. We introduce the linear subspace

$$U := \left\{ \begin{pmatrix} x & y & z \\ \partial\phi/\partial x & \partial\phi/\partial y & \partial\phi/\partial z \end{pmatrix} \,\middle|\, \phi \text{ ternary form of degree } d \right\} \subset W.$$

The action of the group G on W does not restrict to U. In fact, we notice that

$$\begin{pmatrix} 1 & 0 \\ f & a \end{pmatrix} \begin{pmatrix} x & y & z \\ \partial\phi/\partial x & \partial\phi/\partial y & \partial\phi/\partial z \end{pmatrix} \in U$$

if and only if $f = 0$. Let $U^\circ = U \cap W^\circ$ and consider the subgroup

$$H = \left\{ \begin{pmatrix} 1 & 0 \\ 0 & a \end{pmatrix} \,\middle|\, a \in K\backslash\{0\} \right\} \subset G.$$

This yields $\dim(\mathrm{Eig}_{d,\mathrm{sym}}) = \dim U^\circ/H = 3\binom{d+2}{2} - 1 = \frac{1}{2}(d^2+3d)$, as desired. Our configuration spaces Eig_d and $\mathrm{Eig}_{d,\mathrm{sym}}$ are irreducible varieties because they contain the irreducible varieties W°/G and U°/H respectively as dense open subsets. □

6. Real eigenvectors and dynamics

In this section we focus on the real eigenpoints of a tensor A in $(\mathbb{R}^n)^{\otimes d}$. If A is generic then the number of eigenpoints in $\mathbb{P}^{n-1}_{\mathbb{C}}$ equals $((d-1)^n-1)/(d-2)$. Our hope is to show that all of them lie in $\mathbb{P}^{n-1}_{\mathbb{R}}$ for suitably chosen symmetric tensors ϕ. A second question is how many of these real eigenpoints are *robust*, in the sense that they are attracting fixed points of the dynamical system $\nabla\phi : \mathbb{P}^{n-1}_{\mathbb{R}} \dashrightarrow \mathbb{P}^{n-1}_{\mathbb{R}}$. Our results will inform future numerical work along the lines of [**CDN**, Table 4.12].

We begin with a combinatorial construction for the planar case ($n=3$). Consider an arrangement $\mathcal{A}$ of d distinct lines in $\mathbb{P}^2_{\mathbb{R}}$, and let ϕ be the product of d linear forms in x,y,z that define the lines in $\mathcal{A}$. We assume that $\mathcal{A}$ is generic in the sense that no three lines meet in a point. Equivalently, the matroid of $\mathcal{A}$ is a uniform rank 3 matroid on d elements. Such an arrangement $\mathcal{A}$ has $\binom{d}{2}$ *vertices* in $\mathbb{P}^2_{\mathbb{R}}$, and these are the singular points of the reducible curve $\{\phi=0\}$. The complement of $\mathcal{A}$ in $\mathbb{P}^2_{\mathbb{R}}$ has $\binom{d}{2}+1$ connected components, called the *regions* of $\mathcal{A}$.

We are interested in the eigenconfiguration of $\mathcal{A}$, by which we mean the eigenconfiguration of the symmetric tensor ϕ. Theorem 2.1 gives the expected number

$$(6.1)\quad d^2-d+1 \;=\; 1+(d-1)+(d-1)^2 \;=\; 2\binom{d}{2}+1 \;=\; \#\,\text{vertices} + \#\,\text{regions}.$$

The following result shows that this is not just a numerical coincidence.

THEOREM 6.1. *A generic arrangement $\mathcal{A}$ of d lines in $\mathbb{P}^2_{\mathbb{R}}$ has d^2-d+1 complex eigenpoints and they are all real. In addition to the $\binom{d}{2}$ vertices, which are singular eigenpoints, each of the $\binom{d}{2}+1$ regions of $\mathcal{A}$ contains precisely one real eigenpoint.*

PROOF. The singular locus of the curve $\{\phi=0\}$ consists of the vertices of the arrangement $\mathcal{A}$. These are the eigenpoints with eigenvalue 0. Their number is $\binom{d}{2}$.

Let $L_1, L_2, \ldots, L_d$ be the linear forms that define the lines, so $\phi = L_1L_2\cdots L_d$. Consider the following optimization problem on the unit 2-sphere:

$$\text{Maximize } \log|\phi(\mathbf{x})| \;=\; \sum_{i=1}^{d} \log|L_i(x,y,z)| \quad \text{subject to } x^2+y^2+z^2=1.$$

The objective function takes the value $-\infty$ on the d great circles corresponding to $\mathcal{A}$. On each region of $\mathcal{A}$, the objective function takes values in $\mathbb{R}$, and is strictly concave. Hence there exists a unique local maximum $\mathbf{u}^* = (x^*,y^*,z^*)$ in the interior of each region. Such a maximum $\mathbf{u}^*$ is a critical point of the restriction of $\phi(\mathbf{x})$ to the unit 2-sphere. The Lagrange multiplier conditions state that the vector $\mathbf{u}^*$ is parallel to the gradient of ϕ at $\mathbf{u}^*$. This means that $\mathbf{u}^*$ is an eigenvector of ϕ, and hence the pair $\pm\mathbf{u}^*$ defines a real eigenpoint of ϕ in the given region of $\mathbb{P}^2_{\mathbb{R}}$.

We proved that each of the $\binom{d}{2}+1$ regions of $\mathcal{A}$ contains one eigenpoint. In addition, we have the $\binom{d}{2}$ vertices. By Theorem 2.1, the total number of isolated complex eigenpoints cannot exceed $2\binom{d}{2}+1$. This means that there are no eigenpoints in $\mathbb{P}^2_{\mathbb{C}}$ other than those already found. This completes the proof. □

We note that the line arrangement can be perturbed to a situation where the map $\nabla\phi : \mathbb{P}^{n-1}_{\mathbb{R}} \dashrightarrow \mathbb{P}^{n-1}_{\mathbb{R}}$ is regular, i.e. none of the eigenvectors has eigenvalue zero.

COROLLARY 6.2. *There exists a smooth curve of degree d in the real projective plane $\mathbb{P}^2_{\mathbb{R}}$ whose complex eigenconfiguration consists of $d^2 - d + 1$ real points.*

PROOF. The eigenconfiguration of $\phi = L_1 L_2 \cdots L_d$ is 0-dimensional, reduced, and defined over $\mathbb{R}$. By the Implicit Function Theorem, these properties are preserved when ϕ gets perturbed to a generic ternary form ϕ_ϵ that is close to ϕ. □

It is interesting to see what happens when the matroid of $\mathcal{A}$ is not uniform. Here the eigenconfiguration is not reduced. It arises from Theorem 6.1 by degeneration.

EXAMPLE 6.3. Let $d = 6$ and take $\mathcal{A}$ to be the line arrangement defined by

$$\phi \;=\; x \cdot y \cdot z \cdot (x - y) \cdot (x - z) \cdot (y - z).$$

This is the reflection arrangement of type A_4. Its eigenscheme is non-reduced. Each of the 12 regions contains one eigenpoint as before, and the simple vertices $(1:1:0)$, $(1:0:1)$, and $(0:1:1)$ are eigenpoints of multiplicity one. However, each of the triple points $(1:0:0), (0:1:0), (0:0:1), (1:1:1)$ is an eigenpoint of multiplicity 4. This makes sense geometrically: in a nearby generic arrangement, such a vertex splits into three vertices and one new region. We note that the scheme structure at the eigenpoint $(1:0:0)$ is given by the primary ideal $\langle 2yz - z^2, y^2 - z^2 \rangle$.

The concavity argument concerning the optimization problem in the proof of Theorem 6.1 works in arbitrary dimensions, and we record this as a corollary.

COROLLARY 6.4. *Each of the open regions of an arrangement $\mathcal{A}$ of d hyperplanes in $\mathbb{P}^{n-1}_{\mathbb{R}}$ contains precisely one real eigenpoint of $\mathcal{A}$. The number of regions is*

$$\sum_{i=0}^{n-1} \binom{d-1}{i}. \tag{6.2}$$

PROOF. The first part has the same proof as the one for $n = 3$ given above. The formula for the number of regions can be found in [**Sta**, Proposition 2.4]. □

Theorem 6.1 is restricted to $n = 3$ because hyperplane arrangements are singular in codimension 1. Hence the eigenconfiguration of a product of linear forms in $n \geq 4$ variables has components of dimension $n - 3$ in $\mathbb{P}^{n-1}_{\mathbb{R}}$. We conjecture that a fully real eigenconfiguration can be constructed in the vicinity of such a tensor.

CONJECTURE 6.5. *Let ϕ be any product of d nonzero linear forms in $\mathrm{Sym}_1(\mathbb{R}^n)$. Every open neighborhood of ϕ in $\mathrm{Sym}_d(\mathbb{R}^n)$ contains a symmetric tensor ϕ_ϵ such that all $((d-1)^n - 1)/(d-2)$ complex eigenpoints of ϕ_ϵ are real.*

This optimistic conjecture is illustrated by the following variant of Example 1.1.

EXAMPLE 6.6 ($n = d = 4$). The classical Cremona transformation in $\mathbb{P}^3$ is $\nabla\phi$ where $\phi = xyzw$ is the product of the coordinates. The eigenconfiguration of ϕ consists of eight points, one for each sign region in $\mathbb{P}^3$, and the six coordinate lines. The expected number (2.1) of complex eigenpoints is 40. Consider the perturbation

$$\begin{aligned}\phi_\epsilon \;=\; xyzw \,+\, \epsilon\big(&5x^4 + 4x^3y - 2x^2y^2 - 8xy^3 + 7y^4 + 4x^3z + 2x^2yz + 2xy^2z\\ &+2y^3z - 6x^2z^2 + 6xyz^2 + 7y^2z^2 - 8xz^3 + 3yz^3 + 8z^4 - 8x^3w + 2x^2yw\\ &-3xy^2w + 5y^3w + 8x^2zw - 3y^2zw - 5xz^2w - 10yz^2w + 8z^3w - 5x^2w^2\\ &-6xyw^2 - 3y^2w^2 - 6xzw^2 + 3yzw^2 + 3xw^3 + 3yw^3 - 4zw^3 + 3w^4\big).\end{aligned}$$

All 40 complex eigenpoints of this tensor are real, so Conjecture 6.5 holds for ϕ.

REMARK 6.7. Conjecture 6.5 is true for $n \leq 3$. For $n = 3$ this follows from Theorem 6.1: we can take ϕ_ϵ to be any perturbation of the given line arrangement ϕ. For $n = 2$ we take $\phi_\epsilon = \phi$ itself because of the following fact: if a binary form $\phi(x, y)$ is real-rooted then also $x\frac{\partial \phi}{\partial y} - y\frac{\partial \phi}{\partial x}$ is real-rooted. This follows from Corollary 6.4.

We put the lid on this paper with a brief discussion of the dynamical system $\psi : \mathbb{P}^{n-1}_{\mathbb{R}} \dashrightarrow \mathbb{P}^{n-1}_{\mathbb{R}}$ associated with a tensor $A \in (\mathbb{R}^n)^{\otimes d}$. Iterating this map is known as the *tensor power method*, and it is used as a tool in tensor decomposition [**AG**]. This generalizes the power method of numerical linear algebra for computing the eigenvectors of a matrix $A \in (\mathbb{R}^n)^{\otimes 2}$. One starts with some unit vector $\mathbf{v}$ and repeatedly applies the map $\mathbf{v} \mapsto \frac{A\mathbf{v}}{||A\mathbf{v}||}$. For generic inputs A and $\mathbf{v}$, this iteration converges to the eigenvector corresponding to the largest absolute eigenvalue.

Suppose that $\mathbf{u} \in \mathbb{P}^{n-1}_{\mathbb{R}}$ is an eigenpoint of a given tensor $A \in (\mathbb{R}^n)^{\otimes d}$. We say that $\mathbf{u}$ is a *robust eigenpoint* if there exists an open neighborhood $\mathcal{U}$ of $\mathbf{u}$ in $\mathbb{P}^{n-1}_{\mathbb{R}}$ such that, for all starting vectors $\mathbf{v} \in \mathcal{U}$, the iteration of the map ψ converges to $\mathbf{u}$.

EXAMPLE 6.8 (Odeco Tensors). A symmetric tensor is *orthogonally decomposable* (this was abbreviated to *odeco* by Robeva [**Rob**]) if it has the form

$$\phi \;=\; \sum_{i=1}^{n} a_i \mathbf{v}_i^{\otimes d}$$

where $a_1, \ldots, a_n \in \mathbb{R}$ and $\{\mathbf{v}_1, \ldots, \mathbf{v}_n\}$ is an orthogonal basis of $\mathbb{R}^n$. Following [**AG**], the robust eigenpoints of an odeco tensor are the basis vectors $\mathbf{v}_i$, and they can be computed using the tensor power method. Up to an appropriate change of coordinates, the odeco tensors are the Fermat polynomials in (2.4). The robust eigenpoints of $\phi = x_1^d + \cdots + x_n^d$ are the coordinate points $\mathbf{e}_1, \ldots, \mathbf{e}_n$. The region of attraction of the ith eigenpoint $\mathbf{e}_i$ under the iteration of the map $\nabla\phi$ is the set of all points in $\mathbb{R}^{n-1}_{\mathbb{R}}$ whose i^{th} coordinate is largest in absolute value.

Odeco tensors for $n = d = 3$ have three robust eigenvalues. At present we do not know any ternary cubic ϕ with more than three robust eigenpoints. Theorem 6.1 might suggest that products of linear forms are good candidates. However, we ran experiments with random triples of lines in $\mathbb{P}^2_{\mathbb{R}}$, and we observed that the number of robust eigenvalues is usually one and occasionally zero. We never found a factorizable ternary cubic ϕ with two or more robust eigenpoints. The Cremona map in Example 1.1 shows that $\phi = xyz$ is a cubic with zero robust eigenpoints. Here is a similar example that points to the connection with frame theory in [**ORS**].

EXAMPLE 6.9 ($n = d = 3$). We consider the factorizable ternary cubic

$$\phi \;=\; (2x + 2y - z)(2x - y + 2z)(-x + 2y + 2z). \tag{6.3}$$

This equals the frame decomposable tensor seen in [**ORS**, Examples 1.1 and 5.2]:

$$\phi \;=\; \frac{1}{24}\big((-5x+y+z)^3 + (x-5y+z)^3 + (x+y-5z)^3 + (3x+3y+3z)^3\big). \tag{6.4}$$

Its gradient map $\mathbb{P}^2 \dashrightarrow \mathbb{P}^2$ is given by

$$\nabla\phi \;=\; 3\begin{pmatrix} -4x^2 + 4xy + 4xz + 2y^2 + yz + 2z^2 \\ 2x^2 + 4xy + xz - 4y^2 + 4yz + 2z^2 \\ 2x^2 + xy + 4xz + 2y^2 + 4yz - 4z^2 \end{pmatrix}.$$

This has four fixed points and three singular points, for a total of seven eigenpoints:

$$(1:1:-5),\ (1:-5:1),\ (-5:1:1),\ (3:3:3),$$
$$(2:2:-1),\ (2:-1:2),\ (-1:2:2).$$

Note that the pairwise intersections of the lines coincide with the coefficient vectors in (6.3). By plugging $\nabla\phi$ into itself, we verify that the second iterate map equals

$$(\nabla\phi)^{\circ 2} \;=\; \nabla\phi \circ \nabla\phi \;=\; -3^6\phi(x,y,z)\cdot \begin{pmatrix} x & y & z \end{pmatrix}^T.$$

Hence $(\nabla\phi)^{\circ 2}$ is the identity map on all points in $\mathbb{P}^2_{\mathbb{R}}\backslash\{\phi = 0\}$. Every such point lies in a limit cycle of length two. The points on the curve $\{\phi = 0\}$ map to the singular points. We conclude that the ternary cubic ϕ has no robust eigenpoints.

References

[AE] Hirotachi Abo, David Eklund, Thomas Kahle, and Chris Peterson, *Eigenschemes and the Jordan canonical form*, Linear Algebra Appl. **496** (2016), 121–151, DOI 10.1016/j.laa.2015.12.030. MR3464066

[AST] Chris Aholt, Bernd Sturmfels, and Rekha Thomas, *A Hilbert scheme in computer vision*, Canad. J. Math. **65** (2013), no. 5, 961–988, DOI 10.4153/CJM-2012-023-2. MR3095002

[AG] Animashree Anandkumar, Rong Ge, Daniel Hsu, Sham M. Kakade, and Matus Telgarsky, *Tensor decompositions for learning latent variable models*, J. Mach. Learn. Res. **15** (2014), 2773–2832. MR3270750

[BPT] *Semidefinite optimization and convex algebraic geometry*, MOS-SIAM Series on Optimization, vol. 13, Society for Industrial and Applied Mathematics (SIAM), Philadelphia, PA; Mathematical Optimization Society, Philadelphia, PA, 2013. Edited by Grigoriy Blekherman, Pablo A. Parrilo and Rekha R. Thomas. MR3075433

[CS] Dustin Cartwright and Bernd Sturmfels, *The number of eigenvalues of a tensor*, Linear Algebra Appl. **438** (2013), no. 2, 942–952, DOI 10.1016/j.laa.2011.05.040. MR2996375

[CQZ] Kungching Chang, Liqun Qi, and Tan Zhang, *A survey on the spectral theory of nonnegative tensors*, Numer. Linear Algebra Appl. **20** (2013), no. 6, 891–912, DOI 10.1002/nla.1902. MR3141883

[CDN] Chun-Feng Cui, Yu-Hong Dai, and Jiawang Nie, *All real eigenvalues of symmetric tensors*, SIAM J. Matrix Anal. Appl. **35** (2014), no. 4, 1582–1601, DOI 10.1137/140962292. MR3286682

[Ein] Lawrence Ein, *Some stable vector bundles on* $\mathbf{P}^4$ *and* $\mathbf{P}^5$, J. Reine Angew. Math. **337** (1982), 142–153, DOI 10.1515/crll.1982.337.142. MR676047

[Eis] David Eisenbud, *The geometry of syzygies*, Graduate Texts in Mathematics, vol. 229, Springer-Verlag, New York, 2005. A second course in commutative algebra and algebraic geometry. MR2103875

[FS] John Erik Fornæss and Nessim Sibony, *Complex dynamics in higher dimension. I*, Astérisque **222** (1994), 5, 201–231. Complex analytic methods in dynamical systems (Rio de Janeiro, 1992). MR1285389

[Fu1] William Fulton, *Intersection theory*, Ergebnisse der Mathematik und ihrer Grenzgebiete (3) [Results in Mathematics and Related Areas (3)], vol. 2, Springer-Verlag, Berlin, 1984. MR732620

[Fu2] William Fulton, *Algebraic curves. An introduction to algebraic geometry*, W. A. Benjamin, Inc., New York-Amsterdam, 1969. Notes written with the collaboration of Richard Weiss; Mathematics Lecture Notes Series. MR0313252

[GKP] Ronald L. Graham, Donald E. Knuth, and Oren Patashnik, *Concrete mathematics*, 2nd ed., Addison-Wesley Publishing Company, Reading, MA, 1994. A foundation for computer science. MR1397498

[LQY] Guoyin Li, Liqun Qi, and Gaohang Yu, *The Z-eigenvalues of a symmetric tensor and its application to spectral hypergraph theory*, Numer. Linear Algebra Appl. **20** (2013), no. 6, 1001–1029, DOI 10.1002/nla.1877. MR3141890

[Lim] L. H. Lim: Singular values and eigenvalues of tensors: a variational approach, Proceedings of the IEEE International Workshop on *Computational Advances in Multi-Sensor Adaptive Processing* (CAMSAP '05), **1** (2005), pp. 129–132.

[M] M. Maccioni,"Tensor rank and eigenvectors" (PhD diss., in progress), University of Florence, Italy, 2017.

[ORS] Luke Oeding, Elina Robeva, and Bernd Sturmfels, *Decomposing tensors into frames*, Adv. in Appl. Math. **73** (2016), 125–153, DOI 10.1016/j.aam.2015.10.004. MR3433503

[OO] Luke Oeding and Giorgio Ottaviani, *Eigenvectors of tensors and algorithms for Waring decomposition*, J. Symbolic Comput. **54** (2013), 9–35, DOI 10.1016/j.jsc.2012.11.005. MR3032635

[OS] Giorgio Ottaviani and Edoardo Sernesi, *On the hypersurface of Lüroth quartics*, Michigan Math. J. **59** (2010), no. 2, 365–394, DOI 10.1307/mmj/1281531462. MR2677627

[SOS] A. Papachristodoulou, J. Anderson, G. Valmorbida, S. Prajna, P. Seiler and P. Parrilo: SOSTOOLS: Sum of squares optimization toolbox for MATLAB, software available from `http://www.mit.edu/~parrilo/sostools`, 2013.

[PWZ] Marko Petkovšek, Herbert S. Wilf, and Doron Zeilberger, $A = B$, A K Peters, Ltd., Wellesley, MA, 1996. With a foreword by Donald E. Knuth; With a separately available computer disk. MR1379802

[Rob] E. Robeva: Orthogonal decomposition of symmetric tensors, *SIAM J. Matrix Anal. Appl.* **37** (2016), 86–102.

[Sta] Richard P. Stanley, *An introduction to hyperplane arrangements*, Geometric combinatorics, IAS/Park City Math. Ser., vol. 13, Amer. Math. Soc., Providence, RI, 2007, pp. 389–496. MR2383131

[Stu] Bernd Sturmfels, *Solving systems of polynomial equations*, CBMS Regional Conference Series in Mathematics, vol. 97, Published for the Conference Board of the Mathematical Sciences, Washington, DC; by the American Mathematical Society, Providence, RI, 2002. MR1925796

[Qi] Liqun Qi, *Eigenvalues of a real supersymmetric tensor*, J. Symbolic Comput. **40** (2005), no. 6, 1302–1324, DOI 10.1016/j.jsc.2005.05.007. MR2178089

Department of Mathematics, University of Idaho, Moscow, Idaho 83844
E-mail address: `abo@uidaho.edu`

Department of Mathematics, University of California, Berkeley, California 94720
E-mail address: `seigal@berkeley.edu`

Department of Mathematics, University of California, Berkeley, California 94720
E-mail address: `bernd@berkeley.edu`

Contemporary Mathematics
Volume **685**, 2017
http://dx.doi.org/10.1090/conm/685/13712

Sum of squares basis pursuit with linear and second order cone programming

Amir Ali Ahmadi and Georgina Hall

Abstract. We devise a scheme for solving an iterative sequence of linear programs (LPs) or second order cone programs (SOCPs) to approximate the optimal value of any semidefinite program (SDP) or sum of squares (SOS) program. The first LP and SOCP-based bounds in the sequence come from the recent work of Ahmadi and Majumdar on diagonally dominant sum of squares (DSOS) and scaled diagonally dominant sum of squares (SDSOS) polynomials. We then iteratively improve on these bounds by pursuing better bases in which more relevant SOS polynomials admit a DSOS or SDSOS representation. Different interpretations of the procedure from primal and dual perspectives are given. While the approach is applicable to SDP relaxations of general polynomial programs, we apply it to two problems of discrete optimization: the maximum independent set problem and the partition problem. We further show that some completely trivial instances of the partition problem lead to strictly positive polynomials on the boundary of the sum of squares cone and hence make the SOS relaxation fail.

1. Introduction

In recent years, semidefinite programming [**37**] and sum of squares optimization [**22, 31, 33**] have proven to be powerful techniques for tackling a diverse set of problems in applied and computational mathematics. The reason for this, at a high level, is that several fundamental problems arising in discrete and polynomial optimization [**7, 17, 23**] or the theory of dynamical systems [**3, 19, 32**] can be cast as linear optimization problems over the cone of nonnegative polynomials. This observation puts forward the need for efficient conditions on the coefficients $c_\alpha := c_{\alpha_1,\ldots,\alpha_n}$ of a multivariate polynomial

$$p(x) = \sum_\alpha c_{\alpha_1,\ldots,\alpha_n} x_1^{\alpha_1} \ldots x_n^{\alpha_n}$$

that ensure the inequality $p(x) \geq 0$, for all $x \in \mathbb{R}^n$. If p is a quadratic function, $p(x) = x^T Q x + 2c^T x + b$, then nonnegativity of p is equivalent to the $(n+1) \times (n+1)$ symmetric matrix

$$\begin{pmatrix} Q & c \\ c^T & b \end{pmatrix}$$

2010 *Mathematics Subject Classification.* Primary 90C22, 90C06, 90C05.
The authors are partially supported by the AFOSR Young Investigator Program Award and the NSF CAREER Award.

being positive semidefinite and this constraint can be imposed by semidefinite programming. For higher degrees, however, imposing nonnegativity of polynomials is in general an intractable computational task. In fact, even checking if a given quartic polynomial is nonnegative is NP-hard [**13**]. A particularly popular and seemingly powerful sufficient condition for a polynomial p to be nonnegative is for it to decompose as a sum of squares of other polynomials:

$$p(x) = \sum_i q_i^2(x).$$

This condition is attractive for several reasons. From a computational perspective, for fixed-degree polynomials, a sum of squares decomposition can be checked (or imposed as a constraint) by solving a semidefinite program of size polynomial in the number of variables. From a representational perspective, such a decomposition *certifies* nonnegativity of p in terms of an easily verifiable algebraic identity. From a practical perspective, the so-called "sum of squares relaxation" is well-known to produce powerful (often exact) bounds on optimization problems that involve nonnegative polynomials; see, e.g., [**34**]. The reason for this is that constructing examples of nonnegative polynomials that are not sums of squares in relatively low dimensions and degrees seems to be a difficult task[1], especially when additional structure arising from applications is required.

We have recently been interested in leveraging the attractive features of semidefinite programs (SDPs) and sum of squares (SOS) programs, while solving much simpler classes of convex optimization problems, namely *linear programs* (LPs) and *second order cone programs* (SOCPs). Such a research direction can potentially lead to a better understanding of the relative power of different classes of convex relaxations. It also has obvious practical motivations as simpler convex programs come with algorithms that have better scalability and improved numerical conditioning properties. This paper is a step in this research direction. We present a scheme for solving a sequence of LPs or SOCPs that provide increasingly accurate approximations to the optimal value and the optimal solution of a semidefinite (or a sum of squares) program. With the algorithms that we propose, one can use one of many mature LP/SOCP solvers such as [**1, 2, 11**], including simplex-based LP solvers, to obtain reasonable approximations to the optimal values of these more difficult convex optimization problems.

The intuition behind our approach is easy to describe with a contrived example. Suppose we would like to show that the degree-4 polynomial

$$\begin{aligned} p(x) \;=\; & x_1^4 - 6x_1^3x_2 + 2x_1^3x_3 + 6x_1^2x_3^2 + 9x_1^2x_2^2 - 6x_1^2x_2x_3 - 14x_1x_2x_3^2 + 4x_1x_3^3 \\ & +5x_3^4 - 7x_2^2x_3^2 + 16x_2^4 \end{aligned}$$

has a sum of squares decomposition. One way to do this is to attempt to write p as

$$p(x) = z^T(x)Qz(x),$$

where

$$z(x) = (x_1^2, x_1x_2, x_2^2, x_1x_3, x_2x_3, x_3^2)^T \tag{1.1}$$

is the standard (homogeneous) monomial basis of degree 2 and the matrix Q, often called the *Gram matrix*, is symmetric and positive semidefinite. The search for

[1]See [**36**] for explicit examples of nonnegative polynomials that are not sums of squares.

such a Q can be done with semidefinite programming; one feasible solution e.g. is as follows:

$$Q = \begin{pmatrix} 1 & -3 & 0 & 1 & 0 & 2 \\ -3 & 9 & 0 & -3 & 0 & -6 \\ 0 & 0 & 16 & 0 & 0 & -4 \\ 1 & -3 & 0 & 2 & -1 & 2 \\ 0 & 0 & 0 & -1 & 1 & 0 \\ 2 & -6 & 4 & 2 & 0 & 5 \end{pmatrix}.$$

Suppose now that instead of the basis z in (1.1), we pick a different basis

$$\tilde{z}(x) = (2x_1^2 - 6x_1x_2 + 2x_1x_3 + 2x_3^2, \quad x_1x_3 - x_2x_3, \quad x_2^2 - \frac{1}{4}x_3^2)^T. \tag{1.2}$$

With this new basis, we can get a sum of squares decomposition of p by writing it as

$$p(x) = \tilde{z}^T(x) \begin{pmatrix} \frac{1}{2} & 0 & 0 \\ 0 & 1 & 0 \\ 0 & 0 & 4. \end{pmatrix} \tilde{z}(x).$$

In effect, by using a better basis, we have simplified the Gram matrix and made it diagonal. When the Gram matrix is diagonal, its positive semidefiniteness can be imposed as a *linear* constraint (diagonals should be nonnegative).

Of course, the catch here is that we do not have access to the magic basis $\tilde{z}(x)$ in (1.2) a priori. Our goal will hence be to "pursue" this basis (or other good bases) by starting with an arbitrary basis (typically the standard monomial basis), and then iteratively improving it by solving a sequence of LPs or SOCPs and performing some efficient matrix decomposition tasks in the process. Unlike the intentionally simplified example we gave above, we will not ever require our Gram matrices to be diagonal. This requirement is too strong and would frequently lead to our LPs and SOCPs being infeasible. The underlying reason for this is that the cone of diagonal matrices is not full dimensional in the cone of positive semidefinite matrices. Instead, we will be after bases that allow the Gram matrix to be *diagonally dominant* or *scaled diagonally dominant* (see Definition 2.1). The use of these matrices in polynomial optimization has recently been proposed by Ahmadi and Majumdar [**5**, **6**]. We will be building on and improving upon their results in this paper.

1.1. Organization of this paper. The organization of the rest of the paper is as follows. In Section 2, we introduce some notation and briefly review the concepts of "dsos and sdsos polynomials" which are used later as the first step of an iterative algorithm that we propose in Section 3. In this section, we explain how we inner approximate semidefinite (Subsection 3.1) and sum of squares (Subsection 3.2) cones with LP and SOCP-based cones by iteratively changing bases. In Subsection 3.3, we give a different interpretation of our LPs in terms of their corner description as opposed to their facet description. Subsection 3.4 is about duality, which is useful for iteratively outer approximating semidefinite or sum of squares cones.

In Section 4, we apply our algorithms to the Lovász semidefinite relaxation of the maximum stable set problem. It is shown numerically that our LPs and SOCPs converge to the SDP optimal value in very few iterations and outperform some other well-known LP relaxations on a family of randomly generated examples. In Section 5, we consider the partition problem from discrete optimization. As opposed to the stable set problem, the quality of our relaxations here is rather poor. In fact, even the sum of squares relaxation fails on some completely trivial instances. We show this empirically on random instances, and formally prove it on one representative example (Subsection 5.1). The reason for this failure is existence of a certain family of quartic polynomials that are nonnegative but not sums of squares.

2. Preliminaries

We denote the set of real symmetric $n \times n$ matrices by S_n. Given two matrices A and B in S_n, their standard matrix inner product is denoted by $A \cdot B := \sum_{i,j} A_{ij}B_{ij} = \text{Trace}(AB)$. A symmetric matrix A is *positive semidefinite* (psd) if $x^T Ax \geq 0$ for all $x \in \mathbb{R}^n$; this will be denoted by the standard notation $A \succeq 0$, and our notation for the set of $n \times n$ psd matrices is P_n. We say that A is *positive definite* (pd) if $x^T Ax > 0$ for all $x \neq 0$. Any psd matrix A has an upper triangular Cholesky factor $U = \text{chol}(A)$ satisfying $A = U^T U$. When A is pd, the Cholesky factor is unique and has positive diagonal entries. For a cone of matrices in S_n, we define its dual cone $\mathcal{K}^*$ as $\{Y \in S_n : Y \cdot X \geq 0, \ \forall X \in \mathcal{K}\}$.

For a vector variable $x \in \mathbb{R}^n$ and a vector $s \in \mathbb{Z}^n_+$, let a monomial in x be denoted as $x^s = \Pi_{i=1}^n x_i^{s_i}$ which by definition has degree $\sum_{i=1}^n s_i$. A polynomial is said to be *homogeneous* or a *form* if all of its monomials have the same degree. A form $p(x)$ in n variables is nonnegative if $p(x) \geq 0$ for all $x \in \mathbb{R}^n$, or equivalently for all x on the unit sphere in $\mathbb{R}^n$. The set of nonnegative (or positive semidefinite) forms in n variables and degree d is denoted by $PSD_{n,d}$. A form $p(x)$ is a *sum of squares* (sos) if it can be written as $p(x) = \sum_{i=1}^r q_i^2(x)$ for some forms $q_1, \ldots, q_r$. The set of sos forms in n variables and degree d is denoted by $SOS_{n,d}$. We have the obvious inclusion $SOS_{n,d} \subseteq PSD_{n,d}$, which is strict unless $d = 2$, or $n = 2$, or $(n, d) = (3, 4)$ [**20**]. Let $z(x, d)$ be the vector of all monomials of degree exactly d; it is well known that a form p of degree $2d$ is sos if and only if it can be written as $p(x) = z^T(x, d)Qz(x, d)$, for some psd matrix Q [**32, 33**]. An SOS optimization problem is the problem of minimizing a linear function over the intersection of the convex cone $SOS_{n,d}$ with an affine subspace. The previous statement implies that SOS optimization problems can be cast as semidefinite programs.

2.1. DSOS and SDSOS optimization.

In recent work, Ahmadi and Majumdar introduce more scalable alternatives to SOS optimization that they refer to as *DSOS and SDSOS programs* [**5, 6**][2]. Instead of semidefinite programming, these optimization problems can be cast as linear and second order cone programs respectively. Since we will be building on these concepts, we briefly review their relevant aspects to make our paper self-contained.

The idea in [**5, 6**] is to replace the condition that the Gram matrix Q be positive semidefinite with stronger but cheaper conditions in the hope of obtaining more efficient inner approximations to the cone $SOS_{n,d}$. Two such conditions come from the concepts of *diagonally dominant* and *scaled diagonally dominant* matrices in linear algebra. We recall these definitions below.

[2] The work in [**6**] is currently in preparation for submission; the one in [**5**] is a shorter conference version of [**6**] which has already appeared. The presentation of the current paper is meant to be self-contained.

DEFINITION 2.1. A symmetric matrix A is *diagonally dominant* (dd) if $a_{ii} \geq \sum_{j \neq i} |a_{ij}|$ for all i. We say that A is *scaled diagonally dominant* (sdd) if there exists a diagonal matrix D, with positive diagonal entries, which makes DAD diagonally dominant.

We refer to the set of $n \times n$ dd (resp. sdd) matrices as DD_n (resp. SDD_n). The following inclusions are a consequence of Gershgorin's circle theorem [**15**]:

$$DD_n \subseteq SDD_n \subseteq P_n.$$

Whenever it is clear from the context, we may drop the subscript n from our notation. We now use these matrices to introduce the cones of "dsos" and "sdsos" forms which constitute special subsets of the cone of sos forms. We remark that in the interest of brevity, we do not give the original definition of dsos and sdsos polynomials as it appears in [**6**] (as sos polynomials of a particular structure), but rather an equivalent characterization of them that is more useful for our purposes. The equivalence is proven in [**6**].

DEFINITION 2.2 ([**5**,**6**]). Recall that $z(x, d)$ denotes the vector of all monomials of degree exactly d. A form $p(x)$ of degree $2d$ is said to be

- *diagonally-dominant-sum-of-squares* (dsos) if it admits a representation as $p(x) = z^T(x, d)Qz(x, d)$, where Q is a dd matrix.
- *scaled-diagonally-dominant-sum-of-squares* (sdsos) if it admits a representation as $p(x) = z^T(x, d)Qz(x, d)$, where Q is an sdd matrix.

The definitions for non-homogeneous polynomials are exactly the same, except that we replace the vector of monomials of degree exactly d with the vector of monomials of degree $\leq d$. We observe that a quadratic form $x^T Qx$ is dsos/sdsos/sos if and only if the matrix Q is dd/sdd/psd. Let us denote the cone of forms in n variables and degree d that are dsos and sdsos by $DSOS_{n,d}$, $SDSOS_{n,d}$. The following inclusion relations are straightforward:

$$DSOS_{n,d} \subseteq SDSOS_{n,d} \subseteq SOS_{n,d} \subseteq PSD_{n,d}.$$

From the point of view of optimization, our interest in all of these algebraic notions stems from the following theorem.

THEOREM 2.3 ([**5**,**6**]). *For any fixed d, optimization over the cones $DSOS_{n,d}$ (resp. $SDSOS_{n,d}$) can be done with linear programming (resp. second order cone programming) of size polynomial in n.*

The "LP part" of this theorem is not hard to see. The equality $p(x) = z^T(x, d)Qz(x, d)$ gives rise to linear equality constraints between the coefficients of p and the entries of the matrix Q (whose size is $\sim n^{\frac{d}{2}} \times n^{\frac{d}{2}}$ and hence polynomial in n for fixed d). The requirement of diagonal dominance on the matrix Q can also be described by linear inequality constraints on Q. The "SOCP part" of the statement comes from the fact, shown in [**6**], that a matrix A is sdd if and only if it can be expressed as

$$A = \sum_{i<j} M_{2\times 2}^{ij}, \tag{2.1}$$

where each $M_{2\times 2}^{ij}$ is an $n \times n$ symmetric matrix with zeros everywhere except for four entries $M_{ii}, M_{ij}, M_{ji}, M_{jj}$, which must make the 2×2 matrix $\begin{bmatrix} M_{ii} & M_{ij} \\ M_{ji} & M_{jj} \end{bmatrix}$

symmetric and positive semidefinite. These constraints are *rotated quadratic cone* constraints and can be imposed using SOCP [**8, 25**]:

$$M_{ii} \geq 0, \quad \left|\left| \begin{pmatrix} 2M_{ij} \\ M_{ii} - M_{jj} \end{pmatrix} \right|\right| \leq M_{ii} + M_{jj}.$$

We refer to linear optimization problems over the convex cones $DSOS_{n,d}$, $SDSOS_{n,d}$, and $SOS_{n,d}$ as DSOS programs, SDSOS programs, and SOS programs respectively. In general, quality of approximation decreases, while scalability increases, as we go from SOS to SDSOS to DSOS programs. What we present next can be thought of as an iterative procedure for moving from DSOS/SDSOS relaxations towards SOS relaxations without increasing the problem size in each step.

3. Pursuing improved bases

Throughout this section, we consider the standard SDP

$$\begin{aligned} SOS^* := \min_{X \in S_n} \; & C \cdot X \\ \text{s.t. } & A_i \cdot X = b_i, i = 1, \dots, m, \\ & X \succeq 0, \end{aligned} \tag{3.1}$$

which we assume to have an optimal solution. We denote the optimal value by SOS^* since we think of a semidefinite program as a sum of squares program over quadratic forms (recall that $PSD_{n,2} = SOS_{n,2}$). This is so we do not have to introduce additional notation to distinguish between degree-2 and higher degree SOS programs. The main goal of this section is to construct sequences of LPs and SOCPs that generate bounds on the optimal value of (3.1). Section 3.1 focuses on providing upper bounds on (3.1) while Section 3.4 focuses on lower bounds.

3.1. Inner approximations of the psd cone. To obtain upper bounds on (3.1), we need to replace the constraint $X \succeq 0$ by a stronger condition. In other words, we need to provide *inner approximations* to the set of psd matrices.

First, let us define a family of cones

$$DD(U) := \{M \in S_n \mid M = U^T Q U \text{ for some dd matrix } Q\},$$

parametrized by an $n \times n$ matrix U. Optimizing over the set $DD(U)$ is an LP since U is fixed, and the defining constraints are linear in the coefficients of the two unknowns M and Q. Furthermore, the matrices in $DD(U)$ are all psd; i.e., $\forall U$, $DD(U) \subseteq P_n$.

The iteration number k in the sequence of our LPs consists of replacing the condition $X \succeq 0$ by $X \in DD(U_k)$:

$$\begin{aligned} DSOS_k := \min \; & C \cdot X \\ \text{s.t. } & A_i \cdot X = b_i, \; i = 1, \dots, m, \\ & X \in DD(U_k). \end{aligned} \tag{3.2}$$

To define the sequence $\{U_k\}$, we assume that an optimal solution X_k to (3.2) exists for every iteration. As it will become clear shortly, this assumption will be implied simply by assuming that only the first LP in the sequence is feasible. The sequence $\{U_k\}$ is then given recursively by

$$\begin{aligned} U_0 &= I \\ U_{k+1} &= \text{chol}(X_k). \end{aligned} \tag{3.3}$$

Note that the first LP in the sequence optimizes over the set of diagonally dominant matrices as in the work of Ahmadi and Majumdar [**5, 6**]. By defining U_{k+1} as a Cholesky factor of X_k, improvement of the optimal value is guaranteed in each iteration. Indeed, as $X_k = U_{k+1}^T I U_{k+1}$, and the identity matrix I is diagonally dominant, we see that $X_k \in DD(U_{k+1})$ and hence is feasible for iteration $k+1$. This entails that the optimal value at iteration $k+1$ is at least as good as the optimal value at the previous iteration; i.e., $DSOS_{k+1} \leq DSOS_k$. Since the sequence $\{DSOS_k\}$ is lower bounded by SOS^* and monotonic, it must converge to a limit $DSOS^* \geq SOS^*$. We have been unable to formally rule out the possibility that $DSOS^* > SOS^*$. In all of our numerical experiments, convergence to SOS^* happens (i.e., $DSOS^* = SOS^*$), though the speed of convergence seems to be problem dependent (contrast e.g. the results of Section 4 with Section 5). What is easy to show, however, is that if X_k is positive definite[3], then the improvement from step k to $k+1$ is actually *strict.*

THEOREM 3.1. *Let X_k (resp. X_{k+1}) be an optimal solution of iterate k (resp. $k+1$) of (3.2) and assume that X_k is pd and $SOS^* < DSOS_k$. Then,*

$$DSOS_{k+1} < DSOS_k.$$

PROOF. We show that for some $\lambda \in (0,1)$, the matrix $\hat{X} := (1-\lambda)X_k + \lambda X^*$ is feasible to the LP in iteration number $k+1$. We would then have that

$$DSOS_{k+1} \leq C \cdot \hat{X} = (1-\lambda) C \cdot X_k + \lambda C \cdot X^* < DSOS_k,$$

as we have assumed that $C \cdot X^* = SOS^* < DSOS_k = C \cdot X_k$. To show feasibility of $\hat{X}$ to LP number $k+1$, note first that as both X_k and X^* satisfy the affine constraints $A_i \cdot X = b_i$, then $\hat{X}$ must also. Since $X_k = U_{k+1}^T U_{k+1}$ and X_k is pd, U_{k+1} must have positive diagonal entries and is invertible. Let

$$X_{k+1}^* := U_{k+1}^{-T} X^* U_{k+1}^{-1}.$$

For λ small enough the matrix $(1-\lambda)I + \lambda X_{k+1}^*$ will be dd since we know the identity matrix is strictly diagonally dominant. Hence, the matrix

$$\hat{X} = U_{k+1}^T((1-\lambda)I + \lambda X_{k+1}^*)U_{k+1}$$

is feasible to LP number $k+1$. □

A few remarks are in order. First, instead of the Cholesky decomposition, we could have worked with some other decompositions such as the LDL decomposition $X_k = LDL^T$ or the spectral decomposition $X_k = H^T \Lambda H$ (where H has the eigenvectors of X_k as columns). Aside from the efficiency of the Cholesky decomposition, the reason we made this choice is that the decomposition allows us to write X_k as $U^T I U$ and the identity matrix I is at the analytic center of the set of diagonally dominant matrices [**10**, Section 8.5.3]. Second, the reader should see that feasibility of the first LP implies that all future LPs are feasible and lower bounded. While in most applications that we know of the first LP is automatically feasible (see, e.g., the stable set problem in Section 4), sometimes the problem needs to be modified to make this the case. An example where this happens appears in Section 5

[3]This would be the case whenever our inner approximation is not touching the boundary of the psd cone in the direction of the objective. As far as numerical computation is concerned, this is of course always the case.

(see Theorem 5.4), where we apply an SOS relaxation to the partition problem. Alternatively, one can first apply our iterative procedure to a Phase-I problem

$$\begin{aligned} \alpha_k := & \min \alpha \\ & \text{s.t. } A_i \cdot X = b_i, i = 1, \dots, m \\ & X + \alpha I \in DD(U_k), \end{aligned} \tag{3.4}$$

with U_k defined as in (3.3). Indeed, for α large enough, the initial problem in (3.4) (i.e., with $U_0 = I$) is feasible. Thus all subsequent iterations are feasible and continually decrease α. If for some iteration k we get $\alpha_k \leq 0$, then we can start the original LP sequence (3.2) with the matrix U_k obtained from the last iteration of the Phase-I algorithm.

In an analogous fashion, we can construct a sequence of SOCPs that provide upper bounds on SOS^*. This time, we define a family of cones

$$SDD(U) := \{M \in S_n \mid M = U^T Q U, \text{ for some sdd matrix } Q\},$$

parameterized again by an $n \times n$ matrix U. For any U, optimizing over the set $SDD(U)$ is an SOCP and we have $SDD(U) \subseteq P_n$. This leads us to the following iterative SOCP sequence:

$$\begin{aligned} SDSOS_k := & \min C \cdot X \\ & \text{s.t. } A_i \cdot X = b_i, i = 1, \dots, m, \\ & X \in SDD(U_k). \end{aligned} \tag{3.5}$$

Assuming existence of an optimal solution X_k at each iteration, we can once again define the sequence $\{U_k\}$ iteratively as

$$\begin{aligned} U_0 &= I \\ U_{k+1} &= \text{chol}(X_k). \end{aligned}$$

The previous statements concerning strict improvement of the LP sequence as described in Theorem 3.1, as well as its convergence carry through for the SOCP sequence. In our experience, our SOCP bounds converge to the SDP optimal value often faster than our LP bounds do. While it is always true that $SDSOS_0 \leq DSOS_0$ (as $DD \subseteq SDD$), the inequality can occasionally reverse in future iterations.

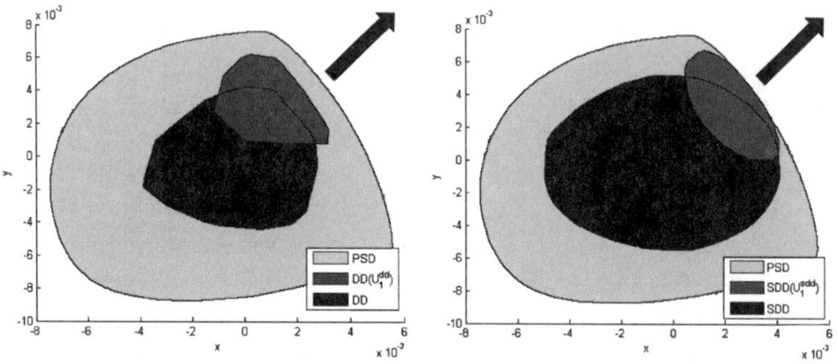

(a) LP inner approximations (b) SOCP inner approximations

FIGURE 1. Improvement after one Cholesky decomposition when maximizing the objective function $x + y$

An illustration of both procedures is given in Figure 1. We generated two random symmetric matrices A and B of size 10×10. The outermost set is the feasible set of an SDP with the constraint $I + xA + yB \succeq 0$. The goal is to maximize the function $x + y$ over this set. The set labeled DD in Figure 1(a) (resp. SDD in Figure 1(b)) consists of the points (x, y) for which $I + xA + yB$ is dd (resp. sdd). Let (x^*_{dd}, y^*_{dd}) (resp. (x^*_{sdd}, y^*_{sdd})) be optimal solutions to the problem of maximizing $x + y$ over these sets. The set labeled $DD(U_1^{dd})$ in Figure 1(a) (resp. $SDD(U_1^{sdd})$ in Figure 1(b)) consists of the points (x, y) for which $I + xA + yB \in DD(U_1^d d)$ (resp. $\in SDD(U_1^{sdd})$) where U_1^{dd} (resp. U_1^{sdd}) corresponds to the Cholesky decomposition of $I + x^*_{dd}A + y^*_{dd}B$ (resp. $I + x^*_{sdd}A + y^*_{sdd}B$). Notice the interesting phenomenon that while the new sets happen to shrink in volume, they expand in the direction that we care about. Already in one iteration, the SOCP gives the perfect bound here.

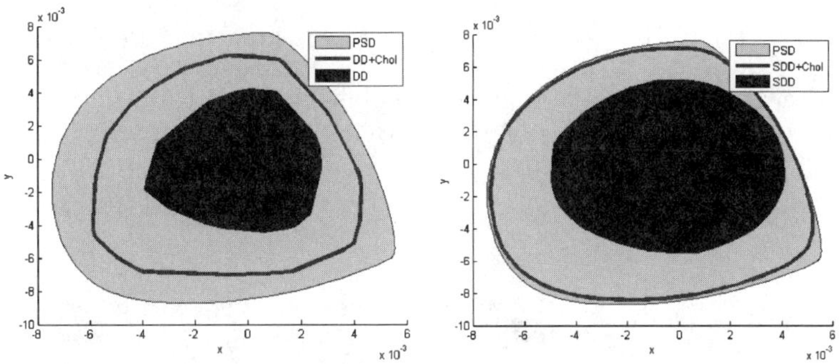

(a) LP inner approximations (b) SOCP inner approximations

FIGURE 2. Improvement in all directions after one Cholesky decomposition

In Figure 2(a), instead of showing the improvement in just the North-East direction, we show it in all directions. This is done by discretizing a large set of directions $d_i = (d_{i,x}, d_{i,y})$ on the unit circle and optimizing along them. More concretely, for each i, we maximize $d_{i,x}x + d_{i,x}y$ over the set $I + xA + yB \in DD_n$. We extract an optimal solution every time and construct a matrix U_{1,d_i} from its Cholesky decomposition. We then maximize in the same direction once again but this time over the set $I + xA + yB \in DD(U_{1,d_i})$. The set of all new optimal solutions is what is plotted with the thick blue line in the figure. We proceed in exactly the same way with our SOCPs to produce Figure 2(b). Notice that both inner approximations after one iteration improve substantially. The SOCP in particular fills up almost the entire spectrahedron.

3.2. Inner approximations to the cone of nonnegative polynomials. A problem domain where inner approximations to semidefinite programs can be useful is in sum of squares programming. This is because the goal of SOS optimization is already to inner approximate the cone of nonnegative polynomials. So by further inner approximating the SOS cone, we will get bounds in the same direction as the SOS bounds.

Let $z(x)$ be the vector of monomials of degree up to d. Define a family of cones of degree-$2d$ polynomials

$$DSOS(U) := \{p \mid p(x) = z^T(x)U^TQUz(x), \text{ for some dd matrix } Q\},$$

parameterized by an $n \times n$ matrix U. We can think of this set as the cone of polynomials that are dsos in the basis $Uz(x)$. If an SOS program has a constraint "p sos", we will replace it iteratively by the constraint $p \in DSOS(U_k)$. The sequence of matrices $\{U_k\}$ is again defined recursively with

$$\begin{aligned} U_0 &= I \\ U_{k+1} &= \text{chol}(U_k^T Q_k U_k), \end{aligned}$$

where Q_k is an optimal Gram matrix of iteration k.

Likewise, let

$$SDSOS(U) := \{p \mid p(x) = z(x)^T U^T Q U z(x), \text{ for some sdd matrix } Q\}.$$

This set can also be viewed as the set of polynomials that are sdsos in the basis $Uz(x)$. To construct a sequence of SOCPs that generate improving bounds on the sos optimal value, we replace the constraint p sos by $p \in SDSOS(U_k)$, where U_k is defined as above.

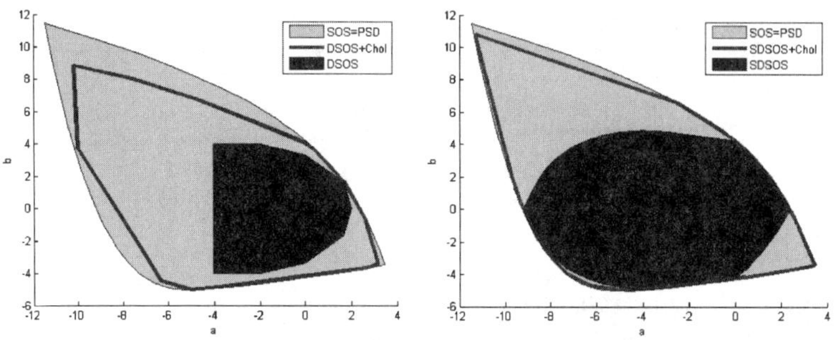

(a) LP inner approximations (b) SOCP inner approximations

FIGURE 3. Improvement in all directions after one Cholesky decomposition

In Figure 3, we consider a parametric family of polynomials

$$p_{a,b}(x_1, x_2) = 2x_1^4 + 2x_2^4 + ax_1^3x_2 + (1-a)x_2^2x_2^2 + bx_1x_2^3.$$

The outermost set in both figures corresponds to the set of pairs (a, b) for which $p_{a,b}$ is sos. As $p_{a,b}$ is a bivariate quartic, this set coincides with the set of (a, b) for which $p_{a,b}$ is nonnegative. The innermost sets in the two subfigures correspond to (a, b) for which $p_{a,b}$ is dsos (resp. sdsos). The thick blue lines illustrate the optimal points achieved when maximizing in all directions over the sets obtained from a single Cholesky decomposition. (The details of the procedure are exactly the same as Figure 2.) Once again, the inner approximations after one iteration improve substantially over the DSOS and SDSOS approximations.

3.3. Extreme-ray interpretation of the change of basis. In this section, we present an alternative but equivalent way of expressing the LP and SOCP-based sequences. This characterization is based on the extreme-ray description of the cone of diagonally dominant/scaled diagonally dominant matrices. It will be particularly useful when we consider outer approximations of the psd cone in Section 3.4.

LEMMA 3.2 (Barker and Carlson [**9**]). *A symmetric matrix M is diagonally dominant if and only if it can be written as*

$$M = \sum_{i=1}^{n^2} \alpha_i v_i v_i^T, \alpha_i \geq 0,$$

where $\{v_i\}$ is the set of all nonzero vectors in $\mathbb{R}^n$ with at most 2 nonzero components, each equal to ± 1.

The vectors v_i are the extreme rays of the DD_n cone. This characterization of the set of diagonally dominant matrices leads to a convenient description of the dual cone:

$$DD_n^* = \{X \in S_n \mid v_i^T X v_i \geq 0, i = 1, \ldots, n^2\}, \tag{3.6}$$

which we will find to be useful in the next subsection. Using Lemma 3.2, we can rewrite the sequence of LPs given in (3.2) as

$$\begin{aligned} DSOS_k := & \min_{X,\alpha_i} C \cdot X \\ & \text{s.t. } A_i \cdot X = b_i, \ i = 1, \ldots, m, \\ & X = \sum_{i=1}^{n^2} \alpha_i (U_k^T v_i)(U_k^T v_i)^T, \\ & \alpha_i \geq 0, i = 1, \ldots, n^2. \end{aligned} \tag{3.7}$$

Let X_k be an optimal solution to the LP in iteration k. The sequence of matrices $\{U_k\}$ is defined just as before:

$$\begin{aligned} U_0 &= I \\ U_{k+1} &= \text{chol}(X_k). \end{aligned}$$

In the first iteration, a linear map is sending (or intuitively "rotating") the extreme rays $\{v_i v_i^T\}$ of the dd cone to a new set of extreme rays $\{(U_1^T v_i)(U_1^T v_i)^T\}$. This procedure keeps repeating itself without ever changing the number of extreme rays.

As the sequence of LPs defined in (3.7) is equivalent to the sequence defined in (3.2), the optimal value of (3.7) improves in each iteration. This can be seen directly: Indeed, X_k is feasible for iteration $k+1$ of (3.7) by taking $\alpha_i = 1$ when v_i has exactly one nonzero entry equal to 1 and $\alpha_i = 0$ otherwise. This automatically implies that $DSOS_{k+1} \leq DSOS_k$. Moreover, the improvement is strict under the assumptions of Theorem 3.1.

The set of scaled diagonally dominant matrices can be described in a similar fashion. In fact, from (2.1), we know that any scaled diagonally dominant matrix M can be written as

$$M = \sum_{i=1}^{\binom{n}{2}} V_i \Lambda_i V_i^T,$$

where V_i is an $n\times 2$ matrix whose columns each contain exactly one nonzero element which is equal to 1, and Λ_i is a 2×2 symmetric psd matrix.

This characterization of SDD_n gives an immediate description of the dual cone

$$SDD_n^* = \left\{X \in S^n \mid V_i^T X V_i \succeq 0, i = 1, \dots, \binom{n}{2}\right\},$$

which will become useful later. Our SOCP sequence in explicit form is then

$$\begin{aligned}
SDSOS_k = \min_{X,\Lambda_i} & \ C \cdot X \\
\text{s.t.} & \ A_i \cdot X = b_i, i = 1, \dots, m, \\
& X = \sum_{i=1}^{\binom{n}{2}} (U_k^T V_i)\Lambda_i (U_k^T V_i)^T, \\
& \Lambda_i \succeq 0.
\end{aligned} \tag{3.8}$$

If X_k is an optimal solution at step k, the matrix sequence $\{U_k\}$ is defined as before:

$$\begin{aligned}
U_0 &= I \\
U_{k+1} &= \text{chol}(X_k).
\end{aligned}$$

The interpretation of (3.8) is similar to that of (3.7).

3.4. Outer approximations of the psd cone. In Section 3.1, we considered inner approximations of the psd cone to obtain upper bounds on (3.1). In many applications, semidefinite programming is used as a "relaxation" to provide outer approximations to some nonconvex sets. This approach is commonly used for relaxing quadratic programs; see, e.g., Section 4, where we consider the problem of finding the largest stable set of a graph. In such scenarios, it does not make sense for us to inner approximate the psd cone: to have a valid relaxation, we need to outer approximate it. This can be easily achieved by working with the dual problems, which we will derive explicitly in this section.

Since $P_n \subseteq DD_n^*$, the first iteration in our LP sequence for outer approximation will be

$$\begin{aligned}
DSOSout_0 := \min_{X} & \ C \cdot X \\
\text{s.t.} & \ A_i \cdot X = b_i, i = 1, \dots, m, \\
& X \in DD_n^*.
\end{aligned}$$

By the description of the dual cone in (3.6), we know this can be equivalently written as

$$\begin{aligned}
DSOSout_0 = \min_{X} & \ C \cdot X \\
\text{s.t.} & \ A_i \cdot X = b_i, \forall i \\
& v_i^T X v_i \geq 0, i = 1, \dots, n^2,
\end{aligned} \tag{3.9}$$

where the v_i's are the extreme rays of the set of diagonally dominant matrices as described in Section 3.3; namely, all vectors with at most two nonzero elements which are either $+1$ or -1. Recall that when we were after inner approximations (Subsection 3.1), the next LP in our sequence was generated by replacing the vectors v_i by $U^T v_i$, where the choice of U was dictated by a Cholesky decomposition of an optimal solution of the previous iterate. In the outer approximation setting, we seemingly do not have access to a psd matrix that would provide us with a Cholesky decomposition. However, we can simply get this from the dual of (3.9)

$$\begin{aligned} DSOSout_0^d := \max_{y,\alpha}\ & b^T y \\ \text{s.t. } & C - \sum_{i=1}^m y_i A_i = \sum_{i=1}^{n^2} \alpha_i v_i v_i^T, \\ & \alpha_i \geq 0, i = 1, \ldots, n^2, \end{aligned}$$

by taking $U_1 = \text{chol}(C - \sum_i y_i^* A_i)$. We then replace v_i by $U_1^T v_i$ in (3.9) to get the next iterate and proceed. In general, the sequence of LPs can be written as

$$\begin{aligned} DSOSout_k = \min_X\ & C \cdot X \\ \text{s.t. } & A_i \cdot X = b_i, i = 1, \ldots, m, \\ & v_i^T U_k X U_k^T v_i \geq 0, \end{aligned}$$

where $\{U_k\}$ is a sequence of matrices defined recursively as

$$\begin{aligned} U_0 &= I \\ U_k &= \text{chol}\left(C - \sum_i y_i^{(k-1)} A_i\right). \end{aligned}$$

The vector y^{k-1} here is an optimal solution to the dual problem at step $k-1$:

$$\begin{aligned} DSOSout_{k-1}^d := \max_{y,\alpha}\ & b^T y \\ \text{s.t. } & C - \sum_{i=1}^m y_i A_i = \sum_{i=1}^{n^2} \alpha_i (U_{k-1}^T v_i)(U_{k-1}^T v_i)^T, \\ & \alpha_i \geq 0, i = 1, \ldots, n^2. \end{aligned}$$

This algorithm again strictly improves the objective value at each iteration. Indeed, from LP strong duality, we have

$$DSOSout_k = DSOSout_k^d$$

and Theorem 3.1 applied to the dual problem states that

$$DSOSout_{k-1}^d < DSOSout_k^d.$$

The sequence of SOCPs for outer approximation can be constructed in an analogous manner:

$$\begin{aligned} SDSOSout_k = \min_X\ & C \cdot X \\ \text{s.t. } & A_i \cdot X = b_i, i = 1, \ldots, m, \\ & V_i^T U_k X U_k^T V_i \succeq 0, i = 1, \ldots, \binom{n}{2}, \end{aligned}$$

where V_i's are $n \times 2$ matrices containing exactly one 1 in each column, and $\{U_k\}$ is a sequence of matrices defined as

$$U_0 = I$$

$$U_k = \text{chol}\left(C - \sum_i y_i^{(k-1)} A_i\right)$$

Here again, the vector $y^{(k-1)}$ is an optimal solution to the dual SOCP at step $k-1$:

$$SDSOSout_{k-1}^d := \max_{y,\Lambda_i} b^T y$$

$$\text{s.t. } C - \sum_{i=1}^m y_i A_i = \sum_{i=1}^{\binom{n}{2}} (U_{k-1}^T v_i)\Lambda_i (U_{k-1}^T v_i)^T,$$

$$\Lambda_i \succeq 0, i = 1, \ldots, \binom{n}{2},$$

where each Λ_i is a 2×2 unknown symmetric matrix.

REMARK 3.3. Let us end with some concluding remarks about our algorithm. There are other ways of improving the DSOS and SDSOS bounds. For example, Ahmadi and Majumdar [**6**, **28**] propose the requirement that $(\sum_{i=1}^n x_i^2)^r p(x)$ be dsos or sdsos as a sufficient condition for nonnegativity of p. As r increases, the quality of approximation improves, although the problem size also increases very quickly. Such hierarchies are actually commonly used in the sum of squares optimization literature. But unlike our approach, they do not take into account a particular objective function and may improve the inner approximation to the PSD cone in directions that we do not care about. Nevertheless, these hierarchies have interesting theoretical implications. Under some assumptions, one can prove that as $r \to \infty$, the underlying convex programs succeed in optimizing over the entire set of nonnegative polynomials; see, e.g., [**6**, **12**, **32**, **35**].

Another approach to improve on the DSOS and SDSOS bounds appears in the recent work in [**4**] with Dash. We show there how ideas from column generation in large-scale integer and linear programming can be used to iteratively improve inner approximations to semidefinite cones. The LPs and SOCPs proposed in that work take the objective function into account and increase the problem size after each iteration by a moderate amount. By contrast, the LPs and SOCPs coming from our Cholesky decompositions in this paper have exactly the same size in each iteration. We should remark however that the LPs from iteration two and onwards are typically more dense than the initial LP (for DSOS) and slower to solve. A worthwhile future research direction would be to systematically compare the performance of the two approaches and to explore customized solvers for the LPs and the SOCPs that arise in our algorithms.

4. The maximum stable set problem

A classic problem in discrete optimization is that of finding the stability number of a graph. The graphs under our consideration in this section are all undirected and unweighted. A *stable set* (or *independent set*) of a graph $G = (V, E)$ is a set of nodes of G no two of which are adjacent. The stability number of G, often denoted by $\alpha(G)$, is the size of its maximum stable set(s). The problem of determining α

has many applications in scheduling (see, e.g., [**16**]) and coding theory [**27**]. As an example, the maximum number of final exams that can be scheduled on the same day at a university without requiring any student to take two exams is given by the stability number of a graph. This graph has courses IDs as nodes and an edge between two nodes if and only if there is at least one student registered in both courses. Unfortunately, the problem of testing whether $\alpha(G)$ is greater than a given integer k is well known to be NP-complete [**21**]. Furthermore, the stability number cannot be approximated within a factor $|V|^{1-\epsilon}$ for any $\epsilon > 0$ unless P=NP [**18**].

A straightforward integer programming formulation of $\alpha(G)$ is given by

$$\begin{aligned} \alpha(G) = \max & \sum_i x_i \\ \text{s.t. } & x_i + x_j \leq 1, \text{ if } \{i,j\} \in E \\ & x_i \in \{0,1\}. \end{aligned}$$

The standard LP relaxation for this problem is obtained by changing the binary constraint $x_i \in \{0,1\}$ to the linear constraint $x_i \in [0,1]$:

$$\begin{aligned} LP := \max & \sum_i x_i \\ \text{s.t. } & x_i + x_j \leq 1, \text{ if } \{i,j\} \in E \\ & x_i \in [0,1]. \end{aligned} \tag{4.1}$$

Solving this LP results in an upper bound on the stability number. The quality of this upper bound can be improved by adding the so-called *clique inequalities*. The set of k-clique inequalities, denoted by C_k, is the set of constraints of the type $x_{i_1} + x_{i_2} + \ldots + x_{i_k} \leq 1$, if $(i_1, \ldots, i_k)$ form a clique (i.e., a complete subgraph) of G. Observe that these inequalities must be satisfied for binary solutions to the above LP, but possibly not for fractional ones. Let us define a family of LPs indexed by k:

$$\begin{aligned} LP^k := \max & \sum_i x_i \\ & x_i \in [0,1] \\ & C_1, \ldots, C_k \text{ are satisfied.} \end{aligned} \tag{4.2}$$

Note that $LP = LP^2$ by construction and $\alpha(G) \leq LP^{k+1} \leq LP^k$ for all k. We will be comparing the bound obtained by some of these well-known LPs with those achieved via the new LPs that we propose further below.

A famous semidefinite programming based upper bound on the stability number is due to Lovász [**27**]:

$$\begin{aligned} \vartheta(G) := \max_X & \ J \cdot X \\ \text{s.t. } & I \cdot X = 1 \\ & X_{ij} = 0, \ \forall \{i,j\} \in E \\ & X \succeq 0, \end{aligned}$$

where J here is the all ones matrix and I is the identity matrix. The optimal value $\vartheta(G)$ is called the Lovász theta number of the graph. We have the following inequalities

$$\alpha(G) \leq \vartheta(G) \leq LP^k, \ \forall k.$$

The fact that $\alpha(G) \leq \vartheta(G)$ is easily seen by noting that if S is a stable set of maximum size and 1_S is its indicator vector, then the rank-one matrix $\frac{1}{|S|}1_S 1_S^T$ is feasible to the SDP and gives the objective value $|S|$. The other inequality states that this SDP-based bound is stronger than the aforementioned LP bound even with all the clique inequalities added (there are exponentially many). A proof can be found e.g. in [**24**, Section 6.5.2].

Our goal here is to obtain LP and SOCP based sequences of upper bounds on the Lovász theta number. To do this, we construct a series of outer approximations of the set of psd matrices as described in Section 3.4. The first bound in the sequence of LPs is given by:

$$\begin{aligned} DSOS_0(G) := &\max_X J \cdot X \\ &\text{s.t. } I \cdot X = 1 \\ &X_{ij} = 0, \ \forall \{i,j\} \in E \\ &X \in DD_n^*. \end{aligned}$$

In view of (3.6), this LP can be equivalently written as

$$\begin{aligned} DSOS_0(G) = &\max_X J \cdot X \\ &\text{s.t. } I \cdot X = 1 \\ &X_{ij} = 0, \ \forall \{i,j\} \in E \\ &v_i^T X v_i \geq 0, i = 1, \ldots, n^2, \end{aligned} \tag{4.3}$$

where v_i is a vector with at most two nonzero entries, each nonzero entry being either $+1$ or -1. This LP is always feasible (e.g., with $X = \frac{1}{n}I$). Furthermore, it is bounded above. Indeed, the last constraints in (4.3) imply in particular that for all i, j, we must have

$$X_{i,j} \leq \frac{1}{2}(X_{ii} + X_{jj}).$$

This, together with the constraint $I \cdot X = 1$, implies that the objective $J \cdot X$ must remain bounded. As a result, the first LP in our iterative sequence will give a finite upper bound on α.

To progress to the next iteration, we will proceed as described in Section 3.4. The new basis for solving the problem is obtained through the dual[4] of (4.3):

$$\begin{aligned} DSOS_0^d(G) := &\max y \\ &\text{s.t. } yI + Y - J = \sum_{i=1}^{n^2} \alpha_i v_i v_i^T \\ &Y_{ij} = 0 \text{ if } i = j \text{ or } \{i,j\} \notin E \\ &\alpha_i \geq 0, i = 1, \ldots, n^2. \end{aligned} \tag{4.4}$$

The second constraint in this problem is equivalent to requiring that $yI + Y - J$ be dd. We can define

$$U_1 = \text{chol}(y_0^* I + Y_0^* - J)$$

[4]The reader should not be confused to see both the primal and the dual as maximization problems. We can make the dual a minimization problem by changing the sign of y.

where (y_1^*, Y_1^*) are optimal solutions to (4.4). We then solve

$$\begin{aligned} DSOS_1(G) := \max_X \ & J \cdot X \\ \text{s.t. } & I \cdot X = 1 \\ & X_{ij} = 0, \ \forall \{i,j\} \in E \\ & v_i^T U_1 X U_1^T v_i \geq 0, i = 1, \ldots, n^2, \end{aligned}$$

to obtain our next iterate. The idea remains exactly the same for a general iterate k: We construct the dual

$$\begin{aligned} DSOS_k^d(G) := \max \ & y \\ \text{s.t. } & yI + Y - J = \sum_{i=1}^{n^2} \alpha_i U_k^T v_i (U_k^T v_i)^T \\ & Y_{ij} = 0 \text{ if } i = j \text{ or } \{i,j\} \notin E \\ & \alpha_i \geq 0, \forall i, \end{aligned}$$

and define

$$U_{k+1} := \text{chol}(y_k^* + Y_k^* - J),$$

where (y_k^*, Y_k^*) is an optimal solution to the dual. The updated primal is then

$$\begin{aligned} DSOS_{k+1}(G) := \max_X \ & J \cdot X \\ \text{s.t. } & I \cdot X = 1 \\ & X_{ij} = 0, \ \forall \{i,j\} \in E \\ & v_i^T U_{k+1} X U_{k+1}^T v_i \geq 0, i = 1, \ldots, n^2. \end{aligned} \tag{4.5}$$

As stated in Section 3.4, the optimal values of (4.5) are guaranteed to strictly improve as a function of k. Note that to get the bounds, we can just work with the dual problems throughout.

An analoguous technique can be used to obtain a sequence of SOCPs. For the initial iterate, instead of requiring that $X \in DD^*$ in (4.3), we require that $X \in SDD^*$. This problem must also be bounded and feasible as

$$P_n \subseteq SDD^* \subseteq DD^*.$$

Then, for a given iterate k, the algorithm consists of solving

$$\begin{aligned} SDSOS_k(G) := \max_X \ & J \cdot X \\ \text{s.t. } & I \cdot X = 1 \\ & X_{ij} = 0, \forall \{i,j\} \in E \\ & V_i^T U_k X U_k^T V_i \succeq 0, i = 1, \ldots, \binom{n}{2}, \end{aligned}$$

where as explained in Section 3.3 each V_i is an $n \times 2$ matrix whose columns contain exactly one nonzero element which is equal to 1. The matrix U_k here is fixed and obtained by first constructing the dual SOCP

$$\begin{aligned} SDSOS_k^d(G) := &\max y \\ &\text{s.t. } yI + Y - J = \sum_{i=1}^{\binom{n}{2}} U_k^T V_i \Lambda_i (U_k^T V_i)^T \\ &Y_{ij} = 0 \text{ if } i = j \text{ or } \{i,j\} \notin E \\ &\Lambda_i \succeq 0, \forall i, \end{aligned}$$

(each Λ_i is a symmetric 2×2 matrix decision variable) and then taking

$$U_k = \text{chol}(y_k^* I + Y_k^* - J).$$

Once again, one can just work with the dual problems to obtain the bounds.

As our first example, we apply both techniques to the problem of finding the stability number of the complement of the Petersen graph (see Figure 4(a)). The exact stability number here is 2 and an example of a maximum stable set is illustrated by the two white nodes in Figure 4(a). The Lovász theta number is 2.5 and has been represented by the continuous line in Figure 4(b). The dashed lines represent the optimal values of the LP and SOCP-based sequences of approximations for 7 iterations. Notice that already within one iteration, the optimal values are within one unit of the true stability number, which is good enough for knowing the exact bound (the stability number is an integer). From the fifth iteration onwards, they differ from the Lovász theta number by only 10^{-2}.

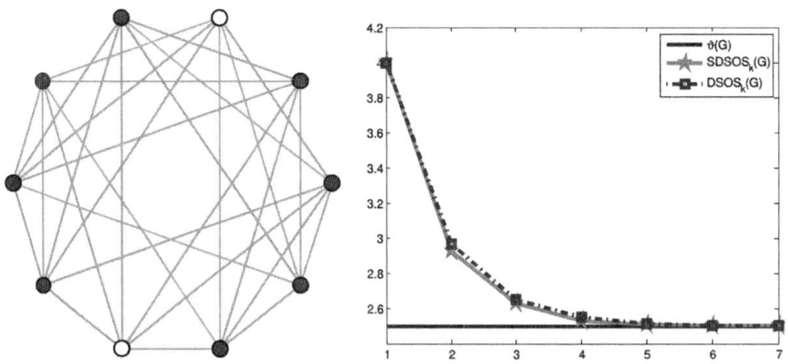

(a) Complement of Petersen graph (b) The Lovász theta number and iterative bounds bounds obtained by LP and SOCP

FIGURE 4. Upper bounding the stability number of the complement of the Petersen graph

Finally, in Table 1, we have generated 100 instances of 20-node Erdös-Rényi graphs with edge probability 0.5. For each instance, we compute the bounds from the Lovász SDP, the standard LP in (4.1), the standard LP with all 3-clique inequalities added (LP^3 in (4.2)), and our LP/SOCP iterative sequences. We focus here on iterations 3,4 and 5 because there is no need to go further. We compare our bounds with the standard LP and the standard LP with 3-clique inequalities because they are LPs of roughly the same size. If any of these bounds are within one unit of the true stable set number, we count this as a success and increment the counter. As can be seen in Table 1, the Lovász theta number is always within a unit of the stable set number, and so are our LP and SOCP sequences ($DSOS_k, SDSOS_k$) after four or at most five iterations. If we look just at the bound after 3 iterations, the success rate of SDSOS is noticeably higher than the success rate of DSOS. Also note that the standard LP with or without the three clique inequalities never succeeds in giving a bound within one unit of $\alpha(G)$.[5]

$\vartheta(G)$	LP	LP^3	$DSOS_3$	$DSOS_4$	$DSOS_5$	$SDSOS_3$	$SDSOS_4$	$SDSOS_5$
100%	0%	0%	14%	83%	100%	69%	100%	100%

TABLE 1. Percentage of instances out of 100 where the bound obtained is less than a unit away from the stability number

5. Partition

The partition problem is arguably the simplest NP-complete problem to state: Given a list of positive integers $a_1, \ldots, a_n$, is it possible to split them into two sets with equal sums? We say that a partition instance is feasible if the answer is yes (e.g., $\{5,2,1,6,3,8,5,4,1,1,10\}$) and infeasible if the answer is no (e.g., $\{47,20,13,15,36,7,46\}$). The partition problem is NP-complete but only weakly. In fact, the problem admits a pseudopolynomial time algorithm based on dynamic programming that can deal with rather large problem sizes efficiently. This algorithm has polynomial running time on instances where the bit size of the integers a_i are bounded by a polynomial in $\log n$ [**14**]. In this section, we investigate the performance and mostly limitations of algebraic techniques for refuting feasibility of partition instances.

Feasibility of a partition instance can always be certified by a short proof (the partition itself). However, unless P=co-NP, we do not expect to always have short certificates of infeasibility. Nevertheless, we can try to look for such a certificate through a sum of squares decomposition. Indeed, given an instance $a := \{a_1, \ldots, a_n\}$, it is not hard to see[6] that the following equivalence holds:

$$(5.1) \quad \begin{bmatrix} a \text{ is an infeasible} \\ \text{partition instance} \end{bmatrix} \Leftrightarrow p_a(x) := \sum_i (x_i^2 - 1)^2 + (\sum_i a_i x_i)^2 > 0, \ \forall x \in \mathbb{R}^n.$$

So if for some $\epsilon > 0$ we could prove that $p_a(x) - \epsilon$ is nonnegative, we would have refuted the feasibility of our partition instance.

DEFINITION 5.1. An instance of partition $a_1, \ldots, a_n$ is said to be *sos-refutable* if there exists $\epsilon > 0$ such that $p_a(x) - \epsilon$ is sos.

[5]All numerical experiments in this paper have been parsed using either SPOT [**29**] or YAMIP [**26**] and solved using the LP/SOCP/SDP solver of MOSEK [**2**].

[6]This equivalence is apparent in view of the zeros of the polynomial on the right hand side of (5.1) corresponding to a feasible partition.

Obviously, any instance of partition that is sos-refutable is infeasible. This suggests that we can consider solving the following semidefinite program

$$\begin{aligned} SOS := &\max\ \epsilon \\ &\text{s.t. } q_a(x) := p_a(x) - \epsilon \text{ is sos} \end{aligned} \tag{5.2}$$

and examining its optimal value. Note that the optimal value of this problem is always greater than or equal to zero as p_a is sos by construction. If the optimal value is positive, we have succeeded in proving infeasibility of the partition instance a.

We would like to define the notions of *dsos-refutable* and *sdsos-refutable* instances analogously by replacing the condition $q_a(x)$ sos by the condition $q_a(x)$ dsos or sdsos. Though (5.2) is guaranteed to always be feasible by taking $\epsilon = 0$, this is not necessarily the case for dsos/sdsos versions of (5.2). For example, the optimization problem

$$\max_{\epsilon}\{\epsilon \mid p_a(x) - \epsilon \text{ dsos}\} \tag{5.3}$$

on the instance $\{1, 2, 2, 1, 1\}$ is infeasible.[7] This is a problem for us as we need the first LP to be feasible to start our iterations. We show, however, that we can get around this issue by modeling the partition problem with homogeneous polynomials.

DEFINITION 5.2. Let p_a be as in (5.1). An instance of partition $a_1, \ldots, a_n$ is said to be *dsos-refutable* (resp. *sdsos-refutable*) if there exists $\epsilon > 0$ such that the quartic form

$$q^h_{a,\epsilon}(x) := p_a\left(\frac{x}{\left(\frac{1}{n}\sum_i x_i^2\right)^{1/2}}\right)\left(\frac{1}{n}\sum_i x_i^2\right)^2 - \epsilon\left(\frac{1}{n}\sum_i x_i^2\right)^2 \tag{5.4}$$

is dsos (resp. sdsos).

Notice that $q^h_{a,\epsilon}$ is indeed a polynomial as it can be equivalently written as

$$\sum_i x_i^4 + \left(\left(\sum_i a_i x_i\right)^2 - 2\sum_i x_i^2\right) \cdot \left(\frac{1}{n}\sum_i x_i^2\right) + (n-\epsilon) \cdot \left(\frac{1}{n}\sum_i x_i^2\right)^2.$$

What we are doing here is homogenizing a polynomial that does not have odd monomials by multiplying its lower degree monomials with appropriate powers of $\sum_i x_i^2$. The next theorem tells us how we can relate nonnegativity of this polynomial to feasibility of partition.

THEOREM 5.3. *A partition instance $a = \{a_1, \ldots, a_n\}$ is infeasible if and only if there exists $\epsilon > 0$ for which the quartic form $q^h_{a,\epsilon}(x)$ defined in (5.4) is nonnegative.*

PROOF. For ease of reference, let us define

$$p^h_a(x) := p_a\left(\frac{x}{\left(\frac{1}{n}\sum_i x_i^2\right)^{1/2}}\right)\left(\frac{1}{n}\sum_i x_i^2\right)^2. \tag{5.5}$$

[7] Under other structures on a polynomial, the same type of problem can arise for sos. For example, consider the Motzkin polynomial **[30]** $M(x_1, x_2) = x_1^2 x_2^4 + x_2^2 x_1^4 - 3x_1^2 x_2^2 + 1$ which is nonnegative everywhere. The problem $\max_\epsilon\{\epsilon \mid M(x) - \epsilon \text{ sos}\}$ is infeasible.

Suppose partition is feasible, i.e, the integers $a_1, \dots, a_n$ can be placed in two sets $\mathcal{S}_1$ and $\mathcal{S}_2$ with equal sums. Let $\bar{x}_i$=1 if a_i is placed in set $\mathcal{S}_1$ and $\bar{x}_i = -1$ if a_i is placed in set $\mathcal{S}_2$. Then $||\bar{x}||_2^2 = n$ and $p_a(\bar{x}) = 0$. This implies that

$$p_a^h(\bar{x}) = p_a(\bar{x}) = 0,$$

and hence having $\epsilon > 0$ would make

$$q_{a,\epsilon}^h(\bar{x}) = -\epsilon < 0.$$

If partition is infeasible, then $p_a(x) > 0$, $\forall x \in \mathbb{R}^n$. In view of (5.5) we see that $p_a^h(x) > 0$ on the sphere $\mathbb{S}$ of radius n. Since p_a^h is continuous, its minimum $\hat{\epsilon}$ on the compact set $\mathbb{S}$ is achieved and must be positive. So we must have

$$q_{a,\hat{\epsilon}}^h(x) = p_a^h(x) - \hat{\epsilon}\left(\frac{1}{n}\sum_i x_i^2\right)^2 \geq 0, \forall x \in \mathbb{S}.$$

By homogeneity, this implies that $q_{a,\hat{\epsilon}}^h$ is nonnegative everywhere. □

Consider now the LP

$$\begin{aligned} &\max_{\epsilon}\ \epsilon \\ &\text{s.t. } q_{a,\epsilon}^h(x) \text{ dsos.} \end{aligned} \tag{5.6}$$

THEOREM 5.4. *The LP in (5.6) is always feasible.*

PROOF. Let $h(x) := \left(\frac{1}{n}\sum_i x_i^2\right)^2$ and recall that $z(x,2)$ denotes the vector of all monomials of degree exactly 2. We can write

$$h(x) = z^T(x,2)Q_h z(x,2)$$

where Q_h is in the strict interior of the DD_n cone (i.e., its entries q_{ij} satisfy $q_{ii} > \sum_j |q_{ij}|, \forall i$). Furthermore, let Q be a symmetric matrix such that $p_a^h(x) = z(x,2)^T Q z(x,2)$. Then

$$q_{a,\epsilon}^h(x) = p_a^h(x) - \epsilon h(x) = z(x,2)^T(Q - \epsilon Q_h)z(x,2).$$

As Q_h is in the strict interior of DD_n, $\exists \lambda > 0$ such that

$$\lambda Q + (1-\lambda)Q_h \text{ is dd.}$$

Taking $\epsilon = -\frac{1-\lambda}{\lambda}$, $Q - \epsilon Q_h$ will be diagonally dominant and $q_{a,\epsilon}^h$ will be dsos. □

As an immediate consequence, the SOCP

$$\begin{aligned} &\max_{\epsilon}\ \epsilon \\ &\text{s.t. } q_{a,\epsilon}^h(x) \text{ sdsos} \end{aligned} \tag{5.7}$$

is also always feasible. We can now define our sequence of LPs and SOCPs as we have guaranteed feasibility of the first iteration. This is done following the strategy and notation of Section 3.2:

$$\begin{aligned} DSOS_k \text{ (resp. } SDSOS_k) := &\max_{\epsilon}\ \epsilon \\ &\text{s.t. } q_{a,\epsilon}^h(x) \in DSOS(U_k) \text{ (resp. } SDSOS(U_k)), \end{aligned} \tag{5.8}$$

where $\{U_k\}$ is a sequence of matrices recursively defined with $U_0 = I$ and U_{k+1} defined as the Cholesky factor of an optimal dd (resp. sdd) Gram matrix of the optimization problem in iteration k.

We illustrate the performance of these LP and SOCP-based bounds on the infeasible partition instance $\{1, 2, 2, 1, 1\}$. The results are in Figure 5. We can use the sum of squares relaxation to refute the feasibility of this instance by either solving (5.2) (the "non-homogenized version") or solving (5.6) with dsos replaced with sos (the "homogenized version"). Both approaches succeed in refuting this partition instance, though the homogenized version gives a slightly better (more positive) optimal value. As a consequence, we only plot the homogeneous bound, denoted by SOS_h, in Figure 5. Notice that the LP and SOCP-based sequences refute the instance from the 6th iteration onwards.

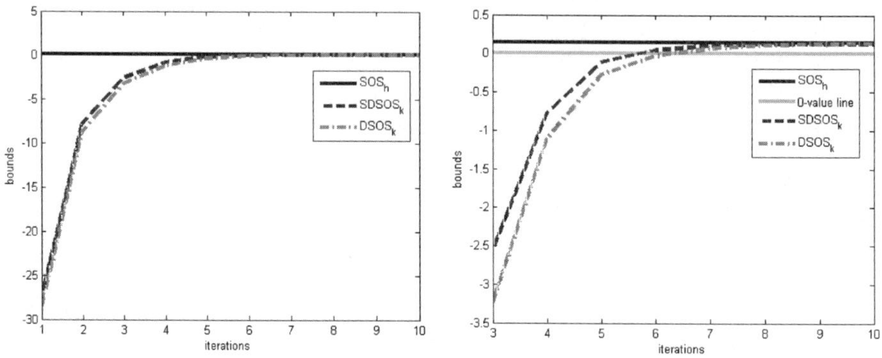

(a) Bounds SOS_h, $DSOS_k$ and $SDSOS_k$

(b) Zoomed-in version of Figure 5(a)

FIGURE 5. Bounds obtained on the {1,2,2,1,1} instance of the partition problem using SDP, as well as the LP/SOCP-based sequences

As our final experiment, we generate 50 infeasible instances of partition with 6 elements randomly generated between 1 and 15. These instances are *trivially infeasible* because we made sure that $a_1 + \cdots + a_6$ is an odd number. In the first column of Table 2, we count the number of successes for sos-refutability (non homogeneous version as defined in Definition 5.1), where a failure is defined as the optimal value of (5.2) being 0 up to numerical precision. The second column corresponds to the number of successes for sos-refutability (homogeneous version). The last 4 columns show the success rate of the LP and SOCP-based sequences as defined in (5.8), after 20 iterations and 40 iterations.

SOS	SOS_h	$DSOS_{20}$	$DSOS_{40}$	$SDSOS_{20}$	$SDSOS_{40}$
56%	56%	12%	16 %	14%	14%

TABLE 2. Rate of success for refutability of infeasible instances of partition

From the experiments, the homogeneous and non-homogeneous versions of (5.2) have the same performance in terms of their ability to refute feasibility. However, we observe that they both fail to refute a large number of completely trivial instances! We prove why this is the case for one representative instance in the next section. The LP and SOCP-based sequences also perform poorly and their convergence is much slower than what we observed for the maximum stable set problem in Section 4.

5.1. Failure of the sum of squares relaxation on trivial partition instances. For complexity reasons, one would expect there to be infeasible instances of partition that are not sos-refutable. What is surprising however is that the sos relaxation is failing on many instances that are totally trivial to refute as the sum of their input integers is odd. We present a proof of this phenomenon on an instance which is arguably the simplest one.[8]

PROPOSITION 5.5. The infeasible partition instance $\{1,1,1,1,1\}$ is not sos-refutable.

PROOF. Let p_a be the polynomial defined in (5.1). To simplify notation, we let $p(x)$ represent $p_a(x)$ for $a = \{1,1,1,1,1\}$. We will show that p is on the boundary of the SOS cone even though we know it is strictly inside the PSD cone. This is done by presenting a dual functional μ that vanishes on p, takes a nonnegative value on all quartic sos polynomials, and a negative value on $p(x) - \epsilon$ for any $\epsilon > 0$. (See Figure 6 for an intuitive illustration of this.)

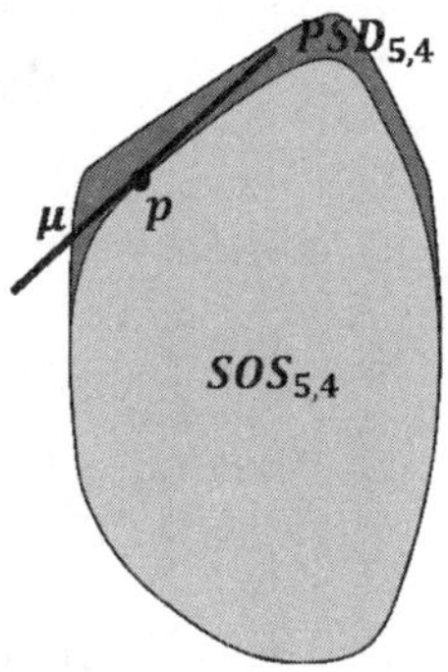

FIGURE 6. The geometric idea behind the proof of Proposition 5.5

The polynomial p when expanded out reads

$$p(x) = n - \sum_i x_i^2 + 2\sum_{i<j} x_i x_j + \sum_i x_i^4. \tag{5.9}$$

Consider the vector of coefficients of p with the ordering as written in (5.9):

(5.10)

$$\vec{p} = \begin{pmatrix} 5 & -1 & -1 & -1 & -1 & -1 & 2 & 2 & 2 & 2 & 2 & 2 & 2 & 2 & 2 & 2 & -1 & -1 & -1 & -1 & -1 \end{pmatrix}.$$

This is a reduced representation of the vector of coefficients of p, in that there are many zeros associated with all other monomials of degree less than or equal to 4, which we are not writing out.

Our goal is to find a vector μ that satisfies

$$\begin{aligned} &\langle \mu, \vec{p} \rangle = 0 \\ &\langle \mu, \vec{q} \rangle \geq 0, \text{ for all } q \text{ sos of degree 4.} \end{aligned} \tag{5.11}$$

If such a μ exists and its first element is nonzero (which by rescaling can then be taken to be 1), then $\langle \mu, \overrightarrow{p - \epsilon} \rangle = \langle \mu, \vec{p} \rangle - \langle \mu, \vec{\epsilon} \rangle = -\epsilon < 0$. This provides us with the required functional that separates $p(x) - \epsilon$ from the set of sos polynomials.

[8] If we were to instead consider the instance [1,1,1], sos would succeed in refuting it.

Selecting the same reduced basis as the one used in (5.10), we take

$$\overrightarrow{\mu_{reduced}} = \begin{pmatrix} 1 & \mathbf{1}_5^T & -\frac{1}{4}\cdot\mathbf{1}_1^T & \mathbf{1}_5^T \end{pmatrix}$$

where $\mathbf{1}_n$ is the all ones vector of size n. The subscript "reduced" denotes the fact that in $\overrightarrow{\mu_{reduced}}$, only the elements of μ needed to verify $\langle \mu, \overrightarrow{p} \rangle = 0$ are presented. Unlike $\overrightarrow{p}$, the entries of μ corresponding to the other monomials are not all zero. This can be seen from the entries of the matrix M that appears further down.

We now show how (5.11) holds. Consider any sos polynomial q of degree less than or equal to 4. We know that it can be written as

$$q(x) = z^T Q z = \mathrm{Tr}\ Q \cdot zz^T,$$

for some $Q \succeq 0$, and a vector of monomials

$$z^T = [1, x_1, x_2, \ldots, x_n, x_1^2, \ldots, x_n^2, x_1x_2, \ldots, x_{n-1}x_n].$$

It is not difficult to see that

$$\langle \mu, \overrightarrow{q} \rangle = \mathrm{Tr}\ Q \cdot (zz^T)|_\mu$$

where by $(zz)^T|_\mu$, we mean a matrix where each monomial in zz^T is replaced with the corresponding element of the vector μ. This yields the matrix

$$M = \begin{pmatrix}
1 & 0 & 0 & 0 & 0 & 0 & 1 & 1 & 1 & 1 & 1 & b & b & b & b & b & b & b & b & b & b \\
0 & 1 & b & b & b & b & 0 & 0 & 0 & 0 & 0 & 0 & 0 & 0 & 0 & 0 & 0 & 0 & 0 & 0 & 0 \\
0 & b & 1 & b & b & b & 0 & 0 & 0 & 0 & 0 & 0 & 0 & 0 & 0 & 0 & 0 & 0 & 0 & 0 & 0 \\
0 & b & b & 1 & b & b & 0 & 0 & 0 & 0 & 0 & 0 & 0 & 0 & 0 & 0 & 0 & 0 & 0 & 0 & 0 \\
0 & b & b & b & 1 & b & 0 & 0 & 0 & 0 & 0 & 0 & 0 & 0 & 0 & 0 & 0 & 0 & 0 & 0 & 0 \\
0 & b & b & b & b & 1 & 0 & 0 & 0 & 0 & 0 & 0 & 0 & 0 & 0 & 0 & 0 & 0 & 0 & 0 & 0 \\
1 & 0 & 0 & 0 & 0 & 0 & 1 & 1 & 1 & 1 & 1 & b & b & b & b & b & b & b & b & b & b \\
1 & 0 & 0 & 0 & 0 & 0 & 1 & 1 & 1 & 1 & 1 & b & b & b & b & b & b & b & b & b & b \\
1 & 0 & 0 & 0 & 0 & 0 & 1 & 1 & 1 & 1 & 1 & b & b & b & b & b & b & b & b & b & b \\
1 & 0 & 0 & 0 & 0 & 0 & 1 & 1 & 1 & 1 & 1 & b & b & b & b & b & b & b & b & b & b \\
1 & 0 & 0 & 0 & 0 & 0 & 1 & 1 & 1 & 1 & 1 & b & b & b & b & b & b & b & b & b & b \\
b & 0 & 0 & 0 & 0 & 0 & b & b & b & b & b & 1 & b & b & b & b & b & b & a & a & a \\
b & 0 & 0 & 0 & 0 & 0 & b & b & b & b & b & b & 1 & b & b & b & a & a & b & b & a \\
b & 0 & 0 & 0 & 0 & 0 & b & b & b & b & b & b & b & 1 & b & a & b & a & b & a & b \\
b & 0 & 0 & 0 & 0 & 0 & b & b & b & b & b & b & b & b & 1 & a & a & b & a & b & b \\
b & 0 & 0 & 0 & 0 & 0 & b & b & b & b & b & b & b & a & a & 1 & b & b & b & b & a \\
b & 0 & 0 & 0 & 0 & 0 & b & b & b & b & b & b & a & b & a & b & 1 & b & b & a & b \\
b & 0 & 0 & 0 & 0 & 0 & b & b & b & b & b & b & a & a & b & b & b & 1 & a & b & b \\
b & 0 & 0 & 0 & 0 & 0 & b & b & b & b & b & a & b & b & a & b & b & a & 1 & b & b \\
b & 0 & 0 & 0 & 0 & 0 & b & b & b & b & b & a & b & a & b & b & a & b & b & 1 & b \\
b & 0 & 0 & 0 & 0 & 0 & b & b & b & b & b & a & a & b & b & a & b & b & b & b & 1
\end{pmatrix},$$

where $a = \frac{3}{8}$ and $b = -\frac{1}{4}$. We can check that $M \succeq 0$. This, together with the fact that $Q \succeq 0$, implies that (5.11) holds.[9] □

---

[9]It can be shown in a similar fashion that $\{1, 1, 1, 1, 1\}$ is not sos-refutable in the homogeneous formulation of (5.4) either.

5.2. Open problems. We showed in the previous subsection that the infeasible partition instance $\{1,1,1,1,1\}$ was not sos-refutable. Many more randomly-generated partition instances that we knew to be infeasible (their sum being odd) also failed to be sos-refutable. This observation motivates the following open problem:

Open Problem 1. Characterize the set of partition instances $\{a_1,\ldots,a_n\}$ that have an odd sum but are not sos-refutable (see Definition 5.1).

Our second open problem has to do with the power of higher order sos relaxations for refuting feasibility of partition instances.

Open Problem 2. For a positive integer r, let us call a partition instance $\{a_1,\ldots,a_n\}$ *r-sos-refutable* if $\exists\epsilon > 0$ such that $(p(x)-\epsilon)(\sum_i x_i^2+1)^r$ is sos. Note that this is also a certificate of infeasibility of the instance. Even though the $\{1,1,1,1,1\}$ instance is not sos-refutable, it is 1-sos-refutable. Furthermore, we have numerically observed that the instance $\{1,1,1,1,1,1,1\}$ (vector of all ones of length 7) is not sos-refutable or 1-sos-refutable, but it is 2-sos-refutable. If we consider the instance consisting of n ones with n odd, and define $\tilde{r}$ to be the minimum r such that $\{1,1,\ldots,1\}$ becomes r-sos-refutable, is it true that $\tilde{r}$ must grow with n?

Acknowledgments

We are grateful to Anirudha Majumdar and Sanjeeb Dash for insightful discussions.

References

[1] *Gurobi optimizer reference manual*, URL: http://www. gurobi. com (2012).

[2] *MOSEK reference manual*, 2013, Version 7. Latest version available at http://www.mosek.com/.

[3] A. A. Ahmadi, *Algebraic relaxations and hardness results in polynomial optimization and Lyapunov analysis*, Ph.D. thesis, Massachusetts Institute of Technology, September 2011, Available at http://aaa.princeton.edu/publications.

[4] A. A. Ahmadi, S. Dash, and G. Hall, *Optimization over structured subsets of positive semidefinite matrices via column generation*, Under Review, 2015.

[5] A. A. Ahmadi and A. Majumdar, *DSOS and SDSOS optimization: LP and SOCP-based alternatives to sum of squares optimization*, Proceedings of the 48th Annual Conference on Information Sciences and Systems, Princeton University, 2014.

[6] ———, *DSOS and SDSOS: more tractable alternatives to sum of squares and semidefinite optimization*, In preparation, 2015.

[7] ———, *Some applications of polynomial optimization in operations research and real-time decision making*, Prepint available at http://arxiv.org/abs/1504.06002. To appear in Optimization Letters, 2015.

[8] F. Alizadeh and D. Goldfarb, *Second-order cone programming*, Math. Program. **95** (2003), no. 1, Ser. B, 3–51, DOI 10.1007/s10107-002-0339-5. ISMP 2000, Part 3 (Atlanta, GA). MR1971381

[9] George Phillip Barker and David Carlson, *Cones of diagonally dominant matrices*, Pacific J. Math. **57** (1975), no. 1, 15–32. MR0404302

[10] Stephen Boyd and Lieven Vandenberghe, *Convex optimization*, Cambridge University Press, Cambridge, 2004. MR2061575

[11] CPLEX, *V12. 4: Users manual for CPLEX*, International Business Machines Corporation **46**, no. 53, 157.

[12] Jesús A. de Loera and Francisco Santos, *An effective version of Pólya's theorem on positive definite forms*, J. Pure Appl. Algebra **108** (1996), no. 3, 231–240, DOI 10.1016/0022-4049(95)00042-9. MR1384003

[13] P. J. C. Dickinson and L. Gijben, *On the computational complexity of membership problems for the completely positive cone and its dual*, Available at `http://www.optimization-online.org/DB-FILE/2011/05/3041.pdf`, 2011.

[14] M. R. Garey and D. S. Johnson, *Computers and Intractability*, vol. 29, Wh Freeman, 2002.

[15] S. A. Gershgorin, *Uber die Abgrenzung der Eigenwerte einer Matrix*, Bulletin de l'Académie des Sciences de l'URSS. Classe des sciences mathématiques et na (1931), no. 6, 749–754.

[16] Martin Charles Golumbic, *Algorithmic graph theory and its applications*, Graph theory, combinatorics and algorithms, Springer, New York, 2005, pp. 41–62, DOI 10.1007/0-387-25036-0_3. MR2177666

[17] Nebojša Gvozdenović and Monique Laurent, *Semidefinite bounds for the stability number of a graph via sums of squares of polynomials*, Math. Program. **110** (2007), no. 1, Ser. B, 145–173, DOI 10.1007/s10107-006-0062-8. MR2306134

[18] J. Håstad, *Clique is hard to approximate within $n^{1-\epsilon}$*, Proceedings of the 37th Annual Symposium on Foundations of Computer Science.

[19] *Positive polynomials in control*, Lecture Notes in Control and Information Sciences, vol. 312, Springer-Verlag, Berlin, 2005. Edited by D. Henrion and A. Garulli. MR2121436

[20] D. Hilbert, *Über die Darstellung Definiter Formen als Summe von Formenquadraten*, Math. Ann. **32** (1888).

[21] Richard M. Karp, *Reducibility among combinatorial problems*, Complexity of computer computations (Proc. Sympos., IBM Thomas J. Watson Res. Center, Yorktown Heights, N.Y., 1972), Plenum, New York, 1972, pp. 85–103. MR0378476

[22] Jean B. Lasserre, *Global optimization with polynomials and the problem of moments*, SIAM J. Optim. **11** (2000/01), no. 3, 796–817, DOI 10.1137/S1052623400366802. MR1814045

[23] Monique Laurent, *Sums of squares, moment matrices and optimization over polynomials*, Emerging applications of algebraic geometry, IMA Vol. Math. Appl., vol. 149, Springer, New York, 2009, pp. 157–270, DOI 10.1007/978-0-387-09686-5_7. MR2500468

[24] M. Laurent and F. Vallentin, *Lecture Notes on Semidefinite Optimization*, 2012.

[25] Miguel Sousa Lobo, Lieven Vandenberghe, Stephen Boyd, and Hervé Lebret, *Applications of second-order cone programming*, Linear Algebra Appl. **284** (1998), no. 1-3, 193–228, DOI 10.1016/S0024-3795(98)10032-0. ILAS Symposium on Fast Algorithms for Control, Signals and Image Processing (Winnipeg, MB, 1997). MR1655138

[26] J. Löfberg, *Yalmip : A toolbox for modeling and optimization in MATLAB*, Proceedings of the CACSD Conference, 2004, Available from `http://control.ee.ethz.ch/~joloef/yalmip.php`.

[27] László Lovász, *On the Shannon capacity of a graph*, IEEE Trans. Inform. Theory **25** (1979), no. 1, 1–7, DOI 10.1109/TIT.1979.1055985. MR514926

[28] A. Majumdar, A. A. Ahmadi, and R. Tedrake, *Control and verification of high-dimensional systems via DSOS and SDSOS optimization*, Proceedings of the 53rd IEEE Conference on Decision and Control, 2014.

[29] A. Megretski, *SPOT: systems polynomial optimization tools*, 2013.

[30] T. S. Motzkin, *The arithmetic-geometric inequality*, Inequalities (Proc. Sympos. Wright-Patterson Air Force Base, Ohio, 1965), Academic Press, New York, 1967, pp. 205–224. MR0223521

[31] Yurii Nesterov, *Squared functional systems and optimization problems*, High performance optimization, Appl. Optim., vol. 33, Kluwer Acad. Publ., Dordrecht, 2000, pp. 405–440, DOI 10.1007/978-1-4757-3216-0_17. MR1748764

[32] P. A. Parrilo, *Structured semidefinite programs and semialgebraic geometry methods in robustness and optimization*, Ph.D. thesis, California Institute of Technology, May 2000.

[33] Pablo A. Parrilo, *Semidefinite programming relaxations for semialgebraic problems*, Math. Program. **96** (2003), no. 2, Ser. B, 293–320, DOI 10.1007/s10107-003-0387-5. Algebraic and geometric methods in discrete optimization. MR1993050

[34] Pablo A. Parrilo and Bernd Sturmfels, *Minimizing polynomial functions*, Algorithmic and quantitative real algebraic geometry (Piscataway, NJ, 2001), DIMACS Ser. Discrete Math. Theoret. Comput. Sci., vol. 60, Amer. Math. Soc., Providence, RI, 2003, pp. 83–99. MR1995016

[35] Bruce Reznick, *Uniform denominators in Hilbert's seventeenth problem*, Math. Z. **220** (1995), no. 1, 75–97, DOI 10.1007/BF02572604. MR1347159

[36] Bruce Reznick, *Some concrete aspects of Hilbert's 17th problem*, Contemporary Mathematics, vol. 253, American Mathematical Society, 2000, pp. 251–272.
[37] Lieven Vandenberghe and Stephen Boyd, *Semidefinite programming*, SIAM Rev. **38** (1996), no. 1, 49–95, DOI 10.1137/1038003. MR1379041

ORFE, Princeton University, Sherrerd Hall, Charlton Street, Princeton, New Jersey 08540

E-mail address: a_a_a@princeton.edu

ORFE, Princeton University, Sherrerd Hall, Charlton Street, Princeton, New Jersey 08540

E-mail address: gh4@princeton.edu

Contemporary Mathematics
Volume **685**, 2017
http://dx.doi.org/10.1090/conm/685/13718

Helly's theorem: New variations and applications

Nina Amenta, Jesús A. De Loera, and Pablo Soberón

ABSTRACT. This survey presents recent Helly-type geometric theorems published since the appearance of the last comprehensive survey, more than ten years ago. We discuss how such theorems continue to be influential in computational geometry and in optimization. The survey contains several open problems.

1. Introduction

Eduard Helly's theorem is now a century old[1] [**Hel23, Rad21**] and it is firmly established as a fundamental result in combinatorial geometry, one with a large number of applications. In its original form it states.

THEOREM 1.1 (Helly's theorem). *Let $\mathcal{F}$ be a finite family of convex sets in $\mathbb{R}^d$. If every $d+1$ or fewer elements of $\mathcal{F}$ intersects, then all the sets in family $\mathcal{F}$ intersect.*

We recommend [**Mat02**] for an in-depth introduction to combinatorial geometry. Since its discovery, Helly's theorem has found a plethora of generalizations, extensions and applications in many areas of mathematics (see [**DGK63, Eck93, Mat02, Wen04**] and the references therein). Our survey focuses on recent advances, roughly after the year 2004, regarding new versions of Helly's theorem and its generalizations. We also discuss applications of Helly theorems in computational geometry and optimization. The developments around this area have been tremendous.

2. The many variations and manifestations of Helly's theorem

One of the reasons for the popularity of Helly's theorem is its versatility; there are many ways to change the framework that yield interesting theorems. Changing the following aspects of Helly's theorem are common ways to get different versions.

- **Convexity of the sets.** The convexity hypothesis on the sets in $\mathcal{F}$ can be replaced by either stronger conditions, such as being axis-parallel boxes or translates of a convex set, or weaker conditions such as conditions on the homology groups of the sets and their intersections.

2010 *Mathematics Subject Classification.* Primary 52B05.

[1]The theorem was stated by Eduard Helly already in 1913, but his participation in the first world war delayed the publication of his proof until 1923. Alas, Johann Radon published a now classical proof a few years earlier too.

- **Quality of intersection as a conclusion.** One can replace the goal of having the whole family intersect by other conditions. This could include containing some integer points, having an intersection of a certain dimension or volume, having a bounded piercing number, etc.
- **Local intersection conditions.** Instead of requiring that all subfamilies of a certain size intersect, one can ask for other combinatorial conditions on families of a fixed size. Some examples are the (p, q) property, fractional conditions, or colorful conditions.

Given a property $\mathcal{P}$, it is standard to call $\mathcal{P}$ a *Helly-type property* if there is an integer h such that for a family $\mathcal{F}$ of sets, *if every h elements of $\mathcal{F}$ satisfy $\mathcal{P}$, so does the entire family $\mathcal{F}$.* The smallest such integer $h(\mathcal{P})$ is the Helly number of property $\mathcal{P}$. Helly's theorem essentially says that *having non-empty intersection* is a Helly-type property for all finite families of convex sets in $\mathbb{R}^d$, with $d+1$ as the Helly number. In this survey we restrict ourselves to Helly-type properties related to geometric intersection of sets, sometimes with additional conditions. In this context, there is a natural way to define Helly numbers for a given family $\mathcal{F}$ of sets, bypassing the specification of a property. Given a family $\mathcal{F}$ of sets, we define the $\mathcal{F}$-helly number $h(\mathcal{F})$ (if it exists) as the smallest integer satisfying the following. For any finite subfamily $\mathcal{G} \subset \mathcal{F}$, if every $h(\mathcal{F})$ sets of $\mathcal{G}$ intersect, all of the sets in $\mathcal{G}$ intersect too. If $h(\mathcal{G})$ is undefined, we simply say $h(\mathcal{G}) = \infty$.

With more than 300 papers published in the last fifteen years with relation to Helly's theorem it is not a surprise that we were forced to restrict ourselves to Helly-type theorems related to geometric intersection of sets. Sadly, we left detailed discussion of many interesting geometric Helly-type result out of this survey (see e.g., [**BJC15**, **Bre14**, **Hol13**, **CIKL13**, **Bre13**, **Kar12**, **Bre12**, **DS11**, **AB11**, **BT11**, **Goa10**, **Far09**, **Eck08**, **BM06b**, **GHP$^+$06**, **ABF06**] and the references therein) and we omitted other very active areas, with less emphasis in geometry, such as Helly theorems over graphs.

In each of the following subsections we will discuss a different family of new variants of Helly's theorem related to the intersections of geometric sets. They include coordinates restrictions and colorful results, as well as quantitative, topological, and fractional versions. We conclude the section with results regarding transversals to convex sets. Many results fall into more than one category, so we have made some arbitrary choices in the presentation. We include a few short proofs to help provide insight, and we establish some new results generalizing or combining existing theorems as we go along.

2.1. Coordinates restrictions and S-Helly numbers. Let us motivate what S-Helly numbers are with an example.

THEOREM 2.1 (Boxes-only Helly Theorem; Halman [**Hal07a**]). *Given a finite set D of at least $2d$ axis parallel boxes in $\mathbb{R}^d$, and a finite set S of points, the boxes in D have a common point in S if every $2d$ of its elements do.*

The result above is highly dependent on the set of points S. Notice we get a radical simplification when $S = \mathbb{R}^d$, namely,

THEOREM 2.2. *Given a finite set D of axis parallel boxes in $\mathbb{R}^d$, the boxes in D have a common point if every two of its elements do.*

There is a wide collection of Helly-type results, where, as in Theorem 2.1, the hypotheses and conclusions of intersection are required to contain points from a specific proper subset S of $\mathbb{R}^d$. A classic result of this sort is Doignon's theorem, in which the point set S is more strictly specified to be $S = \mathbb{Z}^d$. The theorem is a variation of Helly's theorem.

THEOREM 2.3 (Doignon [**Doi73**]). *Given a finite family D of at least 2^d convex sets in $\mathbb{R}^d$, the sets in D have a common point in $\mathbb{Z}^d$ if every 2^d of its elements do.*

Doignon's theorem was rediscovered by David E. Bell [**Bel76**] and Herbert Scarf [**Sca77**], with applications in integer optimization in mind.

We can now define in general the S-Helly numbers:

DEFINITION 2.4. Given a set $S \subset \mathbb{R}^d$, we consider *the S-Helly number*

$$h(S) = h(\{K \cap S : K \subset \mathbb{R}^d \text{ is convex}\}).$$

In other words, it is the smallest positive integer such that the following statement holds. *Given a finite family $\mathcal{F}$ of convex sets in $\mathbb{R}^d$, if the intersection of every $h(S)$ contains a point of S, then there is a point of S in $\cap\mathcal{F}$.* Note that $h(S)$ is the Helly number for the property *"having non-empty intersection over S"*. If no number satisfies the condition above, we say $h(S) = \infty$. We use $h(S)$ to denote the Helly number for set S, since in this context $\mathcal{F}$ is always the family of convex sets.

Helly's original theorem can thus be stated as $h(\mathbb{R}^d) = d + 1$, and Doignon's theorem as $h(\mathbb{Z}^d) = 2^d$. Interestingly, many other sets S give Helly-type theorems as well. For the case of dimension $d = 1$ the reader can prove easily that for any set $S \subset \mathbb{R}$, the Helly number $h(S)$ exists and is at most two, but starting with $\mathbb{R}^2$ the situation becomes considerably more complicated (see for example the results in [**DLLHORP15**]).

Historically, many results about S-Helly numbers were originally discussed in the setting of *abstract convexity*, which focuses on the study of spaces emulating a *convex hull operator* and recreating many results which were known only in Euclidean spaces [**vdV93**]. Many *convexity spaces* can be formed by taking a proper subset $S \subset \mathbb{R}^d$, and defining the convex sets as the intersections of the standard convex sets in $\mathbb{R}^d$ with S, as in Definition 2.4 (see [**DRS81, Hof79, Jam81, KW71**] and the references therein).

In recent years, the excitement around S-Helly numbers has increased due to the rich connections to the theory of optimization, in particular cutting planes (see the end of Subsection 3.4). Motivated by the work of Bell, and Scarf, A. J. Hoffman developed a theory to compute general S-Helly numbers in [**Hof79**]. Hoffmann stated and sketched the following *mixed integer* version of Helly's and Doignon's theorems, which includes them both. Later it was rediscovered by Averkov and Weismantel [**AW12**], who gave a detailed proof:

THEOREM 2.5 (Mixed-Integer Helly theorem [**Hof79, AW12**]). *The Helly number $h(\mathbb{R}^d \times \mathbb{Z}^k)$ equals $(d + 1)2^k$.*

The following methods are useful to bound the S-Helly number of a subset S of $\mathbb{R}^d$. The techniques were studied first by Hoffman [**Hof79**], and later developed by several authors [**AW12, Ave13a, CDS, DLLHORP15**].

We define an *S-vertex-polytope* as the convex hull of points $x_1, x_2, \ldots, x_k \in S$ in convex position such that no other point of S is in $\operatorname{conv}\{x_1, \ldots, x_k\}$. Similarly, an S-*face-polytope* is defined as the intersection of half-spaces $H_1, H_2, \dots, H_k$ such that $\bigcap_i H_i$ has k facets and contains exactly k points of S, one contained in the relative interior of each facet.

A set $S \subset \mathbb{R}^d$ is said to be *discrete* if every point of $\mathbb{R}^d$ has an open neighborhood that contains only a finite number of points of S.

LEMMA 2.6 ([**Ave13a**, Thm. 2.1] [**Hof79**, Prop. 3]). *Assume $S \subset \mathbb{R}^d$ is discrete, then $h(S)$ is equal to the following two numbers, which must be equal:*

(1) *The supremum of the number of faces of an S-face-polytope.*
(2) *The supremum of the number of vertices of an S-vertex-polytope.*

The case when S is not discrete is a bit more delicate. If we denote by $f(S)$ the quantity (1) above, Averkov [**Ave13a**] showed that in general $f(S) \leq h(S)$. Recently, new results involving these two quantities and the S-Helly number $h(S)$ have been obtained by Conforti and Di Summa [**CDS, Ave13a**]. For example, they proved that when S is a closed subset and $f(S)$ is finite, then $h(S) \leq (d+1)f(S)$. They also showed an example where $f(S) = 1$, yet the S-Helly number is infinite (examples appeared in [**DLLHORP15**]). Fortunately, for the calculation of Helly numbers there is a very general result due to A.J. Hoffman characterizing S-Helly numbers.

LEMMA 2.7 (Proposition 2 in [**Hof79**]). *If $S \subset \mathbb{R}^d$, then $h(S)$ can be computed as the supremum of the numbers h such that the following holds: there exists a set $R = \{x_1, \dots, x_h\} \subset S$ such that $\bigcap_i \operatorname{conv}(R \setminus \{x_i\})$ does not intersect S.*

Note that, in particular, the points of R must be in strictly convex position. As a direct application of Lemma 2.7 we have the following proposition.

PROPOSITION 2.8. *If $S_1, S_2 \subseteq \mathbb{R}^d$, then $h(S_1 \cup S_2) \leq h(S_1) + h(S_2)$.*

The Helly number of a product is not always so well-behaved as in the mixed integer case, but Averkov and Weismantel proved the following

PROPOSITION 2.9 ([**AW12**, Thm. 1.1]). *If M is a closed subset of $\mathbb{R}^k$ then $h(\mathbb{R}^d \times M) \leq (d+1)h(M)$.*

Similarly, one can prove a general bound in the case of discrete sets. The result below shows the optimality of Doignon's theorem, for example.

THEOREM 2.10 (Conforti, Di Summa [**CDS**]). *If $S_1, S_2 \subset \mathbb{R}^d$ are discrete sets, then $h(S_1 \times S_2) \geq h(S_1)h(S_2)$.*

It is also natural to study subgroups of $\mathbb{R}^d$. General subgroups and other new examples of sets S with $h(S) < \infty$ are presented in [**DLLHORP15**].

THEOREM 2.11 (De Loera, La Haye, Oliveros, Roldán Pensado [**DLLHORP15**]). *The following bounds on Helly numbers hold*

- *If S is a dense subset of $\mathbb{R}^2$, then $h(S) \leq 4$. This result is sharp.*
- *Let L be a proper sublattice of $\mathbb{Z}^2$. If $S = \mathbb{Z}^2 \setminus L$, then $h(S) \leq 6$. This result is sharp.*
- *Let $L_1, \dots, L_k$ be (possibly translated) sublattices of $\mathbb{Z}^d$. Then the set $S = \mathbb{Z}^d \setminus (L_1 \cup \dots \cup L_k)$ has Helly number $h(S) \leq C_k 2^d$ for some constant C_k depending only on k.*

2.2. Colorful variations. Colorful variations of Helly's theorem come from asking additional combinatorial restrictions to the intersection structure of the family. This result is due to Lovász, and his proof, below, appeared in a paper by Bárány.

THEOREM 2.12 (Colorful Helly theorem; Lovász [**Bár82**]). *Let $\mathcal{F}_1, \mathcal{F}_2, \dots, \mathcal{F}_{d+1}$ be finite families of convex sets in $\mathbb{R}^d$ such that for every choice $K_1 \in \mathcal{F}_1, K_2 \in \mathcal{F}_2, \dots, K_{d+1} \in \mathcal{F}_{d+1}$, the family $\{K_1, \dots, K_{d+1}\}$ has non-empty intersection. Then, there is an index i such that the members of family $\mathcal{F}_i$ have a non-empty intersection.*

Note that if $\mathcal{F}_1 = \mathcal{F}_2 = \ldots = \mathcal{F}_{d+1}$ we obtain Helly's original theorem. The sets $\mathcal{F}_i$ are called *color classes*, which gives the theorem its name.

PROOF. We may assume without loss of generality that the sets are compact. Let v be a direction in $\mathbb{R}^d$. We may choose v so that for every colorful choice of d sets, their intersection attains its v-directional minimum at a single point (a consequence of [**ELR70**], for instance).

Take a colorful choice of d sets such that its v-directional minimum p is maximal (in the direction v). We may assume without loss of generality that the missing color in that d-tuple is $\mathcal{F}_{d+1}$, and denote its sets by $F_1 \in \mathcal{F}_1, \dots, F_d \in \mathcal{F}_d$. We will show that $p \in \cap \mathcal{F}_{d+1}$.

Take any set $F_{d+1} \in \mathcal{F}_{d+1}$. By the condition of the problem, the $(d+1)$-tuple $A = \{F_1, \dots, F_{d+1}\}$ must be intersecting. Let q be a point in its intersection. Notice that $\langle q, v\rangle \geq \langle p, v\rangle$, where $\langle \cdot, \cdot \rangle$ represents the dot product. Let $A_i = A \setminus \{F_i\}$ for $1 \leq i \leq d+1$.

By the definition of p, for each i, the v-directional minimum q_i of $\cap A_i$ satisfies $\langle q_i, v\rangle \leq \langle p, v\rangle$. Since $\cap A_i$ also contains q, by convexity there must be a point $p_i \in \cap A_i$ with $\langle p_i, v\rangle = \langle p, v\rangle$ (notice that $p_{d+1} = p$).

Thus, we have obtained $d+1$ points in the hyperplane $H = \{x : \langle x, v\rangle = \langle p, v\rangle\}$, of dimension $d-1$. By Radon's lemma [**Rad21**], there must be a partition of them into two sets B, C such that $\operatorname{conv} B \cap \operatorname{conv} C \neq \emptyset$. Take any point $p' \in \operatorname{conv} B \cap \operatorname{conv} C$. It is immediate that $p' \in F_i$ for all $1 \leq i \leq d+1$. Thus $p' = p$, which means $p \in F_{d+1}$, as desired. □

Not every Helly-type theorem admits a colorful version. For instance, a colorful version of Theorem 2.2 does not hold even if we permit d colors. A simple counterexample is to take $\mathcal{F}$ to be the set the of facets of a hypercube. We color $\mathcal{F}$ so that each opposite pair of facets carry the same color. Clearly no color class intersects, but every colorful choice intersects in one vertex of the hypercube. There are two interesting questions regarding colorful variations:

- Which Helly-type theorems admit a colorful version?
- Is it possible to get a stronger conclusion than a single color class intersecting?

Sufficient conditions for the existence of colorful version were determined in [**DLLHORP15**], which are essentially the minimal conditions needed to carry out Lovász's original proof. In particular, it follows that any discrete set S with a finite Helly number $h(S)$ admits a Helly-type theorem with $h(S)$ color classes. This extends to some quantitative versions, described in the next subsection [**DLLHRS15**,

DLLHRS15a, DLLHRS15b]. In particular, it shows that Doignon's theorem is colorable.

THEOREM 2.13 (Colorful Doignon; De Loera, La Haye, Oliveros, Roldán-Pensado [**DLLHORP15**]). *Let $\mathcal{F}_1, \mathcal{F}_2, \ldots, F_{2^d}$ be finite families of convex sets in $\mathbb{R}^d$. If the intersection of every colorful choice $F_1 \in \mathcal{F}_1, \ldots, F_{2^d} \in \mathcal{F}_{2^d}$ contains an integer point, there is a color class $\mathcal{F}_i$ such that $\cap\mathcal{F}_i$ contains an integer point.*

There are still plenty of Helly-type theorems that do not fit the criteria in [**DLLHORP15**]. For instance, Karasev presented the following theorem.

THEOREM 2.14 (Karasev [**Kar00**]). *Let K be a convex set in the plane and $\mathcal{F}$ be a family of translates of K. If every pair of sets in $\mathcal{F}$ has a non-empty intersection, then there are three points such that every set in $\mathcal{F}$ contains at least one of them.*

Dol'nikov asked if Theorem 2.14 could be colored.

OPEN PROBLEM 2.15 (Dol'nikov's problem). Let K be a convex set in the plane and $\mathcal{F}_1, \mathcal{F}_2, \mathcal{F}_3$ be finite families of translates of K. If any two translates of K from different families intersect, show that there is an index i and three points in $\mathbb{R}^d$ such that every set of $\mathcal{F}_i$ contains at least one of the points.

The problem above is only solved when K is a triangle, K is centrally symmetric, or when the hypothesis involves four families instead of three [**JCMS15**].

A stronger form of the colorful Helly theorem was shown by Kalai and Meshulam. Their methods work in much more general settings, where the conditions on the color classes are replaced by an arbitrary matroid. In its most direct application to convex sets in $\mathbb{R}^d$, it yields the theorem below. Even though the statement is indeed combinatorially stronger than the original colored version by Lovász, the reader may notice that the proof at the beginning of this subsection actually implies both results. See also the interesting commutative algebra reformulation of colorful variants in [**Flø11**].

THEOREM 2.16 (Kalai, Meshulam [**KM05**]). *Let $\mathcal{F}_1, \mathcal{F}_2, \ldots, \mathcal{F}_{d+1}$ be finite families of convex sets in $\mathbb{R}^d$ such that for every choice $K_1 \in \mathcal{F}_1, K_2 \in \mathcal{F}_2, \ldots, K_{d+1} \in \mathcal{F}_{d+1}$, the family $\{K_1, \ldots, K_{d+1}\}$ has non-empty intersection. Then, there is an index i and a choice of sets $F_j \in \mathcal{F}_j$ for all $j \neq i$ such that the family $\mathcal{F}_i \cup \{F_j : j \neq i\}$ has non-empty intersection.*

If we change the ambient space, colorful results also appear. A subset K of the sphere S^{d-1} is called *strongly convex* if it is contained in an open hemisphere and for every pair of points x, y in K, the shortest arc from x to y is contained in K. In the sphere there are different notions of convexity, which lead to different Helly-type results (see the survey [**Wen04**] and the references therein for details).

THEOREM 2.17 (Holmsen, Pach, Tverberg [**HPT08**]). *Let $C_1, C_2, \ldots, C_{d+1}$ be non-empty, finite collections of closed strongly convex sets on S^{d-1}. If every selection of sets from distinct C_i have a point in common, then for some $1 \leq i < j \leq d+1$ there is a point common to every member of $C_i \cup C_j$.*

This is a consequence of applying duality to the following "very colorful Carathéodory", which also appears in [**ABB$^+$09**]. For more on the interesting topic of colorful Carathéodory theorems see [**Bár82, BO95, BO97**].

THEOREM 2.18 (Very colorful Carathéodory, [**HPT08, ABB$^+$09**]). *Let $C_1, C_2, \ldots, C_{d+1}$ be sets of points in $\mathbb{R}^d$, considered as color classes. If every colorful choice $c_1 \in C_1, \ldots, c_{d+1} \in C_{d+1}$ does not capture the origin (i.e., $0 \notin \operatorname{conv}\{c_1, \ldots, c_{d+1}\}$), then there are two color classes C_i, C_j such that $0 \notin \operatorname{conv}(C_i \cup C_j)$.*

A similar version of Theorem 2.17 holds in $\mathbb{R}^d$ with more generality. Namely, the following new theorem,

THEOREM 2.19. *Let $n > d$ and $\mathcal{F}_1, \ldots, \mathcal{F}_n$ be non-empty, finite collections of closed convex sets in $\mathbb{R}^d$. If every selection of sets, one from each $\mathcal{F}_i$, have a point in common, then there is a set I of $n - d$ indices in $\{1, \ldots, n\}$ such that there is a point in common to every member of $\cup_{i \in I} \mathcal{F}_i$.*

The reader may notice that the original proof by Lovász, described at the beginning of this subsection, proves this result.

2.3. Quantitative variations. More variations of Helly's theorem appear when one wants to quantify the size of the intersection of the sets in the family. The driving principle behind these results is that *given a family $\mathcal{F}$ of convex sets in $\mathbb{R}^d$, if every subfamily of a fixed size has a "large intersection", then the whole family has a "large intersection" too.*

If we consider $\mathcal{C}(\mathbb{R}^d)$ the space of all convex sets of $\mathbb{R}^d$ with the Hausdorff topology, we can measure the size of a convex set by any non-negative monotone function $f : \mathcal{C}(\mathbb{R}^d) \to \mathbb{R}^+ \cup \{\infty\}$. We say f is monotone if $A \subset B$ implies $f(A) \leq f(B)$. The crudest way to define that a convex set K is large is to say that K is large if it is non-empty, corresponding to Helly's theorem. The usual non-quantitative Helly theorems arise when we ask whether $f(K) > 0$ or not. For example, if we are given a subset $S \subset \mathbb{R}^d$, we could define $f(K) = 0$ if $K \cap S = \emptyset$ and $f(K) = 1$ otherwise. This reproduces the S-Helly numbers we mentioned earlier.

Given a discrete set $S \subset \mathbb{R}^d$, one can consider the function $f(K) = |K \cap S|$, which leads to several new Helly-type results. Other notions of size are given by the volume of K or its diameter. We stress that Helly-type theorems based on functions such as $f(K) = \operatorname{vol}(K)$ or $f(K) = \operatorname{diam}(K)$, which vary continuously over the convex sets, yield wildly different results than Helly-type theorems based on functions such as $f(K) = |K \cap S|$, which vary discretely.

Our first example is a quantitative version of Doignon's theorem.

THEOREM 2.20 (Aliev, Bassett, De Loera, Louveaux [**ABDLL**]). *Let d, k be non-negative integers, there exists a constant $c(k, d)$, depending only on k and d, such that if a rational polytope $P_A(b) = \{x : Ax \leq b\}$ in $\mathbb{R}^d$ has exactly k integral solutions, then there is a subset of the inequalities of $Ax \leq b$, of cardinality no more than $c(k, d)$, such that the induced subproblem has exactly the same k integer solutions as $P_A(b)$. Moreover,*

$$c(k, d) \leq \lceil 2(k+1)/3 \rceil (2^d - 2) + 2.$$

Note that the original Doignon-Bell-Scarf theorem is $k = 0$. This theorem implies a version for convex sets.

COROLLARY 2.21. *Let $(X_i)_{i \in \Lambda}$ be a collection of convex sets in $\mathbb{R}^d$, where at least one of these sets is compact. If the intersection of every sub-collection of size at most $c(k-1, d)$ contains at least k integer points, then the intersection of the whole family contains at least k integer points.*

OPEN PROBLEM 2.22. Determine the exact value of the Helly number $c(k,d)$.

The value of $c(k,d)$ by Aliev et al. was shown to be optimal for $k=1$, and not to be optimal starting from $k=3$. Recently Averkov et al. showed that the formula above is tight for $k=2$ [**AMHPW**] and that the Helly number $c(k,d)$ can be used to bound the number of lattice points on the boundary of a strictly convex set in terms of the number of lattice points in the interior. Of course, one can count points of discrete sets, thus one might expect similar quantitative Helly numbers for other discrete sets besides lattices.

DEFINITION 2.23. We say that a discrete set $S \subset \mathbb{R}^d$ has a k-quantitative Helly number $h_k(S)$ if it is the smallest positive integer such that the following statement holds. For every finite family $\mathcal{F}$ of convex sets in $\mathbb{R}^d$, if the intersection of every subfamily of size $h_k(S)$ has at least k points of S, then $\cap\mathcal{F}$ has at least k points of S. If no integer satisfies that condition, we say $h_k(S)=\infty$.

OPEN PROBLEM 2.24. Characterize the pairs (S,k) for which $h_k(S)<\infty$.

For instance, in [**DLLHRS15**], it was shown that the difference of a lattice and the union of some of its sublattices has finite k-quantitative Helly numbers for all k. Moreover, the fact that $h_k(S)<\infty$ is enough to give quantitative versions of other classic results in discrete geometry, such as Tverberg's theorem [**Tve66**].

Historically, the first quantitative Helly-type theorems were presented by Bárány, Katchalski and Pach, and were based on several continuous functions over the convex sets.

THEOREM 2.25 (Large-volume Helly Theorem; Bárány, Katchalski, Pach [**BKP84**, **BKP82**]). *Given a family $\mathcal{F}$ of $n \geq 2d$ convex sets in $\mathbb{R}^d$, if the intersection of every $2d$ members of $\mathcal{F}$ has volume at least one, then the intersection of all the members of $\mathcal{F}$ has volume at least d^{-2d^2}.*

The guarantee on $\mathrm{vol}(\cap\mathcal{F})$ has been improved by Naszódi to d^{-cd} for some absolute constant $c>0$ [**Nas16**], which is optimal up to the value of c. Naszódi's methods have been refined in [**Bra17**]. This begs the question of how much the volume guarantee can be improved if the size of the subfamilies one is willing to check for intersection is increased from $2d$ to a higher cardinality.

THEOREM 2.26 (De Loera, La Haye, Rolnick, Soberón [**DLLHRS15**]). *Let $\varepsilon>0$ and d be a positive integer. There is an integer $n^{\mathrm{vol}}(d,\varepsilon)$ such that the following statement holds. Given a finite family $\mathcal{F}$, if the volume of the intersection of every subfamily of size at most $n^{\mathrm{vol}}(d,\varepsilon)$ is at least one, then*

$$\mathrm{vol}(\cap\mathcal{F}) \geq (1+\varepsilon)^{-1} \sim 1-\varepsilon.$$

Moreover, the optimal value of $n^{\mathrm{vol}}(d,\varepsilon)$ satisfies $n^{\mathrm{vol}}(d,\varepsilon)=\Theta_d(\varepsilon^{-(d-1)/2})$.

The Θ_d notation hides constants which may depend on the dimension. We should emphasize the striking difference between Theorem 2.26 and Corollary 2.21. If we measure our sets by their volume, there is no way to avoid a loss. In other words, if we ask for every subfamily of fixed size to have volume at least one, we cannot guarantee that the intersection of the whole family has volume at least one. However, when counting the number of points of a discrete set S in our convex sets, no such loss appears.

Results such as Theorem 2.26 are closely related to finding good approximations of convex sets by polyhedra. This is a vibrant subfield of convexity; we recommend the surveys [**Bro08**, **Gru93**]. For the sake of obtaining Helly-type theorems, we require a very specific kind of approximation compatible with the function we wish to quantify.

DEFINITION 2.27. Given a function $f : \mathcal{C}(\mathbb{R}^d) \to \mathbb{R}^+ \cup \{\infty\}$ we say that f is compatible with polyhedral approximations of convex sets if for every $\varepsilon > 0$ there is a number $n(f, d, \varepsilon)$, depending only on f, the dimension d, and ε, such that the following holds. For any convex set $K \subset \mathbb{R}^d$ with $f(K) < \infty$, there is a polyhedron P such that

- $K \subset P \subset \mathbb{R}^d$,
- P has at most $n(f, d, \varepsilon)$ facets, and
- $(1 - \varepsilon)f(P) \leq f(K)$.

Let $n(f, d, \varepsilon)$ be the minimum value which satisfies the conditions above.

For example, compatible polyhedral approximations exist for $f(K) = \mathrm{vol}(K)$, for $f(K) = \max\{c^T x : x \in K\}$, and for $f(K) = \int_K g(x)dx$ where $g(x)$ is a non-negative Lipschitz function. In the last example, if we only ask for $g(x)$ to be continuous with respect to the Lebesgue measure, f may not be compatible with polyhedral approximations of convex sets.

In order to see this, consider for each positive integer n the circle $C_n \subset \mathbb{R}^2$ of radius one around the point $(3n, 0)$. Then, we can consider μ_n a uniform probability measure on the set $C_n + B_{1/(2n)}(0)$, where $B_{1/(2n)}(0)$ stands for the ball of radius $1/(2n)$ around the origin. Finally, our function f is simply $f = \sum_{n=1}^{\infty} \mu_n$. This function is continuous but not compatible with polyhedral approximations of convex sets. Indeed, given $\varepsilon > 0$, it is clear that if $K = C_n$, for a polygon $P \supset K$, if $f(P)$ is to satisfy the desired conditions, then P would have to be arbitrarily close to K in Hausdorff distance. This would imply that P has an arbitrarily large number of sides, making hte existence of a universal $n(f, d, \varepsilon)$ impossible.

Using this concept and the properties of such functions f, we can prove the following new theorem. Which is a very general version of Theorem 2.26. The optimality of the bound below is discussed after the proof.

THEOREM 2.28. *Let $d, \varepsilon > 0$ be given. Let $f : \mathcal{C}(\mathbb{R}^d) \to \mathbb{R}^+ \cup \{\infty\}$ be a monotone function which is compatible with polyhedral approximations. Then, for any finite family $\mathcal{F}$ of convex sets in $\mathbb{R}^d$, if every subfamily $\mathcal{F}'$ of size at most $d \cdot n(f, d, \varepsilon)$ satisfies $f(\cap \mathcal{F}') \geq 1$, then the entire family satisfies $f(\cap \mathcal{F}) \geq 1 - \varepsilon$.*

The results mentioned in Section 3 show that the theorem above can be used to approximate the value of the integral of a Lipschitz functions or the volume of the intersection of a family of convex sets. To prove Theorem 2.28 we need the following lemma, which, not surprisingly, is a direct consequence of Helly's theorem.

LEMMA 2.29. *Let $\mathcal{F}$ be a finite family of convex sets in $\mathbb{R}^d$ and H^+ be a closed half-space. If $\cap \mathcal{F} \neq \emptyset$ and $\cap \mathcal{F} \subset H^+$, then there is a subfamily $\mathcal{G} \subset \mathcal{F}$ of size at most d such that $\cap \mathcal{G} \subset H^+$.*

PROOF. Consider $H^- = \mathbb{R}^d \setminus H^+$. Notice that the family $\mathcal{F} \cup \{H^-\}$ has empty intersection and is finite. By Helly's theorem, there must be a subfamily of size at most $d + 1$ which has empty intersection. This subfamily must include H^-, since $\cap \mathcal{F} \neq \emptyset$. The other d elements form the family $\mathcal{G}$ we were looking for. □

PROOF OF THEOREM 2.28. If $f(\cap\mathcal{F}) = \infty$, there is nothing to prove. Otherwise, assume $f(\cap\mathcal{F}) < \infty$. We can construct a polyhedron P as in Definition 2.27 for $K = \cap\mathcal{F}$. Then, for an arbitrary facet of P, consider the half-space determined by that hyperplane which contains $\cap\mathcal{F}$. Using the lemma above, we can find d sets of $\mathcal{F}$ whose intersection is contained in that half-space.

If we repeat this process for each facet of P, we have constructed a family $\mathcal{F}'$ of size at most $d \cdot n(f, d, \varepsilon)$ such that $\cap F' \subset P$. Then, it suffices to notice that

$$f(\cap\mathcal{F}) \geq (1-\varepsilon)f(P) \geq (1-\varepsilon)f(\cap\mathcal{F}') \geq 1-\varepsilon.$$

□

In many cases, one can obtain that the constant $n(f, d, \varepsilon)$ is also a lower bound for the cardinality of the subfamilies to check in order to obtain the conclusion of Theorem 2.28. In other words, in those cases the bound given by Theorem 2.28 is off by a multiplicative factor of at most d.

To do this, it suffices to take a convex set K which does not allow an approximation as in Definition 2.27, e.g., one that cannot be approximated efficiently with a polyhedron with at most $n(f, d, \varepsilon) - 1$ facets, and taking $\mathcal{F}$ to be a sufficiently large family of half-spaces containing K. This idea gives the lower bound of Theorem 2.26, for instance.

Thus, in order to obtain a quantitative Helly-type theorem for other functions, it suffices to prove the existence of efficient polyhedral approximations (i.e., bound $n(f, d, \varepsilon)$). The upper bounds for the volume $n(\text{vol}, d, \varepsilon) \sim n^{\text{vol}}(d, \varepsilon)$ can be found in [**RSW01**], and the lower bounds in [**Gru93**]. A compactness argument over $\mathcal{C}(\mathbb{R}^d)$ in [**DLLHRS15**] shows that $n(\text{diam}, d, \varepsilon) < \infty$. Precise asymptotic bounds $n(\text{diam}, d, \varepsilon) = \Theta_d(\varepsilon^{-(d-1)/2})$ appear in [**Sob16**].

OPEN PROBLEM 2.30. Determine the largest value $r(d)$ such that the following statement holds. For every finite family $\mathcal{F}$ convex sets in $\mathbb{R}^d$, if the intersection of every $2d$ of them has diameter at least one, then $\text{diam}(\cap\mathcal{F}) \geq r(d)$.

Is known that $r(d) \geq d^{-2d}$ [**BKP82**], but the conjecture is $r(d) \sim cd^{-1/2}$ for some constant c. Brazitikos has recently confirmed the conjecture in the case where the sets are centrally symmetric around the origin [**Bra15**], and provided the following result for the non-symmetric case.

THEOREM 2.31 (Brazitikos [**Bra15**]). *There are absolute constants $\alpha > 1, c > 0$ such that the following is true. For any finite family $\mathcal{F}$ of convex bodies in $\mathbb{R}^d$, if the diameter of the intersection of any αd of them has diameter at least one, then the* $\text{diam}(\cap\mathcal{F}) \geq cd^{-3/2}$.

2.4. Topological versions. The convexity in Helly's theorem can be replaced by topological properties on the sets of $\mathcal{F}$ and their intersections. Indeed, convex sets (and their intersections) are contractible sets, so it is to be expected that such conditions should be sufficient to characterize the intersection of finite families. The first theorem of this kind was proved by E. Helly himself.

We say a topological space is a *homology cell* if it is non-empty and its homology groups vanish in all dimensions.

THEOREM 2.32 (Helly [**Hel30**]). *Given a finite family $\mathcal{F}$ of $n \geq d+1$ open homology cells in $\mathbb{R}^d$, if*

- *every subfamily $\mathcal{F}'$ of $\mathcal{F}$ of at most $d+1$ members has non-empty intersection and*
- *for every proper subfamily $\mathcal{F}'$ of $\mathcal{F}$ with non-empty intersection $\cup\mathcal{F}'$ is a homology cell,*

then the intersection of all the members of $\mathcal{F}$ is non-empty.

The intuitive idea explaining this theorem is that if a large finite family $\mathcal{G}$ of sets as above has the property that all proper subfamilies are intersecting but $\mathcal{G}$ is not, then $\cup\mathcal{G}$ is homologically equivalent to a high-dimensional ball. This is impossible if $\cup\mathcal{G}$ is embedded in a low-dimensional Euclidean space.

LEMMA 2.33. *If $\mathcal{G}$ is a finite family of open homology cells in a topological space, such that*

- *every proper subfamily of $\mathcal{G}$ is intersecting and its union is a homology cell,*
- *$\cap\mathcal{G} = \emptyset$, and*
- *$|\mathcal{G}| = n+1$,*

then $\cup\mathcal{G}$ is homologically equivalent to S^{n-1}.

PROOF OF LEMMA 2.33. We proceed by induction on n. If $n = 1$, the result is true. Assume that the result is true for $n = k$ and let us consider a family with the properties above and of size $k+2$.

Let B be a set in $\mathcal{G}$, and let $\mathcal{G}' := \mathcal{G} \setminus \{B\}$. We know $\cap\mathcal{G}' \neq \emptyset$ and $A = \cup\mathcal{G}'$ is a homology cell. The family $\mathcal{G}'_B = \{K \cap B : K \in \mathcal{G}'\}$ satisfies the condition for $n = k$. Thus, $A \cap B = \cup\mathcal{G}'_B$ is homologically equivalent to S^{k-1}. By the Mayer-Vietoris sequence we have that $A \cup B = \cup G$ is homologically equivalent to S^k. Intuitively, $A \cap B$ is homologically equivalent to S^{k-1} and adding the cells A and B we are adding the north and south hemispheres of S^k. □

PROOF OF THEOREM 2.32. If the theorem does not hold, let $\mathcal{G}$ a subfamily of minimal cardinality such that $\cap\mathcal{G} = \emptyset$. Then, by the conditions of the theorem $n = |\mathcal{G}| \geq d+2$ and by the lemma above $\cup\mathcal{G} \sim S^{n-2}$. Since $\cup\mathcal{G}$ is embedded in $\mathbb{R}^d$ and $n-2 \geq d$, this is not possible. □

Of course, being a homology cell is quite a strong condition. The requirements on the sets have been trimmed down. For instance, there is a classical theorem of Breen that extended Helly's theorem:

THEOREM 2.34 (Helly's theorem for star-shaped union; Breen [**Bre90**]). *Let $\mathcal{F}$ be a finite family of compact convex sets in $\mathbb{R}^d$. Then, every subfamily of $\mathcal{F}$ of size at most $d+1$ has starshaped union if and only if $\cap\mathcal{F} \neq \emptyset$.*

To produce a topological generalization of Breen's theorem with a proof similar to that of Theorem 2.32, we only need to ask for conditions on certain homology groups $H_j(\cdot)$ of the subfamilies. Thus:

THEOREM 2.35 (Topological Breen theorem; Montejano [**Mon14**]). *Let $\mathcal{F}$ be a finite family of open subsets of a topological space X. Let d be an integer such that $H_i(U) = 0$ for all $i \geq d$ and every open subset U of X.*

Suppose that $H_{j-2}(\cup\mathcal{F}') = 0$ for any subset $\mathcal{F}' \subset \mathcal{F}$ of size j, $1 \leq j \leq d+1$. Then $\cap\mathcal{F} \neq \emptyset$.

A topological version of Helly's theorem by making requirements on the intersections rather than the unions also holds.

THEOREM 2.36 (Topological Helly theorem; Montejano [**Mon14**]). *Let $\mathcal{F}$ be a finite family of open subsets of a topological space X. Let d be an integer such that $H_i(U) = 0$ for all $i \geq d$ and every open subset U of X.*

Suppose that $H_{d-j}(\cap \mathcal{F}') = 0$ for any subset $\mathcal{F}' \subset \mathcal{F}$ of size j, $1 \leq j \leq d+1$. Then $\cap \mathcal{F} \neq \emptyset$.

This uses fewer conditions than an earlier topological Helly-style theorem by Debrunner [**Deb70**]. Another way to obtain topological versions of Helly's theorem is to work with the *nerve complex* of the family $\mathcal{F}$ instead of $\mathcal{F}$. Given a family of sets $\mathcal{F}$, the nerve complex of $\mathcal{F}$, denoted by $N(\mathcal{F})$, is the simplicial complex formed by adding one vertex for each non-empty element in $\mathcal{F}$ and a face for each intersecting subfamily. It turns out that the nerve complexes of families of convex sets have deep topological properties. A lucid explanation is contained in a recent survey by Tancer [**Tan13**]. It should be noted that some topological versions, such as the colorful topological Helly type results of Kalai and Meshulam [**KM05**], have applications to interesting purely geometric problems. Two examples are applications to geometric versions of Hall's theorem for graph [**HMSM14**] and extensions of the colorful Carathéodory theorem [**HK15**].

One of the key ideas around Helly theorems, as is noted among some proofs in this survey, is to continuously translate a half-space containing the family until it loses a non-trivial intersection. The same idea was used by Wegner to show that nerve complexes of families of convex sets in $\mathbb{R}^d$ are *d-collapsible* [**Weg75**]. In other words, the nerve complex of a finite family of convex sets may be brought down to the empty simplex by iteratively removing faces of dimension at most $d-1$ which are contained in exactly one containment-maximal face. Indeed, the fact that every nerve complex of a finite family of convex sets in $\mathbb{R}^d$ is d-collapsible implies Helly's theorem.

Many Helly-type results hold for families whose nerve complexes are d-collapsible. A stronger condition than being d-collapsible is being *d-Leray*, which means that all homology groups of $N(\mathcal{F})$ of dimension greater than or equal to d are zero (see [**MT09**] where collapsable and Leray complexes are discussed in detailed). By the nerve lemma (see [**Bjö95**]), or as a consequence of being d-collapsible, nerve complexes of families of convex sets in $\mathbb{R}^d$ are d-Leray. In fact, the nerve lemma can be used to provide a very short inductive proof of Helly's theorem. It comes from the observation that a non-intersecting family of convex sets where every $d+1$ of them do intersect has as a nerve the boundary of a $(d+1)$-dimensional simplex, which is impossible if the family is contained in $\mathbb{R}^d$.

We say that a family $\mathcal{G}$ of subsets of $\mathbb{R}^d$ is a *good cover* if the intersection of every subfamily is either empty or contractible. Notice that the family of all convex sets in $\mathbb{R}^d$ is a good cover, and the same application of the nerve lemma shows that the nerve complex of a finite subset of a good cover of $\mathbb{R}^d$ is d-Leray. Kalai and Meshulam showed that good covers are enough to obtain the following result.

THEOREM 2.37 (Kalai, Meshulam [**KM08**]). *Let $\mathcal{G}$ be a good cover of $\mathbb{R}^d$ and $\mathcal{F}$ be a finite family of sets such that the intersection of any subfamily of $\mathcal{F}$ is the union of at most k elements of $\mathcal{G}$. Then, if every $k(d+1)$ pairwise-disjoint elements of $\mathcal{F}$ intersect, so does $\mathcal{F}$.*

This is a generalization of the following result by Amenta.

THEOREM 2.38 (Amenta, [**Ame96**]). *Let $\mathcal{F}$ be a family of sets in $\mathbb{R}^d$ such that the intersection of any members of $\mathcal{F}$ can be expressed as the union of at most k pairwise disjoint compact convex sets. If every $k(d+1)$ or fewer members of $\mathcal{F}$ have a point in common, then there is a point in common to all memebers of $\mathcal{F}$.*

The result above gives a clean solution of a conjecture by Grünbaum and Motzkin [**GM61**]. The conjecture had a more elaborate solution due to Morris [**Mor73**], whose correctness has been put to question. We give a proof of Theorem 2.38 in Section 3. If the disjointness assumption is dropped, larger Helly numbers are needed [**AK95, Mat97**].

There have been two recent of generalizations of this result. The first is a topological generalization by Goaoc, Paták, Patáková, Tancer, and Wagner [**GPP$^+$15**], which further generalizes a previous topological extension [**CdVGG14**]. In this case, we no longer require the family $\mathcal{F}$ to be a good cover of $\mathbb{R}^d$, only bounds on the first $\lceil d/2 \rceil$ Betti numbers over $\mathbb{Z}_2$ of $\cap \mathcal{G}$ are required for each subfamily $\mathcal{G} \subset \mathcal{F}$, which is a significantly weaker condition. This result also generalizes [**Mat97**]. The second generalization is a combinatorial version of Theorem 2.38 by Eckhoff and Nischke. The arguments of their proof rely on fixing and extending some of the arguments by Morris [**Mor73**]. To describe this last result we need the following definitions.

Given a family $\mathcal{F}$ of sets, we say $\mathcal{F}$ is

- *intersectional* if for any finite subfamily $\mathcal{G}$, $\cap \mathcal{G}$ is either in $\mathcal{F}$ or empty, and
- *non-additive* if for any finite subfamily $\mathcal{G}$ of pairwise disjoint subsets of $\mathcal{F}$, if $|\mathcal{G}| \geq 2$ then $\cup \mathcal{G} \notin \mathcal{F}$.

Notice that the family of convex sets in $\mathbb{R}^d$ is intersectional and non-additive.

THEOREM 2.39 (Eckhoff, Nischke [**EN09**]). *Let $\mathcal{C}$ be an intersectional non-additive family. If $\mathcal{F}$ is a finite family of sets such that the intersection of any k or fewer members of $\mathcal{F}$ is a disjoint union of at most k members of $\mathcal{C}$, then $h(\mathcal{F}) \leq k h(\mathcal{C})$.*

Given a good cover $\mathcal{G}$ of $\mathbb{R}^d$, the family of finite intersections of $\mathcal{G}$ is an intersectional non-additive family. Thus, Theorem 2.37 is also implied by Theorem 2.39, which is proved by completely different methods. The results of Eckhoff and Nischke and those of Goaoc et al. do not seem to imply each other.

Topological Helly-numbers are a form of moving to non-convex spaces. Similarly, there are some Helly-type results for curves and for algebraic varieties. Consider for example the following theorem about convex curves.

THEOREM 2.40 (Helly for boundary of convex planar curves, Swanepoel [**Swa03**]). *Given a convex body $K \subset \mathbb{R}^2$, and $\mathcal{F}$ a finite family of homothets (including translates) of the boundary ∂K, if every four elements in $\mathcal{F}$ have a point in common, all of them do.*

One can successfully look at non-linear algebraic versions of Helly's theorem, where the sets in the family $\mathcal{F}$ are algebraic hypersurfaces defined as the zeros of a polynomial. There is indeed a Helly-type theorem for hypersurfaces defined by homogeneous polynomials. This was proved by Motzkin [**Mot55**] and later reproved by Deza and Frankl [**DF87, Fra89**].

THEOREM 2.41 (Motzkin and Deza, Frankl [**Mot55**,**DF87**,**Fra89**]). *Let $f_1, \ldots, f_m$ be a system of homogeneous polynomials in n variables and coefficients over a field. Define $d = \max\{\deg(f_i)\}$. Suppose that every subset of $p = \binom{n+d}{d}$ polynomials $\{f_{i_1}, \ldots, f_{i_p}\} \subset \{f_1, \ldots, f_m\}$ has a solution. Then the entire system of polynomial equations $\{f_1, \ldots, f_m\}$ does as well.*

As an example that resembles Theorem 2.40, take $\mathcal{F} = \{f_1(x,y), \ldots, f_s(x,y)\}$ a family of affine real plane curves of degree at most d. If every $\delta = \binom{d+2}{2}$ of the curves have a common real intersection point, then all curves in $\mathcal{F}$ have a real intersection point. The proof of Theorem 2.41 uses the linear algebraic structure of the vector spaces of polynomials of degree d in n variables. This result was later extended to *quasialgebraic sets* in [**CD14**].

Other "algebraic versions" of Helly's theorem have been considered in recent years. E.g. Tropical geometry is a very active area where the usual arithmetic operations are replaced by the max-plus counterparts, i.e., maximum and addition are the two binary operations in $\mathbb{R}$. In [**GM10**] the authors presented several exciting tropical versions of the classical theorems of convexity (see [**BH04**] for the first tropical version of Helly's theorem and the survey [**NS14**]). Finally, [**BGV13**] shows a Helly-type theorem for finite families of subsets of $\mathbb{R}^n$ which are graphs of monotone maps definable in some o-minimal structure over $\mathbb{R}$. This is a rather powerful framework, since one of the most famous examples of o-minimal structures is that of semi-algebraic sets. On the other hand, graphs of monotone maps are, generally, non-convex, and their intersections, unlike intersections of convex sets, can be topologically complicated.

2.5. Fractional versions and versions with tolerance. The hypothesis statement of many Helly-type theorems requires that all subfamilies of cardinality h satisfy a certain property. In this Subsection we discuss two versions which allow some relaxation of this assumption.

The fractional version of Helly's theorem is obtained by changing the condition that *all* $(d+1)$-tuples are intersecting to the condition that *most* $(d+1)$-tuples are intersecting.

THEOREM 2.42 (Fractional Helly theorem; Katchalski, Liu [**KL79**]). *Given a family $\mathcal{F}$ of n convex sets in $\mathbb{R}^d$ and a parameter $\alpha > 0$, if at least $\alpha\binom{n}{d+1}$ of the $(d+1)$-tuples of $\mathcal{F}$ have a common intersection, then there exists a point contained in at least βn objects of $\mathcal{F}$. Here β is a positive constant depending only on α and d.*

PROOF. We may assume without loss of generality that the sets in $\mathcal{F}$ are compact. Let v be a direction. We consider a v-halfspace to be a set of the form $\{x \in \mathbb{R}^d : \langle x, v\rangle \geq \alpha\}$ for some α. Given an intersecting d-tuple A of $\mathcal{F}$, we can consider H_A to be the containment-minimal v-halfspace such that $(\cap A) \bigcap H_A$ is not empty, which exists by the compactness of the sets. Moreover, v can be chosen so that for each A as above $(\cap A) \bigcap H_A$ consists of a single point p_A.

Given an intersecting $(d+1)$-tuple $B \subset \mathcal{F}$, we can assign to it its d-tuple A for which H_A (its associated containment minimal v-halfspace) is containment-maximal. Helly's theorem immediately gives that $B \cup \{H_A\}$ is intersecting, so all the sets in B contain p_A. A simple counting argument shows that if a positive fraction α of the $(d+1)$-tuples is intersecting, then this process assigns the same

d-tuple A at least βn times for some β depending only on d and α. Thus, at least βn sets contain p_A. □

The proof above only gives $\beta \geq \alpha/(d+1)$, but it is quite versatile. Several of the results below are a consequence of this method with little modifications needed. It was later shown by Kalai [**Kal84**] and Eckhoff [**Eck85**] that $\beta = 1-(1-\alpha)^{1/(d+1)}$, which in particular gives $\beta \to 1$ as $\alpha \to 1$ (Katchalski and Liu showed this limit as well, but not the optimal value for β). Kalai's result has been extended to require weaker topological conditions on the sets [**Hel06, CdVGG12**] giving topological versions of the fractional Helly theorem and the (p, q) theorem. At first glance the fractional Helly theorem shows the robustness of Helly's theorem. In reality it goes further than that. For instance, an integer version of the result above is also possible, as shown below.

THEOREM 2.43 (Alon, Kalai, Matoušek, Meshulam [**AKMM02**]). *Given a family $\mathcal{F}$ of n convex sets in $\mathbb{R}^d$ and a parameter $\alpha > 0$ such that $\alpha\binom{n}{2^d}$ of the 2^d-tuples of $\mathcal{F}$ have an integer point in their intersection, there exists an integer point contained in at least βn sets of $\mathcal{F}$. Here β is a positive constant depending only on α and d*

Again, $\beta \to 1$ as $\alpha \to 1$, showing the robustness of Doignon's theorem.

The technique can be pushed to show the robustness of quantitative Helly theorems for discrete sets S.

THEOREM 2.44 (Rolnick, Soberón [**RS15**]). *Let $S \subset \mathbb{R}^d$ be a discrete set and $k > 0$ an integer such that $h_k(S) < \infty$. Given a parameter $\alpha > 0$, let $\mathcal{F}$ be a family of n convex sets in $\mathbb{R}^d$ such that the intersection of at least $\alpha\binom{n}{h_k(S)}$ of the $h_k(S)$-tuples of $\mathcal{F}$ have at least k points of S in their intersection. Then, there exists a subset of k points of S contained in at least βn sets of F. Here β is a positive constant depending only on α, S and d.*

THEOREM 2.45 (Rolnick, Soberón [**RS15**]). *Let $\varepsilon > 0$ and $d > 0$. There is a constant $n_2(d, \varepsilon)$ such that the following holds. Given a parameter $\alpha > 0$ and a family $\mathcal{F}$ of n convex sets in $\mathbb{R}^d$ such that the intersection of at least $\alpha\binom{n}{n_2(d,\varepsilon)}$ of the $n_2(d, \varepsilon)$-tuples of $\mathcal{F}$ have volume at least one, then there exists a set $C \subset \mathbb{R}^d$ of volume at least $1 - \varepsilon$ contained in at least βn sets of F. Here β is a positive constant depending only on α, ε and d.*

Moreover, $n_2(d, \varepsilon) = O_d\left(\varepsilon^{-(d^2-1)/4}\right)$.

The dependence of $n_2(d, \varepsilon)$ on ε in Theorem 2.45 is not clear. It is possible for this fractional Helly to have a much nicer behavior than its Helly counterpart. For certain functions, such as the diameter, this behavior can be observed [**Sob16**]. Thus, it would be interesting to know the answer to the following problem.

OPEN PROBLEM 2.46. In Theorem 2.45, is it possible to conclude that the resulting set C has volume at least 1 (i.e., can ε be dropped entirely from the statement)?

The first indication that the fractional Helly theorem carries essentially different information than Helly's theorem is the following surprising result of Matoušek and Bárány. It says that for the integer fractional Helly theorem it is not necessary to check the 2^d-tuples in order to get a positive-fraction intersection result, but the

$(d+1)$-tuples are sufficient. Namely,

THEOREM 2.47 (Bárány, Matoušek [**BM03**]). *Given a family $\mathcal{F}$ of n convex sets in $\mathbb{R}^d$ and a parameter $\alpha > 0$ such that $\alpha\binom{n}{d+1}$ of the $(d+1)$-tuples of $\mathcal{F}$ have an integer point in their intersection, then there exists an integer point contained in at least βn sets of $\mathcal{F}$. Here β is a positive constant depending only on α and d.*

In this case, we do not get $\beta \to 1$ as $\alpha \to 1$, since $h(\mathbb{Z}_d) = 2^d$. The proof relies on an argument reducing properties of $\mathbb{Z}^d$-convexity to $\mathbb{R}^d$-convexity [**BM03**, Cor. 2.2], which allows the use of Helly-type theorems for which the Helly number is $d+1$ (in particular, the colorful Helly theorem). The question of which sets $S \subset \mathbb{R}^d$ allow for a fractional Helly theorem that only needs to check $(d+1)$-tuples instead of $h(S)$-tuples was answered by Averkov and Weismantel.

THEOREM 2.48 (Averkov, Weismantel [**AW12**]). *Let $S \subset \mathbb{R}^d$ be a closed set with $h(S) < \infty$. Given a family $\mathcal{F}$ of n convex sets in $\mathbb{R}^d$ and a parameter $\alpha > 0$ such that $\alpha\binom{n}{d+1}$ of the $(d+1)$-tuples of $\mathcal{F}$ have a point of S in their intersection, there exists a point of S contained in at least βn sets of $\mathcal{F}$. Here β is a positive constant depending only on α, S and d.*

It is not known if the same improvements hold for the volumetric and quantitative versions of the fractional Helly theorem. Moreover, since the result of Bárány and Matoušek shows that for fractional versions of Helly's theorem one may drastically weaken the conditions, it is natural to ask the following question.

OPEN PROBLEM 2.49. Determine if the size of the subfamilies one needs to check in Theorem 2.44 can be lowered from $h_k(S)$. In particular, is it possible to obtain a bound that depends on k and d but not on S?

The results of Averkov and Weismantel show that this is the case when $k = 1$. It should be mentioned that additional conditions on the sets allow for improved bounds on β, or relaxed requirements on the size of the subfamilies needed to check. For example the authors of [**BFMP$^+$15**] investigated fractional Helly theorems for finite families of boxes. Their main result is the fractional version of Theorem 2.2:

THEOREM 2.50 (Bárány, Fodor, Martínez-Pérez, Montejano, Oliveros, Pór [**BFMP$^+$15**]). *Let $\mathcal{F}$ be a family of n boxes in $\mathbb{R}^d$, and let $\alpha \in \left(1 - \frac{1}{d}, 1\right]$ be a real number. There exists a real number $\beta(\alpha) > 0$ such that if there are $\alpha\binom{n}{2}$ intersecting pairs in $\mathcal{F}$, then $\mathcal{F}$ contains an intersecting subfamily of size βn.*

An example shows that the bounds for α are best possible.

OPEN PROBLEM 2.51. Is there a fractional version of Theorem 2.14?

Another variation of the fractional Helly theorem is a fractional Breen theorem [**BM06a**]. There, when many $(d+1)$-tuples of a family of convex sets have a star-shaped union, then the conclusion is that many of the sets must have a point in common.

OPEN PROBLEM 2.52. Can one prove similar specialized fractional Helly theorems for restricted families of convex bodies besides boxes?

Fractional Helly theorems work in ambient spaces very different than $\mathbb{R}^d$. Matoušek showed a purely combinatorial version of the fractional Helly theorem, which works on any set system $\mathcal{G}$ of bounded VC-dimension (we define this below, but recommend [**Mat10**] for details). It should be stressed that Matoušek's result gives a large family of fractional Helly theorems for families of sets that do not admit a Helly theorem. Earlier examples of fractional Helly theorems that did not have a Helly theorem were already known (see Theorem 2.64, for instance).

Given a set X and a set system $\mathcal{G}$ on X, we say that the VC-dimension of $\mathcal{G}$ is the largest size of a set A that is *shattered* by $\mathcal{G}$. A set A is shattered if every subset of A can be written as $A \cap G$ for some $G \in \mathcal{G}$. For example, the family of hyperplanes in $\mathbb{R}^d$ has VC-dimension d. Plenty of examples of geometric set systems with bounded VC-dimension can be found in [**Mat10**].

The fractional Helly number given by Matoušek is in terms of the *dual shatter function* of the set system. The dual shatter function of a set system $\mathcal{G}$ is a function $\pi^*_{\mathcal{G}} : \mathbb{N} \to \mathbb{N}$ where $\pi^*_{\mathcal{G}}(m)$ is the maximum number of non-empty fields of the Venn diagram of m subsets of $\mathcal{G}$. In other words, given a set A of m subsets of $\mathcal{G}$, it is the largest possible number of subsets $B \subset A$ such that there are elements contained in every set of B but in no set of $A \setminus B$. It is known that $\pi^*_{\mathcal{G}}(m) = o(m^k)$ for some k if and only if the VC-dimension of $\mathcal{G}$ is bounded.

THEOREM 2.53 (Matoušek [**Mat04**]). *Let $\mathcal{F}$ be a set system whose dual shatter function satisfies $\pi^*_{\mathcal{F}}$ satisfies $\pi^*_{\mathcal{F}}(m) = o(m^k)$. Then, for any finite subfamily $\mathcal{F}' \subset \mathcal{F}$ where $\alpha\binom{|\mathcal{F}'|}{k}$ of the the k-tuples of $\mathcal{F}'$ are intersecting, there is an intersecting subfamily of $\mathcal{F}'$ of size at least $\beta|\mathcal{F}'|$, where $\beta > 0$ only depends on α and k.*

An important corollary of this result is a fractional Helly theorem for semi-algebraic sets. Recall that a semi-algebraic set in $\mathbb{R}^d$ is defined by a Boolean combination of a bounded number of polynomial inequalities of bounded degree. Here the maximum of the number of polynomials used and their degrees is called the *complexity* of the system.

COROLLARY 2.54. *For every fixed B, and $\mathcal{F}$ a finite family of semi-algebraic sets in $\mathbb{R}^d$ of complexity at most B, if at least $\alpha\binom{|\mathcal{F}|}{d+1}$ of the $(d+1)$-tuples of $\mathcal{F}$ intersect, there is a subfamily of $\mathcal{F}$ of size at least $\beta|\mathcal{F}|$ which is intersecting. The constant $\beta > 0$ only depends on $B, d,$ and α.*

OPEN PROBLEM 2.55. Is there an integral version of Corollary 2.54?

The fractional Helly theorem for semi-algebraic sets is not a surprise. The intersection patterns of these families of sets are extremely rich in combinatorial properties (see, for instance, [**CFP$^+$14**, **FPS15**, **APP$^+$05**] and the references therein).

A stronger notion of robustness of Helly's theorem than the one presented by the fractional versions is found in the versions with tolerance. In this case, instead of allowing a positive-fraction subfamily not to be included in the intersecting subfamily, we only allow avoidance of a subfamily of fixed size.

These results are related to the Erdős-Gallai numbers $\eta(d+1, t+1)$, which are constants that come up when studying vertex covers in hypergraphs (see [**EG61**] for a proper definition). The following bound was obtained by Tuza [**Tuz89**].

$$\eta(d+1, t+1) < \binom{d+t+1}{d} + \binom{d+t}{d}.$$

In particular, the inequality above implies that the case $t = 0$. The next theorem directly implies Helly's theorem.

THEOREM 2.56 (Helly with tolerance; Montejano, Oliveros [**MO11**]). *Let $\eta(d+1, t+1)$ be the Erdős-Gallai number for $d+1$ and $t+1$. Given a finite family $\mathcal{F}$ of convex sets in $\mathbb{R}^d$, if out of every $\eta(d+1, t+1)$ of the members of $\mathcal{F}$ there is a point in the intersection of all, with the exception of at most t sets, then there is a point in the intersection of all members of $\mathcal{F}$, with the exception of at most t sets.*

The proof of this theorem is a consequence of combinatorial properties of intersection hypergraphs. Their result is stated in a very general form [**MO11**, Thm. 1.1]. If we apply this to S-Helly numbers, we immediately obtain the following theorem.

THEOREM 2.57 (Helly with tolerance over $S \subset \mathbb{R}^d$). *Let $k > 0$ be a integer and $S \subset \mathbb{R}^d$ be a subset with a finite quantitative Helly number $h_k(S)$. Given a finite family $\mathcal{F}$ of convex sets in $\mathbb{R}^d$, if out of every $\eta(h_k(S), t+1)$ sets in $\mathcal{F}$ there are k points of S contained in all, with the exception of at most t sets, then there is a set of k points of S contained in all members of $\mathcal{F}$, with the exception of at most t.*

2.6. Piercing numbers and (p, q) theorems. When Helly's condition does not hold there may not be a single point in the intersection of all members of a family of convex sets. However, it is still possible that using few points we can intersect all the sets in the family, effectively bounding its *piercing number*. More precisely, given a family $\mathcal{F}$ of non-empty convex sets in $\mathbb{R}^d$, we say that the piercing number of $\mathcal{F}$, denoted by $\pi(\mathcal{F})$ is the minimum number of points needed to intersect every member of $\mathcal{F}$. If no finite set can accomplish this, we say $\pi(\mathcal{F}) = \infty$.

A popular weakening of Helly's condition is the so called (p, q) property. We say that a family $\mathcal{F}$ of convex sets in $\mathbb{R}^d$ has the (p, q) property if *out of every p sets in $\mathcal{F}$, there are q which are intersecting.* A classic conjecture of Hadwiger and Debrunner [**HD57**] asked if this property was enough to bound the piercing number of the family. This was answered positively by Alon and Kleitman.

THEOREM 2.58 ((p, q) theorem; Alon, Kleitman [**AK92**]). *Given $p \geq q \geq d+1$, there is a constant $c = c(p, q, d)$ such that any finite family $\mathcal{F}$ of convex sets in $\mathbb{R}^d$ with the (p, q) property satisfies $\pi(\mathcal{F}) \leq c$.*

We denote by $c(p, q, d)$ the optimal constant that fits the theorem above. A simplified proof can be found in [**AK96, Mat02**]. A survey of Eckhoff collected many results around the (p, q) theorem [**Eck03**]. Even though there are plenty of existence results for (p, q) type problems, finding efficient bounds for $c(p, q, d)$ is extremely difficult. As noted in [**Eck03**], by following the original proof of the (p, q) theorem one obtains $c(4, 3, 2) \leq 276$, which is the first non-trivial case. The conjecture is $c(4, 3, 2) = 3$. For this problem, the best bounds are $3 \leq c(4, 3, 2) \leq 13$ [**KGT01**].

OPEN PROBLEM 2.59. Improve the bounds on $c(4, 3, 2)$.

A key step in the proof the (p, q) theorem is the use of the fractional Helly theorem. The other main ingredient one needs to check is the existence of weak ε-nets for the families of sets in question (for convex sets, see [**ABFK92**]). Once the existence for the appropriate type of ε-nets has been cleared we obtain variations of the (p, q) theorem. Thus, many variations of the fractional Helly theorem imply (p, q) type results.

This yields hyperplane transversal (as Theorem 2.64), quantitative (as theorems 2.44 and 2.45) and bounded VC-dimension (as Theorem 2.53) (p, q) theorems as a consequence, each contained in the corresponding reference. In all these variations, the key modification is either the set system in question or the conditions on the intersection.

However, variations of the (p, q) theorem where the local intersection condition is modified are also possible. For instance, a colorful version of the fractional Helly theorem almost immediately gives the following result.

THEOREM 2.60 (colorful (p, q) theorem; Bárány, Fodor, Montejano, Oliveros, Pór [**BFM$^+$14**]). *Let p, q, d be positive integers with $p \geq q \geq d+1$. Then there is an integer $M(p, q, d)$ such that the following holds. Given finite families $\mathcal{F}_1, \mathcal{F}_2, \ldots, \mathcal{F}_p$ of convex sets such that for every choice $F_1 \in \mathcal{F}_1, \ldots, F_p \in \mathcal{F}_p$ there are q sets of $\{F_1, \ldots, F_p\}$ intersecting, there are $q - d$ indices $i \in \{1, \ldots, p\}$ for which $\pi(\mathcal{F}_i) \leq M(p, q, d)$.*

If one wishes for improved bounds on the piercing number of the family, strengthening the (p, q) condition helps. For example, we can say that a family $\mathcal{F}$ of convex sets has the $(p, q)_r$ property if *out of every p convex sets of $\mathcal{F}$, there are r of the q-tuples which are intersecting.*

THEOREM 2.61 (Montejano, Soberón [**MS11**]). *Let $\mathcal{F}$ be a finite family of convex sets in $\mathbb{R}^d$ and $p \geq q \geq d + 1$ be positive integers. If $\mathcal{F}$ has the $(p, q)_r$ property, $|\mathcal{F}| \geq p$ and*

$$r > \binom{p}{q} - \binom{p+1-d}{q+1-d},$$

then $\pi(\mathcal{F}) \leq p - q + 1$.

Let us sketch a proof of the result above, as it does not rely on a fractional Helly theorem, but rather uses some of the topological tools described earlier to reduce the problem to dimension one.

PROOF. The reader may notice that the condition implies that every $(d-1)$-tuple of $\mathcal{F}$ is intersecting. Let v be a direction such that the intersection of every d-tuple is either empty or has a v-directional minimum at a single point. Let A be the intersecting d-tuple whose v-directional minimum p is maximal. As before, every set $K \in \mathcal{F}$ such that $\{K\} \cup A$ is intersecting contains p.

Let $\mathcal{F}_1$ be the subfamily of $\mathcal{F}$ of sets which contain p, and $\mathcal{F}_2 = \mathcal{F} \backslash \mathcal{F}_1$. Consider the half-space $H = \{x : \langle x, v\rangle < \langle p, v\rangle\}$, and define $\mathcal{F}' = \{K \cap H : K \in \mathcal{F}_2\}$. Similarly, let $A' = \{K \cap H : K \in A\}$. It suffices to show that $\pi(\mathcal{F}') \leq p - q$. If $|\mathcal{F}' \cup A'| \geq p$, it still satisfies the $(p, q)_r$ condition. Otherwise, showing that $\pi(\mathcal{F}) \leq p - q + 1$ can be done by extending $\mathcal{F}_2 \cup A$ to a p-tuple and using a brute-force argument.

The family A' is not intersecting and of size d, so by [**BM02**] there is a hyperplane T with orthogonal projection $p : \mathbb{R}^d \to T$ such that $p(A)$ is still not intersecting. By Lemma 2.33, $\cup(p(A))$ is homologically equivalent to S^{d-2}, which implies that it splits T into two connected components, one of which is bounded. Let x be a point in the bounded component and consider the line $\ell = p^{-1}(x)$. It suffices to show that $\mathcal{F}_\ell = \{K \cap \ell : K \in \mathcal{F}'\}$ has the $(p - d, q - d + 1)$-property, as then we would be able to pierce it with $p - q$ points, since $c(p', q', 1) = p' - q' + 1$ for $p' \geq q' \geq 2$.

By the construction of ℓ, every convex set K that intersect every $(d-1)$-tuple of A has a point of ℓ. However, a purely combinatorial argument shows that for any $(p-d)$-tuple B in $\mathcal{F}'$, since $B \cup A'$ has the $(p,q)_r$ property with r as in the theorem, there must be a $(q-d+1)$-tuple C of B such that the union of C and any $(d-1)$-tuple of A' is intersecting, as desired. □

2.7. Transversal theorems. Many Helly-type theorems arise when one changes the conditions that the sets intersect to that of having a transversal by an affine subspace. These tantalizing variations often required additional conditions on the family. Notice that we can think of transversality theorems as intersection theorems. One way is to observe that if a set of convex sets in $\mathbb{R}^d$ is pierced through by a linear subspace L of dimension k, then it means that, under the projection to the orthogonal complement of L, all the projected convex sets in $\mathbb{R}^{d-k}$ must intersect. Another way is to replace each convex set by the space of potential transversals to it, as a subset of a Grassmanian manifold. Thus, a Helly-type transversality theorem for a family of convex set is equivalent to a Helly-type theorem for the intersection of this new family. In the following theorems about hyperplane transversals, the assumption of convexity on $\mathcal{F}$ can be replaced by connectedness without a problem.

It is clear that a piercing or transversal line crosses the sets in the family in some order when read from left to right. When the sets are disjoint one can assume the sets are ordered by such a line. It is intuitively clear such linear ordering of sets is a necessary, although not sufficient, condition for the existence of a transversal. Now we are ready to state the first ever transversal theorem, given by Hadwiger in 1957.

THEOREM 2.62 (Hadwiger [**Had57**]). *Let $\mathcal{F}$ be a finite family of pairwise disjoint connected sets in the plane, and λ be a linear ordering of the sets. If every three elements of $\mathcal{F}$ have a line transversal that intersects them consistently with λ, then there is a line transversal to the whole family.*

Precise conditions for the existence of hyperplane transversals were determined by Pollack and Wenger. Given a family $\mathcal{F}$ of connected sets in $\mathbb{R}^d$, we say there is a *consistent k-ordering of $\mathcal{F}$* if there is a function $f : \mathcal{F} \to \mathbb{R}^k$ such that for any two subfamilies $A, B \subset \mathcal{F}$ with $|A| + |B| = k+2$ and $A \cap B = \emptyset$, if $\operatorname{conv}\{f(a) : a \in A\} \cap \operatorname{conv}\{f(b) : b \in B\} \neq \emptyset$, then $\operatorname{conv}(\cup A) \cap \operatorname{conv}(\cup B) \neq \emptyset$. In other words, the family has all the Radon partitions of a family of points in $\mathbb{R}^k$.

THEOREM 2.63 (Pollack, Wenger [**PW90**]). *A finite family of connected sets in $\mathbb{R}^d$ has a hyperplane transversal if and only if there is a consistent k-ordering of $\mathcal{F}$ for some $0 \le k \le d-1$.*

The case $k=1, d=2$ implies Hadwiger's theorem, and the pairwise disjointness is not needed. Since a consistent k-ordering only needs to be checked for the $(k+2)$-tuples, this constitutes a Helly-type theorem. It should be emphasized that although the condition of k-ordering is necessary for a Helly-type theorem, it is not for the fractional version.

THEOREM 2.64 (Alon, Kalai [**AK95**]). *Given a family $\mathcal{F}$ of n connected sets in $\mathbb{R}^d$ and a parameter $\alpha > 0$ such that $\alpha\binom{n}{d+1}$ of the $(d+1)$-tuples of $\mathcal{F}$ have a hyperplane transversal, then there is a hyperplane transversal to at least βn sets of $\mathcal{F}$. Here β is a positive constant depending only on α and d.*

Pollack and Wenger's result can be colored. If we have a colored family $\mathcal{F}$ of connected sets in $\mathbb{R}^d$, we say there is a *consistent colored k-ordering of $\mathcal{F}$* if

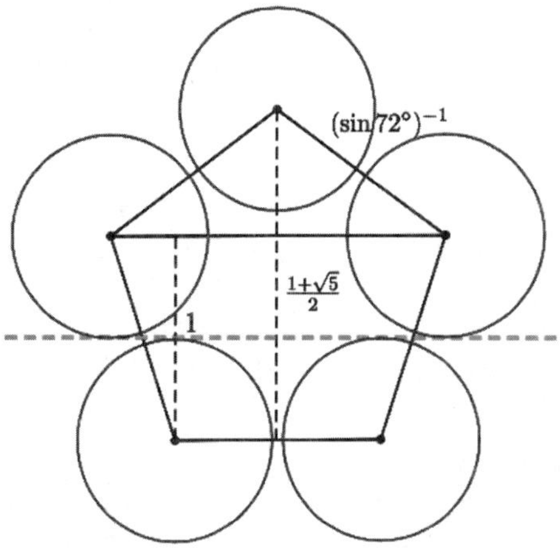

FIGURE 1. Construction showing that the constant in Problem 2.67 cannot be improved.

there is a function $f : \mathcal{F} \to \mathbb{R}^k$ such that for any two subfamilies $A, B \subset \mathcal{F}$ with $|A| + |B| = k+2$, $A \cap B = \emptyset$ and $A \cup B$ rainbow (i.e., not repeating any color), if $\operatorname{conv}\{f(a) : a \in A\} \cap \operatorname{conv}\{f(b) : b \in B\} \neq \emptyset$, then $\operatorname{conv}(\cup A) \cap \operatorname{conv}(\cup B) \neq \emptyset$. In other words, the family has all the colored Radon partitions of a colored family of points in $\mathbb{R}^k$.

THEOREM 2.65 (Holmsen, Roldán-Pensado [**HRP**]). *For $d \geq 1$ and $d-2 \leq k \leq d-1$, there is number $r(d,k)$ such that the following statement holds. Given a family $\mathcal{F}$ of connected sets in $\mathbb{R}^d$, colored with r colors, if there is a consistent colored k-ordering of $\mathcal{F}$, then there is a color that admits a hyperplane transversal to all its sets.*

This extends a result of Arocha, Bracho and Montejano [**ABM08**], that showed $r(2,1) = 3$ (i.e., a colorful version of Hadwiger's theorem). Holmsen and Roldán-Pensado obtained the general bound $r(d, d-1) \leq 2d^2 + 3$ and a new proof of $r(2,1) = 3$. However, the optimal number of colors needed is still unknown.

OPEN PROBLEM 2.66 ([**HRP**]). Determine if $r(d, d-1) = d+1$ for all d.

The problem above has a positive answer if the family has some additional separation properties on the sets [**MO08**]. Other colorful versions of transversal results are contained in [**Mon13**]. If one wishes to work without the k-ordering condition, then it is possible to obtain results by relaxing the conclusion of the theorem.

Given a convex body K in the plane, $\lambda > 0$ and family $\mathcal{F} = \{x_1 + K, \ldots, x_n + K\}$ of translates of K, we consider $\lambda\mathcal{F} = \{x_1 + \lambda K, \ldots, x_n + \lambda K\}$. A classic question of Grünbaum [**Grü58**] is to find the minimal $\lambda = \lambda(K, s)$ such that if every s elements of $\mathcal{F}$ have a line transversal, $\lambda\mathcal{F}$ has a line transversal.

For instance if B is a disk, it is known that $\lambda(B,3) \geq \frac{1+\sqrt{5}}{2}$ (see Figure 1) but the exact value is still open.

OPEN PROBLEM 2.67. Prove that $\lambda(B,3) = \frac{1+\sqrt{5}}{2}$.

It is known that $\lambda(B,4) = \frac{1+\sqrt{5}}{2}$ [**JC07**] and $\lambda(B,3) < 1.79005$ [**JCRP11**]. Additionally, if we ask the balls to be pairwise disjoint, then there is a Helly-type theorem for line transversals, with the Helly number being five [**Tve89**]. This result extends to higher dimensions.

THEOREM 2.68 (Cheong, Goaoc, Holmsen, Petitjean [**CGHP08**]). *Let $\mathcal{F}$ be a family of at least $4d-1$ pairwise disjoint unit balls in $\mathbb{R}^d$. If every $4d-1$ have a line transversal, then $\mathcal{F}$ has a line transversal.*

3. Computational applications of Helly-type theorems

Helly's theorem has many applications, sometimes in unexpected situations, from voting theory [**BNS+10**] to problems for clustering incomplete data points [**GLS10**], covering problems [**DDGG09**], and recently in a new algorithmic approach to the interpolation of data by differentiable functions [**Fef05**]. Once more, for sake of space and coherence of narrative, we mostly discuss optimization algorithms where Helly numbers and Helly-type theorems play a central role.

Helly's theorem plays a basic role in the theory of convex optimization algorithms. Given n convex constraints (their feasible solutions being convex sets) it tells us that checking all subfamilies of size $d+1$ for empty intersection (a common solution) would either produce a subset of $d+1$ constraints that certifies that the entire set has empty intersection (no common solution), or that some common solution exists. Checking for the intersection (or lack thereof) among every subset of $d+1$ constraints is not a very efficient algorithm to solve convex programs; surely we can do much better. But the basic idea of checking only small subfamilies of constraints to determine a property of the entire family set system shows the way towards developing randomized optimization algorithms that leverage other Helly-type theorems.

3.1. Frameworks and algorithms.

Linear optimization is perhaps the central problem in the theory of optimization, and it is often used as a subroutine to solve much more complicated problems which include non-linear or integrality constraints. For an introduction to linear programming see [**Sch86**]. A *Linear Program* consists of a pair $(\mathcal{H},\omega)$, where $\mathcal{H}$ is a family of n linear halfspace constraints in $\mathbb{R}^d$, and ω is a linear function on $\mathbb{R}^d$. The output of a linear programming algorithm should be either a point in the intersection $\cap\mathcal{H}$ minimizing ω, if the intersection is non-empty, or a set of $d+1$ constraints certifying that the intersection is empty.

Helly's theorem describes the combinatorics of linear halfspace intersections, and it plays a natural role in combinatorial random sampling algorithms that were developed mainly to solve the linear programming problem. Recall, a combinatorial algorithm accesses the input only through primitive operations which produce discrete results such as Boolean value or a subset of an input set, rather than floating point numbers. Its running time, when given as a function of the number of primitive operations, does not depend on the size of the coefficients. Two significant combinatorial algorithms are Clarkson's random sampling algorithm, which uses the solutions of small subproblems of size $O(d^2)$, and the classic simplex algorithm from linear optimization.

Clarkson observed that his algorithm could be applied to other problems such as integer programming and smallest enclosing ball. Sharir and Welzl [**SW92**] introduced an abstract framework, the LP-type problems, to describe the requirements for these algorithms to work. Other work in this direction applying randomization and abstractions of linear programming include [**Gär95**, **BSV03**, **Hal07a**]. Last, but not least, in 2008, Gärtner, Matoušek, Rüst and Škovroň [**GMRŠ08**] established a hierarchy of algorithmic frameworks. Starting with the most concrete definition, and progressing to the more abstract case, they propose the following three levels of abstraction:

DEFINITION 3.1. A *concrete LP-type problem* is a triple $(X, \mathcal{H}, \preceq)$, where X is a set linearly ordered by $\preceq$, $\mathcal{H}$ is a finite multiset whose elements are subsets of X, and for any $\mathcal{G} \subseteq \mathcal{H}$, if $\cap\mathcal{G} \neq \emptyset$, then $\cap\mathcal{G}$ has a unique minimum element with respect to $\preceq$. We call this unique minimum element $\omega(G)$ (this is in relation to the more abstract notion below).

DEFINITION 3.2. Consider a pair $(\mathcal{H}, \omega)$, where $\mathcal{H}$ is a finite set of constraints, and ω is a function $\omega : \mathcal{F} \in 2^{\mathcal{H}} \to \Lambda$, where Λ is a linearly ordered set including an element ∞. We say h *violates* F when $\omega(F + h) > \omega(F)$. The pair $(\mathcal{H}, \omega)$ is an *LP-type problem* if it satisfies:

Monotonicity: $\omega(G + h) \geq \omega(G)$ holds for all $G \subseteq \mathcal{H}$ and $h \in \mathcal{H}$, and
Locality: For all $F \subseteq G \subseteq \mathcal{H}$, such that $\omega(F) = \omega(G)$, h violates F if and only if h violates G.

DEFINITION 3.3. A *violator space* is a pair $(\mathcal{H}, \mathsf{V})$, where $\mathcal{H}$ is a finite set of constraints and V a mapping $2^{\mathcal{H}} \to 2^{\mathcal{H}}$ from subsets of $\mathcal{H}$ to their sets of violators, such that the following two axioms hold:

Consistency: $G \cap \mathsf{V}(G) = \emptyset$ holds for all $G \subseteq \mathcal{H}$, and
Locality: $\mathsf{V}(G) = \mathsf{V}(F)$ holds for all $F \subseteq G \subseteq \mathcal{H}$ such that $G \cap \mathsf{V}(F) = \emptyset$.

The definition of LP-type problems abstracts away the ground set X used in concrete LP-type problem and the definition of violator spaces is a further abstraction, taking away the objective function ω. These abstract frameworks are useful because we can often show that a computational challenge has an algorithmic solution, by proving the axioms are satisfied for an LP-type problem, perhaps without showing that its constraints define convex, quasi-convex, or even connected sets. This is often done by using a Helly-type theorem, as we see in the next section. We summarize here some of the results of Gärtner, Matoušek, Rüst and Škovroň on the relationship between the levels of abstraction:

THEOREM 3.4 (Gärtner, Matoušek, Rüst, Škovroň [**GMRŠ08**]). *i) For every LP-type problem, there is a concrete LP-type problem whose bases achieve the same values of ω (so in this sense all LP-type problems are concrete).*
ii) All LP-type problems are violator spaces, but not all violator spaces are LP-type problems.

The simplex method pivots from one basic feasible solution to an adjacent one. That is, it visits vertices of a polyhedron by moving along its one-dimensional faces. The following abstractions replace these notions. A *basis* of $G \in 2^{\mathcal{H}}$ is a minimal cardinality subset $B \subseteq G$ such that the solution of B is the solution of G. For LP-type problems, we define:

DEFINITION 3.5. A *basis* $B \subseteq \mathcal{H}$ is a subset such that $\omega(B - h) < \omega(B)$ holds for all $h \in B$. For $G \subseteq \mathcal{H}$, a basis of G is a minimal subset B of G with $\omega(B) = \omega(G)$.

DEFINITION 3.6. The size of a largest basis of an LP-type problem is called its *combinatorial dimension* and denoted by δ.

For example, the combinatorial dimension of linear programs in d variables is $d+1$. Similar definitions can be made for violator spaces but we omit them here. As with the simplex algorithm, the solution found by a combinatorial algorithm for an LP-type problems is a basis for the input set $\mathcal{H}$ of constraints.

The following observation links the combinatorial dimension of an LP-type problem with a Helly-type theorem about its constraint set $\mathcal{H}$:

LEMMA 3.7 (Amenta [**Ame94b**]). *A concrete LP-type problem of combinatorial dimension δ is always associated with a Helly-type theorem, in which the Helly number is $\delta+1$: a finite set $\mathcal{F} \in 2^{\mathcal{H}}$ of constraints has non-empty intersection if and only if every subset $\mathcal{G} \subseteq \mathcal{F}$ of size $\delta+1$ has non-empty intersection.*

Now let us describe the combinatorial primitives required in these frameworks. The only combinatorial primitive operation required with violator spaces is the *violator test primitive*, which reports whether $h \in \mathcal{H}$ violates $G \subset \mathcal{H}$. Simplex-like combinatorial algorithms for LP-type problems might also use a *basis computation primitive*, corresponding to the pivot operation in linear programming: given a subset $B+h \subset \mathcal{H}$ of constraints, where B is a basis and h is a violator of B, produce a new basis $B' \subseteq B+h$. Basis computations can always be accomplished via violator tests, in time exponential in δ, but for some LP-type problems they can be done more quickly. A simplex algorithm steps from basis to adjacent basis using basis computations; they avoid cycling. Clarkson's algorithm [**Cla95**], however, can compute a basis for violator spaces in which a series of basis computation steps might form a cycle. Acyclic violator spaces are precisely LP-type problems [**GMRŠ08**].

Finally, we consider the running times of combinatorial LP algorithms. Clarkson stated the following result about linear programs and integer linear programs (ILPs):

THEOREM 3.8 (Clarkson [**Cla95**]). *Given an $n \times d$ matrix A, a vector $\mathbf{b} \in \mathbb{R}^n$ and the integer program $\min\{c^T\mathbf{x} : A\mathbf{x} \leq \mathbf{b}, \mathbf{x} \in \mathbb{Z}^d, \mathbf{0} \leq \mathbf{x} \leq \mathbf{u}\}$, one can find a solution to this problem in a expected number of steps of order $O(d^2 n \log(n)) + d\log(n)O(d^{d/2})$. While the algorithm is in general exponential, it gives the best complexity for solving integer linear programs when the number of variables d is fixed.*

More generally, in terms of violator spaces, we have:

THEOREM 3.9. [**GMRŠ08**, Theorem 27] *Using Clarkson's algorithm, a basis of $\mathcal{H}$ in a violator space $(\mathcal{H}, \mathsf{V})$ can be found using an expected $O\left(\delta n + \delta^{O(\delta)}\right)$ calls to the violator test primitive.*

The Ph.D thesis [**Sko07**] shows that violator spaces give the most general framework in which Clarkson's algorithm converges to a solution within this time bound.

Simple analysis [**Sei91, SW92**] shows that the randomized dual-simplex algorithm (also known as random facet) solves any LP-type problems in time using $O(n)$ violator tests and basis computations (where n is the number of constraints), with an exponential upper bound on the running time in terms of δ. This often gives an efficient algorithm when δ is a constant, to certify whether a family $\mathcal{F}$ of input objects has some Helly-type property. A sophisticated analysis of a combined algorithm, using the randomized simplex algorithm to solve the small subproblems in

Clarkson's algorithm, provided the first sub-exponential bound for a combinatorial linear programming algorithm. We say that the running time is sub-exponential in δ if its running time is exponential in some function in $o(\delta)$, such as $\sqrt{\delta}$.

THEOREM 3.10. [**MSW96,Kal92**] *The combined algorithm requires* $O(e^{O(\delta \lg n)})$ *basis computations and* $O(ne^{O(\delta \lg n)})$ *violation tests*

When both primitive operations can be performed in sub-exponential time we get an overall sub-exponential time algorithm.

Violator spaces are a powerful general framework. Recently in [**DLPS15**], the authors prove that problems about the solvability of systems of non-linear polynomial equations can be approached via violator spaces and Clarkson's algorithm can be adapted to this situation. The Helly numbers in question correspond to those discussed in Theorem 2.41.

3.2. Using LP-type problems to prove Helly-type theorems. As we have seen above, the Helly numbers, in the form of a combinatorial dimension, are important measures of the scalability of various algorithms (both of LP-type problems and violator spaces). Amenta [**Ame94b**] gave several constructions for objective functions that could be used to form LP-type problems from existing Helly-type theorems. Often, either lexicographic ordering or a strictly convex function such as distance from the origin can be used to construct an operator $\preceq$ in the definition of a concrete LP-type problem, above. Using this idea, many of the Helly-type theorems in this survey lead easily to concrete LP-type problems, implying that they can be solved in linear time in fixed dimension. For instance, consider to Theorem 2.5, which leads to:

THEOREM 3.11. *Let $\mathcal{D}$ be a finite multiset of convex sets in $\mathbb{Z}^{d-k} \times \mathbb{R}^k$, and let $\preceq$ be a lexicographic order on $\mathbb{Z}^{d-k} \times \mathbb{R}^k$. Then finding the minimum point of $\cap\mathcal{D}$, if one exists, is a concrete LP-type problem of combinatorial dimension at most $\delta = 2^{d-k}(k+1) - 1$.*

PROOF. The fact that it is a concrete LP-type problem follows immediately Definition 3.1. To establish the combinatorial dimension, consider any $\mathcal{G} \subseteq \mathcal{F}$ such that $\cap\mathcal{G}$ is non-empty with unique lexicographic minimum point $x_G \in X$. Consider any basis $\mathcal{B} \subseteq \mathcal{G}$, and augment $\mathcal{B}$ with the open convex set $C_G = \{x \succ x_G\} \cap \mathbb{Z}^{d-k} \times \mathbb{R}^k$. Then $\cap(B \cup C_G) = \emptyset$ while any of its proper subsets are non-empty, so $|B| \leq 2^{d-k}(k+1) - 1$. □

In a similar vein, De Loera et al. showed that every S-Helly number, and the corresponding S-Helly theorem, allow one to use Clarkson's algorithm, by directly constructing a violator space.

THEOREM 3.12 ([**DHOR15**]). *Let $S \subseteq \mathbb{R}^d$ be a closed set with a finite Helly number $h(S)$. Assume that the violation test primitive can be computed. Using Clarkson's algorithm, one can minimize $c^T x$, where c is a linear function, subject to $f_i(x)0$, f_i convex constraint for all $i = 1, 2, \ldots, n$, and for $x \in S$, can be found using an expected $O\left(\delta n + h(S)^{O(h(S))}\right)$ calls to the violator test primitive. Thus the algorithm is linear in n and exponential in $h(S)$.*

What is perhaps more surprising is that one can use algorithms that solve LP-type problems to prove Helly-type theorems. The following theorem, which we mentioned in Subsection 2.4 (Theorem 2.38), makes use of Lemma 3.7:

THEOREM 3.13 (Amenta [**Ame96**]). *Let $(X, \mathcal{H}, \preceq)$ be a concrete LP-type problem of combinatorial dimension δ with the property that $\preceq$ is a total order on the points of X. Let $\mathcal{F}$ be a family of subsets of X such that, for every $\mathcal{G} \subseteq \mathcal{F}$ with $\cap\mathcal{G} \neq \emptyset$, the intersection $\cap\mathcal{G}$ is the disjoint union of at most r elements of $\mathcal{H}$. Then $(X, \mathcal{F}, \preceq)$ is a concrete LP-type problem of combinatorial dimension at most $r(\delta+1)-1$.*

PROOF. We sketch the proof. Once again, it is clear that $(X, \mathcal{F}, \preceq)$ is a concrete LP-type problem, so we just have to establish the combinatorial dimension. We will count the constraints in any basis B such that $\cap B \neq \emptyset$, as follows. The point x achieving $\omega(B)$ is contained in one connected component C_h for each $h \in B$. As the unique minimum of $\cap C_h$, it corresponds to a basis consisting at most δ elements of $\mathcal{H}$. We move the elements of B corresponding to those basis elements from B into a holding set S. Next, we remove the constraint $h \in B$ such that $\omega(S+B-h)$ is maximal; h must be unique since the point x' achieving $\omega(S+B-h) \in h'$, for all $h' \in S+B$, but $x' \notin h$. We remove h from B. The point x' must lie in a different connected component of $S+B-h$ from x, since x is the minimum point in its connected component and all the constraints of $\mathcal{H}$ determining x are held in S. Now we repeat; we count the $\leq d$ constraints of B adjacent to x', move those to S, and then look for another remaining element of B to remove. We continue this process until B is empty; all elements have either been removed or moved into S. The final set $\cap S$ consists of at most r elements of $\mathcal{H}$, with the minimum of each is determined by at most d constraints, so the size of S is at most rd. At most $r-1$ elements were removed, each accounting for a new connected component of $\cap S$, so the combinatorial dimension is $r(\delta+1)-1$. □

Theorem 3.13 gives an easy proof of the theorem first conjectured by Grünbaum and Motzkin [**GM61**] that we mentioned this back in Subsection 2.4.

THEOREM 3.14 (Morris, Amenta [**Mor73**],[**Ame94b**]). *Let $\mathcal{F}$ be a family of sets in $\mathbb{R}^d$, such that the common intersect of any non-empty finite sub-family of $\mathcal{F}$ is the disjoint union of at most r closed convex sets. Then $\mathcal{F}$ has a Helly number of at most $r(d+1)$.*

We can apply Theorem 3.13 to give new Helly-type results in other situations, for instance with Theorem 3.11, above:

THEOREM 3.15. *Let $\mathcal{D}$ be the family of convex sets in $\mathbb{Z}^{d-k} \times \mathbb{R}^k$, and let $\preceq$ be a lexicographic order on $\mathbb{Z}^{d-k} \times \mathbb{R}^k$. Let $\mathcal{F}$ be a finite family of subsets of X such that, for every $\mathcal{G} \subseteq \mathcal{F}$ with $\cap\mathcal{G} \neq \emptyset$, the intersection $\cap\mathcal{G}$ is the disjoint union of at most r elements of $\mathcal{D}$. Then $\mathcal{F}$ has Helly number at most $r(2^{d-k}(k+1))$.*

OPEN PROBLEM 3.16. Are there S-Helly, integral, or mixed-integer versions of more general Helly-type theorems with algorithmic proofs?

3.3. Other Applications. In addition to linear programming, there are a few other useful sub-exponential convex programming problems in higher dimensional spaces:

THEOREM 3.17 (Gärtner [**Gär95**]). *There is a sub-exponential basis computation primitive for the following LP-type problems:*

- *Minimum enclosing ball: Given n points in $\mathbb{R}^d$, find the ball of minimum radius enclosing all the points.*
- *Polytope distance: Given two polytopes in $\mathbb{R}^d$, specified by their n vertices, find the distance between them, or report that they intersect.*
- *Maximum margin separator: Find the linear separator between two sets of n points in $\mathbb{R}^d$ that maximizes the distance between the separating hyperplane and either set, or report that they cannot be separated. This problem is relevant to Support Vector Machine classifiers* [**BDW01**]*, and it corresponds to Kirchberger's Theorem.*

All these geometric problems lead to essentially the same computations.

Halman [**Hal07b**] showed that several kinds of stochastic games can be formulated as LP-type problems and solved in sub-exponential time. In a major breakthrough, work based on properties of the Markov decision process has recently led to a new sub-exponential *lower* bound for the dual-simplex and other randomized combinatorial algorithms for linear programming [**FHZ14**]; earlier, Matoušek had shown that there were LP-type problems that required expected subexponential time [**Mat94**].

Open Problem 3.18. What other natural problems can be solved in expected sub-exponential time by LP-type algorithms?

Many geometric problems can be solved in time linear in the number of constraints when the dimension is fixed. Doignon's theorem implies that the combinatorial dimension of integer programming is $2^d - 1$. Violator tests, or the small sub-problems in Clarkson's algorithm can be solved using Lenstra's algorithm for integer programming in fixed dimension [**Len83**] (as pointed out by Clarkson). The complexity of Lenstra's algorithm is $O(ms + s^2)$, where s is the bit-complexity of the input and m is the number of input constraints; this is not a combinatorial algorithm, but polynomial in m. So overall this gives a linear-time algorithm for integer programming in fixed dimension. An more efficient algorithm for the small subproblems was given by Eisenbrand [**Eis03**].

Often LP-type problems can also be cast as fixed-dimensional linear or convex programming problems, in which $\mathcal{F}$ corresponds to a set of convex constraints and the objective function ω is a convex or lexicographic function. For instance, finding a line transversal of a family $K_1, \ldots, K_n$ of closed intervals parallel to the y-axis in the plane can be formalized as a linear programming problem in the two-dimensional space of lines. N. Megiddo generalized this by formulating line transversals of boxes in any fixed dimension as a constant number of linear programs [**Meg96**].

More interesting LP-type problems, corresponding to Helly-type theorems in which the sets in $\mathcal{F}$ are not necessarily convex, are often defined using a set-up analogous to quasi-convex programming. A good example is:

- Line transversal of disjoint convex disks in the plane: Given a set P of disk centers the plane, with minimum distance one between each pair of points, determine the smallest radius $r < 1/2$ such that the family of congruent disks of radius r centered at the points of P has a line transversal [**Ame94b**], or report that the set has no transversal.

Using the theorem that the Helly number for line transversals of unit disks in the plane is five [**Tve89**], we find that the combinatorial dimension of this problem is also five, even though the dimension of the set of lines in the plane is two, and the set of transversals is not necessarily connected. Here is a general set-up, in which the constraints grow as a function of a parameter λ.

DEFINITION 3.19. A *parameterized LP-type problem* is a concrete LP-type problem $(X, \mathcal{H}, \preceq)$ in which the ground set $X = (\lambda', X')$, the linear order $\preceq = \lambda'$, and the level sets are nested. That is, a level set $h_\lambda = \{x' : (\lambda', x') \in \mathcal{H} \text{ and } \lambda' = \lambda\}$ is contained in the level set h_μ for all $\mu \geq \lambda$.

THEOREM 3.20 ([**Ame94b**], Theorem 5.1). *Let $(X, \mathcal{H}, \preceq)$ be a parameterized LP-type problem. When the minimum point $\omega(G)$ in a subset of constraints $G \in H$ always exists and is achieved at a unique point, and the Helly number of the level sets h_λ for $h \in \mathcal{H}$ is δ, then the combinatorial dimension is also δ.*

This is a generalization of quasi-convex programming, in which the level sets h_λ are convex. Note that with this construction the Helly number and the combinatorial dimension are identical, unlike in the more straightforward definition of concrete LP-type problems in which the constraints themselves have Helly-type property, rather than their level sets. Some useful parameterized LP-type problems include:

- Given two convex polygons in the plane, find their Hausdorff distance [**Ame94a**] under translation; this is relevant to computer vision [**BJ97**].
- Covering with three squares (rectilinear 3-center problem): Given a set of n points in the plane, find the smallest three congruent squares that cover all of the points [**SW96**].
- Given a set of n spheres, all contained in the unit ball in $\mathbb{R}^d$, find the Möbius transformation of the unit ball that maximizes the minimum radius of the transformed spheres [**BE01**]. This has applications to graph layout on the sphere or on the hyperbolic plane.

Finding a line transversal of balls in $\mathbb{R}^3$ can also be done in time linear in the number of constraints by Theorem 2.68. That result applies to a more general family of sets, so-called "pairwise inflatable" balls. It improves a result of Holmsen, Katchalski and Lewis [**HKL03**]. The lower bound for this Helly number is $2d-1$ [**CGH12**].

OPEN PROBLEM 3.21. Is there an LP-type problem of combinatorial dimension $2d-2$ for finding a line transversal of disjoint unit balls in $\mathbb{R}^3$, which could be used to improve this Helly number?

In general, the LP-type problems are associated with "standard" LP-type problems. As we have seen in this survey, there are many related families of combinatorial geometric theorems.

OPEN PROBLEM 3.22. Are there non-trivial computational applications of fractional or volumetric Helly-type theorems, Hadwiger-type theorems, or of (p, q) theorems?

Not all computational applications of Helly's theorem involve LP-type algorithms.

THEOREM 3.23 (Centerpoint theorem [**Neu45, Rad47**]). *Let P be a set of n points in $\mathbb{R}^d$. There exists a point $p \in \mathbb{R}^d$ such that every closed halfspace with boundary passing through p contains at least $\lceil \frac{n}{d+1} \rceil$ of the points in P.*

Such a point p is called a *centerpoint*, and it can be seen as a generalization of the median to data in higher-dimensional Euclidean space. The *Tukey depth* of an arbitrary point p is the minimum number of points contained in a closed halfspace with boundary passing through p; a center point has Tukey depth $n/(d+1)$, but depending on the distribution the Tukey depth of a point might be as high as $n/2$. Finding or approximating center points or points of maximal Tukey depth is a difficult computational challenge in statistics, since, naively, the number of half-spaces one needs to consider is $O(n^d)$. Using the dependencies between these half spaces leads to a more efficient algorithm [**Cha04**] for points of maximal Tukey depth with running time $O(n^{d-1})$. Finding approximate center points in sub exponential time [**CEM$^+$96, MS09, MW13**] is very useful in partitioning problems for efficient parallel computation, and of recent interest in statistics [**CANR08**]. The *ham-sandwich theorem* states that for any d finite sets of points in $\mathbb{R}^d$ there is a hyperplane that bisects all of the sets at once, i.e., has at most half of the points on each side. The computational complexity of this problem and its connections to other problems in convexity is discussed in [**KTW11**].

3.4. Chance-constrained and S-optimization. In all the above applications and situations one has a randomized algorithm to solve a deterministic geometric optimization problem using discrete data (e.g., finitely many data points). But, in some situations, the data points are uncertain, governed by a probability distribution, or part of a continuously parametrized set of constraints. Consider for example the following twist on the problem of smallest-radius enclosing ball of finitely many points: we are given points $(u_1, u_2, \ldots, u_d) \in \mathbb{R}^d$, belonging to an unknown measurable set. Our goal is to find the center x of a ball of smallest radius R that contains a "large proportion" of those points, i.e., we are willing to allow some of the points to be outside the ball. The data points are uncertain, the bad news is we may not know explicitly the probability measure. We are trying to make a decision for the center x and the radius R, that guarantees a large proportion of the constraints are satisfied, but we allow some of the constraints not to be satisfied. In our example, there is a distance constraint per possible point, but not all the distance conditions have to be satisfied at once, we only expect this with high probability. The optimization problem is a classical example of a *chance-constrained convex optimization* problem.

$$\begin{aligned} &\min \quad R \\ &\text{subject to} \quad \Pr\left[\left\{\sqrt{\sum_{1}^{d}(x_i - u_i)^2} - R \leq 0\right\}\right] \geq 1 - \epsilon, x \in \mathbb{R}^{d+1}. \end{aligned}$$

In principle there are infinitely many constraints to consider, but it is it possible to get a reasonable solution, with only small proportion of constraints not satisfied, using partial deterministic convex optimization problem consisting of N samples. In our example, by selecting N out of the given points and carrying a computation for those, we can get a candidate center and radius. The induced ball leaves only a small proportion of the points out.

Indeed, and idea presented by Calafiore and Campi [**CC05, CC06**] (one where Helly's theorem plays a significant role) is to consider only a finite number of sampled instances of the uncertainty affecting the system, and to solve the corresponding standard convex problem. For continuous real-valued variables, the number of scenarios N that need be considered is reasonably small and that the solution of the scenario problem has generalization properties, i.e., it satisfies with high probability also unseen scenarios. They proved an efficient bound on the sample size N needed for the partial convex optimization problem. The size of N increases slowly with the required probabilistic levels of success.

More formally, the original family of convex optimization problems has convex constraints which are parameterized $\{\min g(x) \text{ subject to } f(x,w) \leq 0, \quad w \in \Delta\}$. Here $g(x)$ is a convex objective function and f is convex function in the variables x and parameterized using a measurable set Δ from which the variables w take their values. Given a *tolerance of risk* ϵ one wishes to find an optimal solution for

$$\begin{aligned} CCP(\epsilon) = \min \quad & g(x) \\ \text{subject to} \quad & Pr[\{w : f(x,w) \leq 0\}] \geq 1-\epsilon, \\ & x \in K \text{ convex set}, \\ & x \in \mathbb{R}^d. \end{aligned}$$

The goal is to find a solution that is optimal among all solutions that satisfy all but "a few constraints". In the Calafiori and Campi algorithm, one collects N random independent identically distributed samples $w^1, \ldots, w^N$, and constructs a *sampled convex program*

$$\begin{aligned} SCP(N) = \min \quad & g(x) \\ \text{subject to} \quad & f(x,w^i) \leq 0, \quad i = 1,2,\ldots,N, \\ & x \in K \text{ convex set}, \\ & x \in \mathbb{R}^d. \end{aligned}$$

This leads to two natural questions.

- How is the solution of $SCP(N)$ related to the solution for $CCP(\epsilon)$?
- How many samples need to be drawn in order to guarantee that the resulting randomized optimal solution from $SCP(N)$ violates only a "small portion" of the full set of constraints?

Given $x \in \mathbb{R}^d$, the *probability of violation* of x is $V(x) := Pr[\{w \in \Delta : f(x,w) > 0\}]$. A solution x with small associated $V(x)$ is feasible for "most" of the problem instances. For example, if we take the uniform probability density (with respect to Lebesgue's measure), then $V(x)$ is just the volume of the set of parameters w for which $f(x,w) \leq 0$ is violated. Given $\epsilon \in [0,1]$, if for $\mathbf{x}$ we have $\Pr[f(x,w) \leq 0] \geq 1-\epsilon$, we say it is an *$\epsilon$-level feasible solution.* This is precisely when $V(x) < \epsilon$. If x^N is the optimal solution of $SCP(N)$. $V(x^N)$ is a random variable in the space Δ^N with product probability measure $Pr \times Pr \times \cdots \times Pr = Pr^N$.

THEOREM 3.24 (Feasibility of x^N). *Let $0 < \epsilon \leq 1$ (tolerance) , $0 < \delta < 1$ (lack of confidence). Then, either $CCP(\epsilon)$ is infeasible or when N is sufficiently large. With probability at least $1-\delta$, the optimal solution x^N of $SCP(N)$ is a feasible solution of $CCP(\epsilon)$ with $V(x^N) \leq \epsilon$. Here sufficiently large means:*

$$N \geq \frac{2d}{\epsilon} \ln\left(\frac{1}{\epsilon}\right) + \frac{2}{\epsilon} \ln\left(\frac{1}{\delta}\right) + 2d$$

THEOREM 3.25 (Optimality gap of x^N). *Let $0 < \epsilon \leq 1$ (tolerance) , $0 < \delta < 1$ (lack of confidence). Define $\epsilon_1 = 1 - (1-\delta)^{1/N}$. Then*

(1) *Let J^ϵ be the optimal objective value of $CCP(\epsilon)$ and J^N the optimal value of $SCP(N)$. With probability at least $1-\delta$, if $SCP(N)$ is feasible, it holds that $J^\epsilon \leq J^N$.*
(2) *$\epsilon_1 < \epsilon$ and if $CCP(\epsilon_1)$ is feasible, then with probability at least $1-\delta$ we have $J^N \leq J^{\epsilon_1}$.*

As it turns out, it was proved in [**DHOR15**] that the results of Calafiore and Campi for real-valued variables can be extended in greater generality. In Section 2.1 we have discussed S-Helly theorems for different proper subsets of $\mathbb{R}^d$. The authors of [**DHOR15**] introduced the notion of *S-optimization*, a natural generalization of continuous, integer, and mixed-integer optimization:

DEFINITION 3.26. Given $S \subset \mathbb{R}^d$, the optimization problem with equations and inequalities and variables taking values on S,

$$\begin{aligned} \max \quad & f(x) \\ \text{subject to} \quad & g_i(x) \leq 0, \quad i = 1, 2, \ldots, n, \\ & h_j(x) = 0, \quad j = 1, 2, \ldots, m, \\ & x \in S, \end{aligned}$$

will be called an *S-optimization problem.*

Clearly when $S = \mathbb{R}^d$ the S-optimization problem is the usual continuous optimization problem, $S = \mathbb{Z}^d$ is just integer optimization, and $S = \mathbb{Z}^k \times \mathbb{R}^{d-k}$ is the case of mixed-integer optimization. When only linear constraints are present this is an *S-linear program.* When all constraints are convex we call this an *S-convex program.*

To motivate the study of S-optimization for unusual sets S, below is an example that shows the modeling power of using sophisticated $S \subset \mathbb{R}^d$ (typically discrete sets).

EXAMPLE 3.27. Given a graph $G = (V, E)$, we reformulate the classic graph K-coloring query as the solvability of the following linear system of modular inequations: For all (i, j) in $E(G)$ consider the inequations $c_i \not\equiv c_j \mod K$. This is a system on $|V|$ variables and it has a solution if and only if the graph is K-colorable. Note that the set of points $c = (c_1, \ldots, c_{|V|})$ with $c_i \equiv c_j \mod K$ is a lattice, which we call $L_{i,j}$. Therefore, solving our system of inequalities is equivalent to finding a $c \in S = \mathbb{Z}^{|V|} \setminus (\bigcup_{i,j} L_{i,j})$. Consequently, the problem of deciding k-colorability is equivalent to the problem of finding a solution to an S-linear system of equations, where the variables take values on S, the set difference of a lattice and a union of several sublattices.

THEOREM 3.28. *Let $S \subseteq \mathbb{R}^d$ be a set with a finite Helly number $h(S)$. Let $0 < \epsilon \leq 1$ (tolerance), $0 < \delta < 1$ (distrust) be chosen parameters. Let $f(x, w)$ be a convex function in x and measurable in w. Suppose there is an optimal value x_* of* *the linear minimization chance-constrained problem*

$$\begin{aligned} CCP(\epsilon) = \min \quad & c^T x \\ \textit{subject to} \quad & Pr[f(x,w) > 0] < \epsilon, \\ & x \in K \textit{ convex set}, \\ & x \in S. \end{aligned}$$

Then from a sufficiently large random sample of N different i.i.d values for w (specifically, $w^1, w^2, \ldots, w^N$), x_* can be δ-approximated by x_N, the optimal solution of the convex optimization problem

$$\begin{aligned} SCP(N) = \min \quad & c^T x \\ \textit{subject to} \quad & f(x, w^i) \leq 0, \quad i = 1, 2, \ldots, N, \\ & x \in K \ \textit{convex set}, \\ & x \in S. \end{aligned}$$

More precisely, if x_N exists and the size of the sample $N \geq \frac{2(h(S)-1)}{\epsilon} \ln(1/\epsilon) + \frac{2}{\epsilon} \ln(1/\delta) + 2(h(S)-1)$, then the undesirable event of high-infeasibility $V(x_N) > \epsilon$ has probability less than δ of occurring.

Recently in [**Gär15**], Gärtner investigated how to use LP-type problems and violators spaces in the context of chance-constrained optimization. Given a nondegenerate LP-type problem of combinatorial dimension δ, he considered sampling, with the subsequent removal of the k constraints that lead to the best improvement of objective function.

To conclude we wish to discuss yet another way in which Helly numbers and the optimal Helly configurations (i.e., families that show the Helly number is smallest possible already) appear in deterministic algorithms for mixed integer optimization.

The theory presented in Subsection 2.1 directly influences the modern methods for solving mixed integer programming problems. The most popular technique to solve mixed-integer optimization problems is *branch-and-cut* (see the book [**CCZ14**] for an introduction). Branch-and-cut is based on three main ingredients: relaxations of the problem (typically to continuous linear optimization), clever enumeration, and an appropriate tightening of the formulation by adding cuts. Namely, tightening the formulation is done by adding new valid inequalities called *cutting planes*. Commercial software uses a variety of cutting plane techniques, but the most successful is the Gomory mixed-integer cuts introduced in [**Gom58**].

Extending Gomory's technique has led to a large variety of work which relates to S-optimization and the construction of *maximal S-free sets*. These are the S-facet polyhedra from Lemma 2.6. In 2007, Andersen, Louveaux, Weismantel and Wolsey (see [**ALW10**] for the fundamental association to geometry) put forward a new idea for the generation of cutting planes. Essentially, they showed that, besides non-negativity constraints, the facet-defining inequalities are associated with *splits* (a region between two parallel lines), triangles and quadrilaterals whose interior does not contain an integer point. This allows one to derive valid inequalities by exploiting the combined effect of two rows, instead of a single row. The extension of this model to any dimension (i.e., number of integer variables) was pioneered by Borozan and Cornuéjols [**BC09**] and by Basu, Conforti, Cornuéjols and Zambelli

[**BCCZ10**]. As described in Subsection 2.1, the maximal S-free sets can be used to compute the S-Helly numbers, which bring a connection to integer optimization.

Once more one gets the striking result that the facet-defining inequalities of an integer opimization problem are naturally associated with full-dimensional convex sets whose interior do not contain an integer point. Furthermore these sets are polyhedra. Lovász stated earlier in [**Lov89**] that maximal convex sets whose interior does not contain an integer point are polyhedra, but the first complete proof appears in [**BCCZ10**], and an alternate proof can be found in [**Ave13b**].

These proofs use the simultaneous approximation theorem of Dirichlet and Minkowski's convex body theorem. Important results from number theory, convex geometry and Helly numbers are essential to get a computable formula for the cut-generating function. The details are explained in the recent excellent survey by [**BCDS15**]. This application to cut generation methods has stimulated several direct progress in the combinatorial geometry around Helly's theorem (see e.g., recent work in [**DLLHORP15**]). Once more geometry comes to the help of optimization.

Acknowlegements

We thank the participants of the Mathematical Research Community "Algebraic and Geometric Methods in Discrete Applied Mathematics" for their enthusiasm and support for this book project. The second author is grateful for the support received through an NSA grant. The first author gratefully acknowledges the support of NSF-IIS-1117663 and NSF-IIS-0964357.

References

[AB11] J. L. Arocha and J. Bracho, *A Helly type theorem for abstract projective geometries*, Discrete Comput. Geom. **45** (2011), no. 2, 223–229, DOI 10.1007/s00454-010-9287-7. MR2765526

[ABB+09] J. L. Arocha, I. Bárány, J. Bracho, R. Fabila, and L. Montejano, *Very colorful theorems*, Discrete Comput. Geom. **42** (2009), no. 2, 142–154, DOI 10.1007/s00454-009-9180-4. MR2519872

[ABDLL] I. Aliev, R. Bassett, J. A. De Loera, and Q. Louveaux, *A Quantitative Doignon-Bell-Scarf Theorem*, to appear in Combinatorica, preprint available in Math ArXiv 1405.2480v1.

[ABF06] G. Ambrus, A. Bezdek, and F. Fodor, *A Helly-type transversal theorem for n-dimensional unit balls*, Arch. Math. (Basel) **86** (2006), no. 5, 470–480, DOI 10.1007/s00013-005-1446-3. MR2229364

[ABFK92] N. Alon, I. Bárány, Z. Füredi, and D. J. Kleitman, *Point selections and weak ϵ-nets for convex hulls*, Combin. Probab. Comput. **1** (1992), no. 3, 189–200, DOI 10.1017/S0963548300000225. MR1208800

[ABM08] J. L. Arocha, J. Bracho, and L. Montejano, *A colorful theorem on transversal lines to plane convex sets*, Combinatorica **28** (2008), no. 4, 379–384, DOI 10.1007/s00493-008-2385-y. MR2452840

[AK92] N. Alon and D. J. Kleitman, *Piercing convex sets and the Hadwiger-Debrunner (p,q)-problem*, Adv. Math. **96** (1992), no. 1, 103–112, DOI 10.1016/0001-8708(92)90052-M. MR1185788

[AK95] N. Alon and G. Kalai, *Bounding the piercing number*, Discrete Comput. Geom. **13** (1995), no. 3-4, 245–256, DOI 10.1007/BF02574042. MR1318775

[AK96] N. Alon and D. J. Kleitman, *A purely combinatorial proof of the Hadwiger Debrunner (p,q) conjecture*, Electron. J. Combin. **4** (1997), no. 2, Research Paper 1, approx. 8 pp. (electronic). The Wilf Festschrift (Philadelphia, PA, 1996). MR1444148

[AKMM02] N. Alon, G. Kalai, J. Matoušek, and R. Meshulam, *Transversal numbers for hypergraphs arising in geometry*, Adv. in Appl. Math. **29** (2002), no. 1, 79–101, DOI 10.1016/S0196-8858(02)00003-9. MR1921545

[ALW10] K. Andersen, Q. Louveaux, and R. Weismantel, *An analysis of mixed integer linear sets based on lattice point free convex sets*, Math. Oper. Res. **35** (2010), no. 1, 233–256, DOI 10.1287/moor.1090.0439. MR2676765

[Ame94a] N. Amenta, *Bounded boxes, Hausdorff distance, and a new proof of an interesting helly-type theorem*, Proceedings of the tenth annual symposium on Computational geometry, ACM, 1994, pp. 340–347.

[Ame94b] N. Amenta, *Helly-type theorems and generalized linear programming*, Discrete Comput. Geom. **12** (1994), no. 3, 241–261, DOI 10.1007/BF02574379. ACM Symposium on Computational Geometry (San Diego, CA, 1993). MR1298910

[Ame96] N. Amenta, *A short proof of an interesting Helly-type theorem*, Discrete Comput. Geom. **15** (1996), no. 4, 423–427, DOI 10.1007/BF02711517. MR1384884

[APP+05] N. Alon, J. Pach, R. Pinchasi, R. Radoičić, and M. Sharir, *Crossing patterns of semi-algebraic sets*, J. Combin. Theory Ser. A **111** (2005), no. 2, 310–326, DOI 10.1016/j.jcta.2004.12.008. MR2156215

[Ave13a] G. Averkov, *On maximal S-free sets and the Helly number for the family of S-convex sets*, SIAM J. Discrete Math. **27** (2013), no. 3, 1610–1624, DOI 10.1137/110850463. MR3106473

[Ave13b] G. Averkov, *A proof of Lovász's theorem on maximal lattice-free sets*, Beitr. Algebra Geom. **54** (2013), no. 1, 105–109, DOI 10.1007/s13366-012-0092-8. MR3027668

[AMHPW] Gennadiy Averkov, Bernardo González Merino, Matthias Henze, Ingo Paschke, Stefan Weltge. *Tight bounds on discrete quantitative Helly numbers.* arXiv:1602.07839

[AW12] G. Averkov and R. Weismantel, *Transversal numbers over subsets of linear spaces*, Adv. Geom. **12** (2012), no. 1, 19–28, DOI 10.1515/advgeom.2011.028. MR2911157

[Bár82] I. Bárány, *A generalization of Carathéodory's theorem*, Discrete Math. **40** (1982), no. 2-3, 141–152, DOI 10.1016/0012-365X(82)90115-7. MR676720

[BC09] V. Borozan and G. Cornuéjols, *Minimal valid inequalities for integer constraints*, Math. Oper. Res. **34** (2009), no. 3, 538–546, DOI 10.1287/moor.1080.0370. MR2555335

[BCCZ10] A. Basu, M. Conforti, G. Cornuéjols, and G. Zambelli, *Maximal lattice-free convex sets in linear subspaces*, Math. Oper. Res. **35** (2010), no. 3, 704–720, DOI 10.1287/moor.1100.0461. MR2724071

[BCDS15] A. Basu, M. Conforti, and M. Di Summa, *A geometric approach to cut-generating functions*, Math. Program. **151** (2015), no. 1, Ser. B, 153–189, DOI 10.1007/s10107-015-0890-5. MR3347553

[BDW01] J. L Balcazar, Y. Dai, and O. Watanabe, *Provably fast training algorithms for support vector machines*, Data Mining, 2001. ICDM 2001, Proceedings IEEE International Conference on, IEEE, 2001, pp. 43–50.

[BE01] M. Bern and D. Eppstein, *Optimal Möbius transformations for information visualization and meshing*, Algorithms and data structures (Providence, RI, 2001), Lecture Notes in Comput. Sci., vol. 2125, Springer, Berlin, 2001, pp. 14–25, DOI 10.1007/3-540-44634-6_3. MR1936397

[Bel76] D. E. Bell, *A theorem concerning the integer lattice*, Studies in Appl. Math. **56** (1976/77), no. 2, 187–188. MR0462617

[BFM+14] I. Bárány, F. Fodor, L. Montejano, D. Oliveros, and A. Pór, *Colourful and fractional (p, q)-theorems*, Discrete Comput. Geom. **51** (2014), no. 3, 628–642, DOI 10.1007/s00454-013-9559-0. MR3201247

[BFMP+15] I. Bárány, F. Fodor, A. Martínez-Pérez, L. Montejano, D. Oliveros, and A. Pór, *A fractional Helly theorem for boxes*, Comput. Geom. **48** (2015), no. 3, 221–224, DOI 10.1016/j.comgeo.2014.09.007. MR3276400

[BGV13] S. Basu, A. Gabrielov, and N. Vorobjov, *A Helly-type theorem for semi-monotone sets and monotone maps*, Discrete Comput. Geom. **50** (2013), no. 4, 857–864, DOI 10.1007/s00454-013-9540-y. MR3138139

[BH04] W. Briec and C. Horvath, $\mathbb{B}$*-convexity*, Optimization **53** (2004), no. 2, 103–127, DOI 10.1080/02331930410001695283. MR2058292

[BJ97] R. Basri and D. W. Jacobs, *Recognition using region correspondences*, Int. J. Comput. Vis. **25** (1997), no. 2, 145–166.

[BJC15] I. Bárány and J. Jerónimo-Castro, *Helly type theorems for the sum of vectors in a normed plane*, Linear Algebra Appl. **469** (2015), 39–50, DOI 10.1016/j.laa.2014.11.019. MR3299055

[Bjö95] A. Björner, *Topological methods*, Handbook of combinatorics, Vol. 1, 2, Elsevier, Amsterdam, 1995, pp. 1819–1872. MR1373690

[BKP82] I. Bárány, M. Katchalski, and J. Pach, *Quantitative Helly-type theorems*, Proc. Amer. Math. Soc. **86** (1982), no. 1, 109–114, DOI 10.2307/2044407. MR663877

[BKP84] I. Bárány, M. Katchalski, and J. Pach, *Helly's theorem with volumes*, Amer. Math. Monthly **91** (1984), no. 6, 362–365, DOI 10.2307/2322144. MR750523

[BM02] J. Bracho and L. Montejano, *Helly-type theorems on the homology of the space of transversals*, Discrete Comput. Geom. **27** (2002), no. 3, 387–393, DOI 10.1007/s00454-001-0076-1. MR1921561

[BM03] I. Bárány and J. Matoušek, *A fractional Helly theorem for convex lattice sets*, Adv. Math. **174** (2003), no. 2, 227–235, DOI 10.1016/S0001-8708(02)00037-3. MR1963693

[BM06a] I. Bárány and J. Matoušek, *Berge's theorem, fractional Helly, and art galleries*, Discrete Math. **306** (2006), no. 19-20, 2303–2313, DOI 10.1016/j.disc.2005.12.028. MR2261903

[BM06b] V. Boltyanski and H. Martini, *Jung's theorem for a pair of Minkowski spaces*, Adv. Geom. **6** (2006), no. 4, 645–650, DOI 10.1515/ADVGEOM.2006.037. MR2267041

[BNS+10] D. E. Berg, S. Norine, F. E. Su, R. Thomas, and P. Wollan, *Voting in agreeable societies*, Amer. Math. Monthly **117** (2010), no. 1, 27–39, DOI 10.4169/000298910X474961. MR2599465

[BO95] I. Bárány and S. Onn, *Carathéodory's theorem, colourful and applicable*, Intuitive geometry (Budapest, 1995), Bolyai Soc. Math. Stud., vol. 6, János Bolyai Math. Soc., Budapest, 1997, pp. 11–21. MR1470753

[BO97] I. Bárány and S. Onn, *Colourful linear programming and its relatives*, Math. Oper. Res. **22** (1997), no. 3, 550–567, DOI 10.1287/moor.22.3.550. MR1467385

[Bra17] S. Brazitikos, *Brascamp-Lieb inequality and quantitative versions of Helly's theorem*, Mathematika, **63**(2017), no. 1, 272–291. DOI: 10/1112/S00255793136000255.

[Bra15] ———, *Fractional Helly theorem for the diameter of convex sets*, to appear in Discrete and Computational Geometry. **[math.MG]** (2015).

[Bre90] M. Breen, *Starshaped unions and nonempty intersections of convex sets in* $\mathbf{R}^d$, Proc. Amer. Math. Soc. **108** (1990), no. 3, 817–820, DOI 10.2307/2047806. MR990413

[Bre12] M. Breen, *Helly-type theorems for intersections of sets starshaped via orthogonally convex paths*, Aequationes Math. **84** (2012), no. 3, 207–217, DOI 10.1007/s00010-012-0151-0. MR2996413

[Bre13] M. Breen, *Dual Helly-type theorems for unions of sets starshaped via staircase paths*, Ars Combin. **110** (2013), 129–141. MR3100317

[Bre14] M. Breen, *A Helly-type theorem for intersections of orthogonally starshaped sets in* $\mathbb{R}^d$, Period. Math. Hungar. **68** (2014), no. 1, 45–53, DOI 10.1007/s10998-014-0014-7. MR3189144

[Bro08] E. M. Bronshteĭn, *Approximation of convex sets by polyhedra* (Russian, with Russian summary), Sovrem. Mat. Fundam. Napravl. **22** (2007), 5–37, DOI 10.1007/s10958-008-9144-x; English transl., J. Math. Sci. (N. Y.) **153** (2008), no. 6, 727–762. MR2336506

[BSV03] H. Björklund, S. Sandberg, and S. Vorobyov, *A discrete subexponential algorithm for parity games*, STACS 2003, Lecture Notes in Comput. Sci., vol. 2607, Springer, Berlin, 2003, pp. 663–674, DOI 10.1007/3-540-36494-3_58. MR2066792

[BT11] G. Bacsó and Z. Tuza, *Optimal guard sets and the Helly property*, European J. Combin. **32** (2011), no. 1, 28–32, DOI 10.1016/j.ejc.2010.08.001. MR2727457

[CANR08] J. A. Cuesta-Albertos and A. Nieto-Reyes, *The random Tukey depth*, Comput. Statist. Data Anal. **52** (2008), no. 11, 4979–4988, DOI 10.1016/j.csda.2008.04.021. MR2526207

[CC05] G. Calafiore and M. C. Campi, *Uncertain convex programs: randomized solutions and confidence levels*, Math. Program. **102** (2005), no. 1, Ser. A, 25–46, DOI 10.1007/s10107-003-0499-y. MR2115479

[CC06] G. C. Calafiore and M. C. Campi, *The scenario approach to robust control design*, IEEE Trans. Automat. Control **51** (2006), no. 5, 742–753, DOI 10.1109/TAC.2006.875041. MR2232597

[CCZ14] M. Conforti, G. Cornuéjols, and G. Zambelli, *Integer programming*, Graduate Texts in Mathematics, vol. 271, Springer, Cham, 2014. MR3237726

[CD14] G. R. Chelnokov and V. L. Dol'nikov, *On transversals of quasialgebraic families of sets*, J. Combin. Theory Ser. A **125** (2014), 194–213, DOI 10.1016/j.jcta.2014.03.002. MR3207472

[CDS] M. Conforti and M. Di Summa, *On maximal S-free convex sets and the parameters* $f(S), h(S)$, to appear in SIAM Journal on Discrete Mathematics.

[CdVGG12] É. Colin de Verdière, G. Ginot, and X. Goaoc, *Multinerves and Helly numbers of acyclic families*, Computational geometry (SCG'12), ACM, New York, 2012, pp. 209–217, DOI 10.1145/2261250.2261282. MR3024716

[CdVGG14] É. Colin de Verdière, G. Ginot, and X. Goaoc, *Helly numbers of acyclic families*, Adv. Math. **253** (2014), 163–193, DOI 10.1016/j.aim.2013.11.004. MR3148550

[CEM+96] K. L. Clarkson, D. Eppstein, G. L. Miller, C. Sturtivant, and S.-H. Teng, *Approximating center points with iterative Radon points*, Internat. J. Comput. Geom. Appl. **6** (1996), no. 3, 357–377, DOI 10.1142/S021819599600023X. ACM Symposium on Computational Geometry (San Diego, CA, 1993). MR1409651

[CFP+14] D. Conlon, J. Fox, J. Pach, B. Sudakov, and A. Suk, *Ramsey-type results for semi-algebraic relations*, Trans. Amer. Math. Soc. **366** (2014), no. 9, 5043–5065, DOI 10.1090/S0002-9947-2014-06179-5. MR3217709

[CGH12] O. Cheong, X. Goaoc, and A. Holmsen, *Lower bounds to Helly numbers of line transversals to disjoint congruent balls*, Israel J. Math. **190** (2012), 213–228, DOI 10.1007/s11856-011-0179-1. MR2956239

[CGHP08] O. Cheong, X. Goaoc, A. Holmsen, and S. Petitjean, *Helly-type theorems for line transversals to disjoint unit balls*, Discrete Comput. Geom. **39** (2008), no. 1-3, 194–212, DOI 10.1007/s00454-007-9022-1. MR2383759

[Cha04] T. M. Chan, *An optimal randomized algorithm for maximum Tukey depth*, Proceedings of the Fifteenth Annual ACM-SIAM Symposium on Discrete Algorithms, ACM, New York, 2004, pp. 430–436. MR2291081

[CIKL13] J. Cardinal, H. Ito, M. Korman, and S. Langerman, *Helly numbers of polyominoes*, Graphs Combin. **29** (2013), no. 5, 1221–1234, DOI 10.1007/s00373-012-1203-x. MR3091383

[Cla95] K. L. Clarkson, *Las Vegas algorithms for linear and integer programming when the dimension is small*, J. Assoc. Comput. Mach. **42** (1995), no. 2, 488–499, DOI 10.1145/201019.201036. MR1409744

[DDGG09] J. Demouth, O. Devillers, M. Glisse, and X. Goaoc, *Helly-type theorems for approximate covering*, Discrete Comput. Geom. **42** (2009), no. 3, 379–398, DOI 10.1007/s00454-009-9167-1. MR2519886

[Deb70] H. E. Debrunner, *Helly type theorems derived from basic singular homology*, Amer. Math. Monthly **77** (1970), 375–380. MR0261443

[DF87] M. Deza and P. Frankl, *A Helly type theorem for hypersurfaces*, J. Combin. Theory Ser. A **45** (1987), no. 1, 27–30, DOI 10.1016/0097-3165(87)90043-4. MR883890

[DGK63] L. Danzer, B. Grünbaum, and V. Klee, *Helly's theorem and its relatives*, Proc. Sympos. Pure Math., Vol. VII, Amer. Math. Soc., Providence, R.I., 1963, pp. 101–180. MR0157289

[DHOR15] J. A. DeLorea, R. N. La Haye, D. Oliveros, and E. Roldán-Pensado. *Beyond Chance-Constrained Convex Mixed-Integer Optimization: A Generalized Calafiore-Campi Algorithm and the notion of S-optimization.* Preprint available at arXiv: 1504.00076. (2015).

[DLLHORP15] J. A. De Loera, R. N. La Haye, D. Oliveros, and E. Roldán-Pensado, *Helly numbers of algebraic subsets R^d and an extension of Doignon's theorem*, arXiv:1508.02380 **[math.MG]** (2015). To appear in Advances in Geometry.

[DLLHRS15] J. A. De Loera, R. N. La Haye, D. Rolnick, and P Soberón, *Quantitative Tverberg, Helly, & Carathéodory theorems*, arXiv:1503.06116 **[math.MG]** (2015).

[DLLHRS15a] J. A. De Lorea, R. N. La Haye, D. Rolnick, and P. Soberón. *Quantitative Tverberg theorems over lattices and other discrete sets.* To appear in Discrete and Compt. Geometry.

[DLLHRS15b] J. A. De Lorea, R. N. La Haye, D. Rolnick, and P. Soberón. *Quantitative combinatorial geometry for continuous parameters.* To appear in Discrete and Compt. Geometry.

[DLPS15] J. A. De Loera, S. Petrovic, and D. Stasi, *Random sampling in computational algebra: Helly numbers and violator spaces*, Journal of Symbolic Computation, **77**, pp. 1–15 (2016).

[Doi73] J.-P. Doignon, *Convexity in cristallographical lattices*, J. Geometry **3** (1973), 71–85. MR0387090

[DRS81] J.-P. Doignon, J. R. Reay, and G. Sierksma, *A Tverberg-type generalization of the Helly number of a convexity space*, J. Geom. **16** (1981), no. 2, 117–125, DOI 10.1007/BF01917580. MR642260

[DS11] M. Domokos and E. Szabó, *Helly dimension of algebraic groups*, J. Lond. Math. Soc. (2) **84** (2011), no. 1, 19–34, DOI 10.1112/jlms/jdq101. MR2819688

[Eck85] J. Eckhoff, *An upper-bound theorem for families of convex sets*, Geom. Dedicata **19** (1985), no. 2, 217–227, DOI 10.1007/BF00181472. MR809468

[Eck93] J. Eckhoff, *Helly, Radon, and Carathéodory type theorems*, Handbook of convex geometry, Vol. A, B, North-Holland, Amsterdam, 1993, pp. 389–448. MR1242986

[Eck03] J. Eckhoff, *A survey of the Hadwiger-Debrunner (p,q)-problem*, Discrete and computational geometry, Algorithms Combin., vol. 25, Springer, Berlin, 2003, pp. 347–377, DOI 10.1007/978-3-642-55566-4_16. MR2038482

[Eck08] J. Eckhoff, *Common transversals in the plane: the fractional perspective*, European J. Combin. **29** (2008), no. 8, 1872–1880, DOI 10.1016/j.ejc.2008.01.020. MR2463163

[EG61] P. Erdős and T. Gallai, *On the minimal number of vertices representing the edges of a graph.* (English, with Russian summary), Magyar Tud. Akad. Mat. Kutató Int. Közl. **6** (1961), 181–203. MR0144332

[Eis03] F. Eisenbrand, *Fast integer programming in fixed dimension*, Algorithms—ESA 2003, Lecture Notes in Comput. Sci., vol. 2832, Springer, Berlin, 2003, pp. 196–207, DOI 10.1007/978-3-540-39658-1_20. MR2085453

[ELR70] G. Ewald, D. G. Larman, and C. A. Rogers, *The directions of the line segments and of the r-dimensional balls on the boundary of a convex body in Euclidean space*, Mathematika **17** (1970), 1–20. MR0270271

[EN09] J. Eckhoff and K.-P. Nischke, *Morris's pigeonhole principle and the Helly theorem for unions of convex sets*, Bull. Lond. Math. Soc. **41** (2009), no. 4, 577–588, DOI 10.1112/blms/bdp024. MR2521353

[Far09] B. Farb, *Group actions and Helly's theorem*, Adv. Math. **222** (2009), no. 5, 1574–1588, DOI 10.1016/j.aim.2009.06.004. MR2555905

[Fef05] C. L. Fefferman, *A sharp form of Whitney's extension theorem*, Ann. of Math. (2) **161** (2005), no. 1, 509–577, DOI 10.4007/annals.2005.161.509. MR2150391

[FHZ14] O. Friedmann, T. Dueholm Hansen, and U. Zwick, *Random-facet and random-bland require subexponential time even for shortest paths*, arXiv:1410.7530 **[cs.DS]** (2014).

[Flø11] G. Fløystad, *The colorful Helly theorem and colorful resolutions of ideals*, J. Pure Appl. Algebra **215** (2011), no. 6, 1255–1262, DOI 10.1016/j.jpaa.2010.08.009. MR2769230

[FPS15] J. Fox, J. Pach, and A. Suk, *Density and regularity theorems for semi-algebraic hypergraphs*, SODA '15: Proceedings of the Twenty-Sixth Annual ACM-SIAM Symposium on Discrete Algorithms, SIAM, January 2015, pp. 1517–1530.

[Fra89] P. Frankl, *Helly-type theorems for varieties*, European J. Combin. **10** (1989), no. 3, 243–245, DOI 10.1016/S0195-6698(89)80058-7. MR1029170

[Gär95] B. Gärtner, *A subexponential algorithm for abstract optimization problems*, SIAM J. Comput. **24** (1995), no. 5, 1018–1035, DOI 10.1137/S0097539793250287. MR1350756

[Gär15] B. Gärtner, *Sampling with removal in LP-type problems*, J. Comput. Geom. **6** (2015), no. 2, 93–112. MR3315835

[GHP+06] J. E. Goodman, A. Holmsen, R. Pollack, K. Ranestad, and F. Sottile, *Cremona convexity, frame convexity and a theorem of Santaló*, Adv. Geom. **6** (2006), no. 2, 301–321, DOI 10.1515/ADVGEOM.2006.018. MR2243303

[GLS10] J. Gao, M. Langberg, and L. J. Schulman, *Clustering lines in high-dimensional space: classification of incomplete data*, ACM Trans. Algorithms **7** (2010), no. 1, Art. 8, 26, DOI 10.1145/1868237.1868246. MR2747125

[GM61] B. Grünbaum and T. S. Motzkin, *On components in some families of sets*, Proc. Amer. Math. Soc. **12** (1961), 607–613. MR0159262

[GM10] S. Gaubert and F. Meunier, *Carathéodory, Helly and the others in the max-plus world*, Discrete Comput. Geom. **43** (2010), no. 3, 648–662, DOI 10.1007/s00454-009-9207-x. MR2587842

[GMRŠ08] B. Gärtner, J. Matoušek, L. Rüst, and P. Škovroň, *Violator spaces: structure and algorithms*, Discrete Appl. Math. **156** (2008), no. 11, 2124–2141, DOI 10.1016/j.dam.2007.08.048. MR2437006

[Goa10] X. Goaoc, *Some discrete properties of the space of line transversals to disjoint balls*, Nonlinear computational geometry, IMA Vol. Math. Appl., vol. 151, Springer, New York, 2010, pp. 51–83, DOI 10.1007/978-1-4419-0999-2_3. MR2560555

[Gom58] R. E. Gomory, *Outline of an algorithm for integer solutions to linear programs*, Bull. Amer. Math. Soc. **64** (1958), 275–278. MR0102437

[GPP+15] X. Goaoc, P. Paták, Z. Patáková, M. Tancer, and U. Wagner, *Bounding Helly numbers via Betti numbers*, 31st International Symposium on Computational Geometry, LIPIcs. Leibniz Int. Proc. Inform., vol. 34, Schloss Dagstuhl. Leibniz-Zent. Inform., Wadern, 2015, pp. 507–521. MR3392802

[Grü58] B. Grünbaum, *On common transversals*, Arch. Math. **9** (1958), 465–469. MR0102780

[Gru93] P. M. Gruber, *Aspects of approximation of convex bodies*, Handbook of convex geometry, Vol. A, B, North-Holland, Amsterdam, 1993, pp. 319–345. MR1242984

[Had57] H. Hadwiger, *Ueber Eibereiche mit gemeinsamer Treffgeraden* (German), Portugal. Math. **16** (1957), 23–29. MR0099015

[Hal07a] N. Halman, *Discrete and lexicographic Helly-type theorems*, Discrete Comput. Geom. **39** (2008), no. 4, 690–719, DOI 10.1007/s00454-007-9028-8. MR2413154

[Hal07b] N. Halman, *Simple stochastic games, parity games, mean payoff games and discounted payoff games are all LP-type problems*, Algorithmica **49** (2007), no. 1, 37–50, DOI 10.1007/s00453-007-0175-3. MR2344393

[HD57] H. Hadwiger and H. Debrunner, *Über eine Variante zum Hellyschen Satz* (German), Arch. Math. (Basel) **8** (1957), 309–313. MR0092987

[Hel23] E. Helly, *Über mengen konvexer körper mit gemeinschaftlichen punkte.*, Jahresbericht der Deutschen Mathematiker-Vereinigung **32** (1923), 175–176.

[Hel30] E. Helly, *Über Systeme von abgeschlossenen Mengen mit gemeinschaftlichen Punkten* (German), Monatsh. Math. Phys. **37** (1930), no. 1, 281–302, DOI 10.1007/BF01696777. MR1549795

[Hel06] S. Hell, *Tverberg-type theorems and the fractional helly property*, Ph.D. thesis, Technische Universität Berlin, 2006, https://opus4.kobv.de/opus4-tuberlin/frontdoor/index/index/docId/1371.

[HK15] A. Holmsen and R. Karasev, *Colorful theorems for strong convextity*, arXiv:1509.08783 **[math.CO]** (2015). To appear in the Proc. Amer. Math. Soc.

[HKL03] A. Holmsen, M. Katchalski, and T. Lewis, *A Helly-type theorem for line transversals to disjoint unit balls*, Discrete Comput. Geom. **29** (2003), no. 4, 595–602, DOI 10.1007/s00454-002-0793-0. MR1976608

[HMSM14] A. F. Holmsen, L. Martinez-Sandoval, and L. Montejano, *A geometric Hall-type theorem*, Proc. Amer. Math. Soc. **144** (2016), no. 2, 503–511, DOI 10.1090/proc12733. MR3430829

[Hof79] A. J. Hoffman, *Binding constraints and Helly numbers*, Second International Conference on Combinatorial Mathematics (New York, 1978), Ann. New York Acad. Sci., vol. 319, New York Acad. Sci., New York, 1979, pp. 284–288. MR556036

[Hol13] A. F. Holmsen, *Geometric transversal theory: $T(3)$-families in the plane*, Geometry—intuitive, discrete, and convex, Bolyai Soc. Math. Stud., vol. 24, János Bolyai Math. Soc., Budapest, 2013, pp. 187–203, DOI 10.1007/978-3-642-41498-5_7. MR3204559

[HPT08] A. F. Holmsen, J. Pach, and H. Tverberg, *Points surrounding the origin*, Combinatorica **28** (2008), no. 6, 633–644, DOI 10.1007/s00493-008-2427-5. MR2488743

[HRP] A. F. Holmsen and E. Roldán-Pensado, *The colored Hadwiger transversal theorem in R^d*, Combinatorica, 36:417(2016). DOI:10.1007/s00493-014-3192-2.

[Jam81] R. E. Jamison-Waldner, *Partition numbers for trees and ordered sets*, Pacific J. Math. **96** (1981), no. 1, 115–140. MR634767

[JC07] J. Jerónimo Castro, *Line transversals to translates of unit discs*, Discrete Comput. Geom. **37** (2007), no. 3, 409–417, DOI 10.1007/s00454-006-1281-8. MR2301526

[JCMS15] J. Jerónimo-Castro, A. Magazinov, and P Soberón, *On a problem by Dol'nikov*, Discrete Math. **338** (2015), no. 9.

[JCRP11] J. Jerónimo-Castro and E. Roldán-Pensado, *Line transversals to translates of a convex body*, Discrete Comput. Geom. **45** (2011), no. 2, 329–339, DOI 10.1007/s00454-010-9293-9. MR2765533

[Kal84] G. Kalai, *Intersection patterns of convex sets*, Israel J. Math. **48** (1984), no. 2-3, 161–174, DOI 10.1007/BF02761162. MR770699

[Kal92] G. Kalai, *A subexponential randomized simplex algorithm*, Proceedings of the twenty-fourth annual ACM symposium on Theory of computing, ACM, 1992, pp. 475–482.

[Kar00] R. N. Karasev, *Transversals for families of translates of a two-dimensional convex compact set*, Discrete Comput. Geom. **24** (2000), no. 2-3, 345–353, DOI 10.1007/s004540010040. The Branko Grünbaum birthday issue. MR1758055

[Kar12] R. N. Karasev, *Analogues of the central point theorem for families with d-intersection property in $\mathbb{R}^d$*, Combinatorica **32** (2012), no. 6, 689–702, DOI 10.1007/s00493-012-2603-5. MR3063156

[KGT01] D. J. Kleitman, A. Gyárfás, and G. Tóth, *Convex sets in the plane with three of every four meeting*, Combinatorica **21** (2001), no. 2, 221–232, DOI 10.1007/s004930100020. Paul Erdős and his mathematics (Budapest, 1999). MR1832447

[KL79] M. Katchalski and A. Liu, *A problem of geometry in $\mathbf{R}^n$*, Proc. Amer. Math. Soc. **75** (1979), no. 2, 284–288, DOI 10.2307/2042758. MR532152

[KM05] G. Kalai and R. Meshulam, *A topological colorful Helly theorem*, Adv. Math. **191** (2005), no. 2, 305–311, DOI 10.1016/j.aim.2004.03.009. MR2103215

[KM08] G. Kalai and R. Meshulam, *Leray numbers of projections and a topological Helly-type theorem*, J. Topol. **1** (2008), no. 3, 551–556, DOI 10.1112/jtopol/jtn010. MR2417442

[KTW11] C. Knauer, H. R. Tiwary, and D. Werner, *On the computational complexity of ham-sandwich cuts, Helly sets, and related problems*, 28th International Symposium on Theoretical Aspects of Computer Science, LIPIcs. Leibniz Int. Proc. Inform., vol. 9, Schloss Dagstuhl. Leibniz-Zent. Inform., Wadern, 2011, pp. 649–660. MR2853467

[KW71] D. C. Kay and E. W. Womble, *Axiomatic convexity theory and relationships between the Carathéodory, Helly, and Radon numbers*, Pacific J. Math. **38** (1971), 471–485. MR0310766

[Len83] H. W. Lenstra Jr., *Integer programming with a fixed number of variables*, Math. Oper. Res. **8** (1983), no. 4, 538–548, DOI 10.1287/moor.8.4.538. MR727410

[Lov89] L. Lovász, *Geometry of numbers and integer programming*, Mathematical programming (Tokyo, 1988), Math. Appl. (Japanese Ser.), vol. 6, SCIPRESS, Tokyo, 1989, pp. 177–201. MR1114315

[Mat94] J. Matoušek, *Lower bounds for a subexponential optimization algorithm*, Random Structures Algorithms **5** (1994), no. 4, 591–607, DOI 10.1002/rsa.3240050408. MR1293081

[Mat97] J. Matoušek, *A Helly-type theorem for unions of convex sets*, Discrete Comput. Geom. **18** (1997), no. 1, 1–12, DOI 10.1007/PL00009305. MR1453439

[Mat02] J. Matoušek, *Lectures on discrete geometry*, Graduate Texts in Mathematics, vol. 212, Springer-Verlag, New York, 2002. MR1899299

[Mat04] J. Matoušek, *Bounded VC-dimension implies a fractional Helly theorem*, Discrete Comput. Geom. **31** (2004), no. 2, 251–255, DOI 10.1007/s00454-003-2859-z. MR2060639

[Mat10] J. Matoušek, *Geometric discrepancy*, Algorithms and Combinatorics, vol. 18, Springer-Verlag, Berlin, 1999. An illustrated guide. MR1697825

[Meg96] N. Megiddo, *Program product for displaying a line passing through a plurality of boxes*, July 2 1996, US Patent 5,533,178.

[MO08] L. Montejano and D. Oliveros, *Colourful transversal theorems*, Contrib. Discrete Math. **3** (2008), no. 2, 60–75. MR2458134

[MO11] L. Montejano and D. Oliveros, *Tolerance in Helly-type theorems*, Discrete Comput. Geom. **45** (2011), no. 2, 348–357, DOI 10.1007/s00454-010-9296-6. MR2765535

[Mon13] L. Montejano, *Transversals, topology and colorful geometric results*, Geometry—intuitive, discrete, and convex, Bolyai Soc. Math. Stud., vol. 24, János Bolyai Math. Soc., Budapest, 2013, pp. 205–218, DOI 10.1007/978-3-642-41498-5_8. MR3204560

[Mon14] L. Montejano, *A new topological Helly theorem and some transversal results*, Discrete Comput. Geom. **52** (2014), no. 2, 390–398, DOI 10.1007/s00454-014-9613-6. MR3249387

[Mor73] H. C. Morris, *TWO PIGEON-HOLE PRINCIPLES AND UNIONS OF CONVEXLY-DISJOINT SETS*, ProQuest LLC, Ann Arbor, MI, 1974. Thesis (Ph.D.)–California Institute of Technology. MR2624077

[Mot55] T. S. Motzkin, *A proof of Hilbert's Nullstellensatz*, Math. Z. **63** (1955), 341–344. MR0074388

[MS09] G. L. Miller and D. R Sheehy, *Approximate center points with proofs*, Proceedings of the twenty-fifth annual symposium on Computational geometry, ACM, 2009, pp. 153–158.

[MS11] L. Montejano and P. Soberón, *Piercing numbers for balanced and unbalanced families*, Discrete Comput. Geom. **45** (2011), no. 2, 358–364, DOI 10.1007/s00454-010-9295-7. MR2765536

[MSW96] J. Matoušek, M. Sharir, and E. Welzl, *A subexponential bound for linear programming*, Algorithmica **16** (1996), no. 4-5, 498–516.

[MT09] J. Matoušek and M. Tancer, *Dimension gaps between representability and collapsibility*, Discrete Comput. Geom. **42** (2009), no. 4, 631–639, DOI 10.1007/s00454-008-9091-9. MR2556459

[MW13] W. Mulzer and D. Werner, *Approximating Tverberg points in linear time for any fixed dimension*, Discrete Comput. Geom. **50** (2013), no. 2, 520–535, DOI 10.1007/s00454-013-9528-7. MR3090531

[Nas16] M. Naszódi, *Proof of a conjecture of Bárány, Katchalski and Pach*, Discrete Comput. Geom. **55** (2016), no. 1, 243–248, DOI 10.1007/s00454-015-9753-3. MR3439267

[Neu45] B. H. Neumann, *On an invariant of plane regions and mass distributions*, J. London Math. Soc. **20** (1945), 226–237. MR0017564

[NS14] V. Nitica and S. Sergeev, *Tropical convexity over max-min semiring*, Tropical and idempotent mathematics and applications, Contemp. Math., vol. 616, Amer. Math. Soc., Providence, RI, 2014, pp. 241–260, DOI 10.1090/conm/616/12301. MR3221336

[PW90] R. Pollack and R. Wenger, *Necessary and sufficient conditions for hyperplane transversals*, Combinatorica **10** (1990), no. 3, 307–311, DOI 10.1007/BF02122783. MR1092546

[Rad21] J. Radon, *Mengen konvexer Körper, die einen gemeinsamen Punkt enthalten* (German), Math. Ann. **83** (1921), no. 1-2, 113–115, DOI 10.1007/BF01464231. MR1512002

[Rad47] R. Rado, *A theorem on general measure*, J. London Math. Soc. **21** (1946), 291–300 (1947). MR0021962

[RS15] D. Rolnick and P. Soberón, *Quantitative (p, q) theorems in combinatorial geometry*, arXiv:1504.01642 **[math.MG]** (2015).

[RSW01] S. Reisner, C. Schütt, and E. Werner, *Dropping a vertex or a facet from a convex polytope*, Forum Math. **13** (2001), no. 3, 359–378, DOI 10.1515/form.2001.012. MR1831090

[Sca77] H. E. Scarf, *An observation on the structure of production sets with indivisibilities*, Proc. Nat. Acad. Sci. U.S.A. **74** (1977), no. 9, 3637–3641. MR0452678

[Sch86] A. Schrijver, *Theory of linear and integer programming*, Wiley-Interscience Series in Discrete Mathematics, John Wiley & Sons, Ltd., Chichester, 1986. A Wiley-Interscience Publication. MR874114

[Sei91] R. Seidel, *Small-dimensional linear programming and convex hulls made easy*, Discrete Comput. Geom. **6** (1991), no. 5, 423–434, DOI 10.1007/BF02574699. MR1115100

[Sko07] P. Skovron, *Abstract models of optimization problems*, Ph.D. thesis, Charles University, Prague, 2007.

[Sob16] P. Soberón, *Helly-type theorems for the diameter*, to appear Bulletin of the London Mathematical Society (2016).

[SW92] M. Sharir and E. Welzl, *A combinatorial bound for linear programming and related problems*, STACS 92 (Cachan, 1992), Lecture Notes in Comput. Sci., vol. 577, Springer, Berlin, 1992, pp. 569–579. MR1255620

[SW96] Micha Sharir and Emo Welzl, *Rectilinear and polygonal p-piercing and p-center problems*, Proceedings of the twelfth annual symposium on Computational geometry, ACM, 1996, pp. 122–132.

[Swa03] K. J. Swanepoel, *Helly-type theorems for homothets of planar convex curves*, Proc. Amer. Math. Soc. **131** (2003), no. 3, 921–932 (electronic), DOI 10.1090/S0002-9939-02-06722-9. MR1937431

[Tan13] M. Tancer, *Intersection patterns of convex sets via simplicial complexes: a survey*, Thirty essays on geometric graph theory, Springer, New York, 2013, pp. 521–540, DOI 10.1007/978-1-4614-0110-0_28. MR3205172

[Tuz89] Z. Tuza, *Minimum number of elements representing a set system of given rank*, J. Combin. Theory Ser. A **52** (1989), no. 1, 84–89, DOI 10.1016/0097-3165(89)90064-2. MR1008162

[Tve66] H. Tverberg, *A generalization of Radon's theorem*, J. London Math. Soc. **41** (1966), 123–128. MR0187147

[Tve89] H. Tverberg, *Proof of Grünbaum's conjecture on common transversals for translates*, Discrete Comput. Geom. **4** (1989), no. 3, 191–203, DOI 10.1007/BF02187722. MR988742

[vdV93] M. L. J. van de Vel, *Theory of convex structures*, North-Holland Mathematical Library, vol. 50, North-Holland Publishing Co., Amsterdam, 1993. MR1234493

[Weg75] G. Wegner, *d-collapsing and nerves of families of convex sets*, Arch. Math. (Basel) **26** (1975), 317–321. MR0375333

[Wen04] R. Wenger, *Helly-type theorems and geometric transversals*, Handbook of discrete and computational geometry, CRC Press Ser. Discrete Math. Appl., CRC, Boca Raton, FL, 1997, pp. 63–82. MR1730160

Department of Computer Science, University of California, Davis, One Shields Avenue, Davis, California 95616

E-mail address: amenta@cs.ucdavis.edu

Department of Mathematics, University of California, Davis, One Shields Avenue, Davis, California 95616

E-mail address: deloera@math.ucdavis.edu

Department of Mathematics, Northeastern University, 360 Huntington Ave., Boston, Massachusetts 02115

E-mail address: p.soberonbravo@neu.edu

Contemporary Mathematics
Volume **685**, 2017
http://dx.doi.org/10.1090/conm/685/13752

Representation theory of the symmetric group in voting theory and game theory

Karl-Dieter Crisman and Michael E. Orrison

Abstract. This paper is a survey of some of the ways in which the representation theory of the symmetric group has been used in voting theory and game theory. In particular, we use permutation representations that arise from the action of the symmetric group on tabloids to describe, for example, a surprising relationship between the Borda count and Kemeny rule in voting. We also explain a powerful representation-theoretic approach to working with linear symmetric solution concepts in cooperative game theory. Along the way, we discuss new research questions that arise within and because of the representation-theoretic framework we are using.

1. Introduction

Symmetry arises in many ways in voting theory and game theory. In this paper, we highlight one such appearance by surveying some of the ways in which the *representation theory* of the symmetric group has been used in these fields. Our primary goal is to show how certain well-understood permutation representations of the symmetric group can be used to make sense of foundational ideas in voting theory and game theory, and how using these representations can in turn help researchers formulate novel questions and meaningful generalizations.

Among other examples, we use the representation theory of the symmetric group to describe a surprising relationship between the Borda count and the Kemeny rule in voting. We also use it to construct and make sense of certain infinite families of solution concepts in cooperative game theory. Along the way, we describe much of the representation-theoretic framework used in voting theory and game theory, and we discuss several new research questions that arise therein.

The permutation representations on which we focus are those associated to the usual action of the symmetric group S_n on simple combinatorial objects called tabloids. Representation theory experts will have no trouble seeing that we only scratch the surface on what could be done with these representations in voting theory and game theory. Another goal for this paper is therefore to encourage readers to contribute to these fields by going well beyond what we present here.

As a brief introduction to the role that symmetry can play in voting, consider a situation in which voters are asked to choose their favorite candidate, A, B, or C. If we end with an outcome of $(13, 4, 7)$ for A, B, C, respectively, then it is

2010 *Mathematics Subject Classification.* Primary 91B12, 91A12, 20C30.

natural to say this is essentially the same outcome as $(9,0,3)$, because there are four votes for each candidate that "cancel" each other. We can go further, though, by writing $(13,4,7) = (8,8,8) + (5,-4,-1)$ and noting that the first vector on the right captures how many voters there were, while the second vector captures the different levels of support the candidates received, which in the end is what really matters.

Interestingly, such decompositions arise by having the symmetric group S_3 act on the candidates by permuting their labels, which in turn induces an action on the set of outcomes. Furthermore, by having the symmetric group act on the labels of candidates in much more complicated voting situations, similar decompositions of voting data arise, and we are able to use what we know about the resulting subspaces of voting data to say something worthwhile (as we will demonstrate in Section 3) about how information is used by different kinds of voting procedures.

In the next section, we introduce most of the notation we will use for the rest of the paper, and we motivate the use of tabloids for indexing the kind of data with which we will be working. Although we do not assume our readers will be familiar with the ideas we will be presenting from voting theory and game theory, we will assume that our readers have a basic working knowledge of the representation theory of finite groups (see, for example, [**14**, **39**]).

Finally, we encourage interested readers to view the ideas presented in this paper within the larger framework of harmonic analysis on finite groups [**3**, **4**, **8**, **43**]. In our opinion, doing so makes it much easier to see how the representation theory of the symmetric group (and other finite groups) has been applied outside of voting theory and game theory, in fields, for example, such as statistics [**2**, **8**, **9**, **26**] and machine learning [**12**, **24**, **25**]. Doing so will also reveal that the approach we are taking in this paper has already found success in a variety of other settings.

2. Background

In this section, we define tabloids and their associated permutation representations of the symmetric group. We also describe how tabloids can be used to index the kind of voting and game-theoretic data with which we will be working. Good references for the material in this section are [**13**] and [**37**].

Let n be a positive integer. A *composition* of n is a list $\lambda = (\lambda_1, \ldots, \lambda_m)$ of positive integers whose sum is n. If it is also the case that $\lambda_1 \geq \cdots \geq \lambda_m$, then we say λ is a *partition* of n. For example, $(2,3,1,3)$ is a composition of 9, and $(4,2,2,1)$ is a partition of 9. If λ is a composition, then we will denote by $\overline{\lambda}$ the partition obtained by reordering the numbers in λ so that they form a non-increasing list. For example, if $\lambda = (2,3,1,3)$, then $\overline{\lambda} = (3,3,2,1)$.

If λ is a composition of n, then the *Young diagram of shape* λ is the left-justified array of boxes that has λ_i boxes in its ith row (see Figure 1). If we fill these boxes with the numbers $1, \ldots, n$ without repetition, then we create a *Young tableau of shape* λ. Two tableaux of shape $\lambda = (\lambda_1, \ldots, \lambda_m)$ are then said to be *row equivalent* if they have the same set of λ_1 numbers in the first row, the same set of λ_2 numbers in the second row, and so on. An equivalence class of tableaux is then called a *tabloid of shape* λ.

We often denote a tabloid by first forming a representative tableau and then removing the vertical dividers within each row (see Figure 2). For convenience, we will usually choose the representative tableau whose entries in each row are in

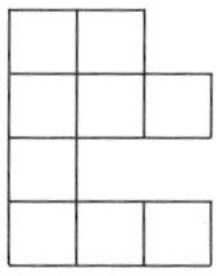

FIGURE 1. The Young diagram of shape $(2, 3, 1, 3)$.

ascending order, and we will read the entries from top to bottom when putting such representatives and their associated tabloids in lexicographic order.

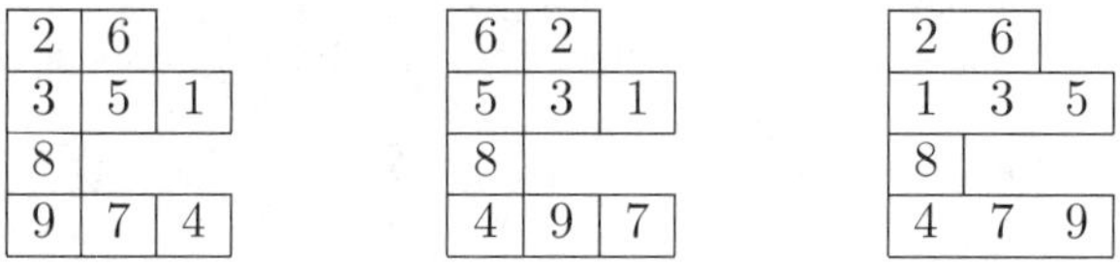

FIGURE 2. Two row equivalent tableaux and their tabloid.

Suppose λ is a composition of n. Let X^λ denote the set of tabloids of shape λ, and let M^λ denote the vector space of real-valued functions defined on X^λ. If $x \in X^\lambda$, then $\mathbf{f}_x$ will denote its associated indicator function with the property that $\mathbf{f}_x(x) = 1$ and $\mathbf{f}_x(y) = 0$ for all $y \neq x$. Note that the set of such indicator functions forms an orthonormal basis for M^λ with the respect to the usual inner product $\langle \cdot, \cdot \rangle$ on M^λ, which is defined by setting

$$\langle \mathbf{f}, \mathbf{g} \rangle = \sum_{x \in X^\lambda} \mathbf{f}(x)\mathbf{g}(x)$$

for all $\mathbf{f}, \mathbf{g} \in M^\lambda$. We will call this basis of indicator functions the *usual basis* of M^λ, and when necessary we will assume it has been ordered with respect to the lexicographic ordering of the associated tabloids in X^λ.

In this paper, we will be interested in a variety of real-valued functions defined on X^λ for different choices of λ. We will also be interested in certain linear transformations defined on M^λ.

For example, consider an election involving n candidates labeled 1 through n. If we ask the voters to tell us their top favorite λ_1 candidates, then their next top favorite λ_2 candidates, and so on, then we can interpret each of their responses as a tabloid $x \in X^\lambda$. For example, if $\lambda = (2, 3, 1, 3)$, and a voter chooses the tabloid in Figure 2, then we will take that to mean candidates 2 and 6 are her top two favorite candidates, but we do not know if the voter prefers one of these two candidates over the other. The function $\mathbf{f} \in M^\lambda$ in which we are then interested is simply the one for which $\mathbf{f}(x)$ is the total number of voters who chose tabloid $x \in X^\lambda$.

We will mostly focus on three kinds of compositions in this paper. The first composition is the all-ones partition $(1, \ldots, 1)$. The tabloids in $X^{(1,\ldots,1)}$ naturally correspond to all of the possible permutations of the numbers $1, \ldots, n$. Therefore, $M^{(1,\ldots,1)}$ is helpful when describing voting situations in which voters are asked to provide a full ranking of candidates. The second kind of composition has the form $(k, n-k)$ where $1 \leq k \leq n-1$. In this case, we will use the tabloids in $X^{(k,n-k)}$ to index k-element subsets of $\{1, \ldots, n\}$, where each k-element subset corresponds to the tabloid whose top row contains all of its elements. Finally, the third composition is (n). The set $X^{(n)}$ consists of a single, one-rowed tabloid which we will use to index the entire set $\{1, \ldots, n\}$.

As an example of how we will use $M^{(1,\ldots,1)}$ and $M^{(k,n-k)}$ in this paper, in the next section we will consider the voting situation in which voters have been asked to fully rank a set of n candidates. We will then use that information, which we will view as an element of $M^{(1,\ldots,1)}$, to assign points to each of the individual candidates. The result may therefore be viewed as an element of $M^{(1,n-1)}$. Furthermore, we will do all of this using a linear transformation from $M^{(1,\ldots,1)}$ to $M^{(1,n-1)}$.

Now it is natural in voting to insist that a voting method not depend on how we label the candidates. Fortunately, we may easily express this property using a group action. More specifically, note that the symmetric group S_n acts naturally on X^λ where if $\sigma \in S_n$ and $x \in X^\lambda$, then $\sigma \cdot x$ is the tabloid one gets by applying the permutation σ to each entry of x.

The action of S_n on X^λ extends to an action of S_n on M^λ where if $\sigma \in S_n$, $\mathbf{f} \in M^\lambda$, and $x \in X^\lambda$, then

$$(\sigma \cdot \mathbf{f})(x) = \mathbf{f}(\sigma^{-1} \cdot x).$$

In other words, M^λ may be viewed as a module over the group algebra $\mathbb{R}S_n$. For example, $M^{(n)}$ corresponds to the trivial $\mathbb{R}S_n$-module. This is because $X^{(n)}$ consists of a single tabloid, and every permutation in S_n fixes this tabloid.

In the voting example above, we can view the action of S_n on $M^{(1,\ldots,1)}$ and $M^{(1,n-1)}$ as the result of changing the labels on the candidates. When the outcome of a voting method does not depend on the labels that have been assigned to the candidates, we say that the voting method is *neutral* (see, for example, [**30**, **42**], among many other references for general voting theory concepts). Insisting that a voting method based on a linear transformation $T : M^{(1,\ldots,1)} \to M^{(1,n-1)}$ be neutral then simply becomes the requirement that T must be an $\mathbb{R}S_n$-module homomorphism. We will focus only on voting methods that are neutral in this paper.

Fortunately, the representation theory of M^λ is well-understood. The irreducible $\mathbb{R}S_n$-modules are parametrized by the partitions of n, and we denote the irreducible module corresponding to the partition μ by S^μ. These are the well-known *Specht modules* where, for example, $S^{(n)}$ corresponds to the one-dimensional trivial $\mathbb{R}S_n$-module.

If λ is a partition of n (and not just a composition), then M^λ is isomorphic to a direct sum of Specht modules

$$M^\lambda \cong \bigoplus_\mu \kappa_{\mu\lambda} S^\mu$$

where the $\kappa_{\mu\lambda}$ are *Kostka numbers* and are used here record the multiplicity of each Specht module in M^λ (see, for example, Sections 2.3 and 2.11 of [**37**]).

Note that if λ is a composition of n and not necessarily a partition, then the module M^λ is easily seen to be isomorphic to $M^{\overline{\lambda}}$ by simply reordering the rows of the tabloids in X^λ. We may therefore just as easily work with compositions as with partitions of n when dealing with real-valued functions (e.g., voting data) defined on sets (e.g., rankings of candidates) that are indexed by tabloids.

The modules $M^{(1,\ldots,1)}$, $M^{(k,n-k)}$, and $M^{(n)}$ have particularly straightforward decompositions in terms of Specht modules. For example, $M^{(1,\ldots,1)}$ is isomorphic to the regular $\mathbb{R}S_n$-module, and thus

$$M^{(1,\ldots,1)} \cong \bigoplus_{\mu} (\dim S^{\mu}) S^{\mu}.$$

If $1 \le k \le n/2$ (so that $(n-k,k)$ is a partition of n), then

$$M^{(k,n-k)} \cong M^{(n-k,k)} \cong S^{(n)} \oplus S^{(n-1,1)} \oplus S^{(n-2,2)} \oplus \cdots \oplus S^{(n-k,k)}.$$

Lastly, because $M^{(n)}$ corresponds to the trivial $\mathbb{R}S_n$-module, we have that $M^{(n)} \cong S^{(n)}$ (see Chapter 7 of [**8**] or Section 2.11 of [**37**]).

Before leaving this section, we introduce some specialized notation and terminology that we will use later in the paper.

First, note that $M^{(1,n-1)}$ is a direct sum of two irreducible submodules. These submodules are easy to describe (see also p. 39 in [**37**]). One of the submodules, which we will denote by U_0, is isomorphic to $S^{(n)}$ and consists of all of the constant functions in $M^{(1,n-1)}$:

$$U_0 = \left\{ \mathbf{f} \in M^{(1,n-1)} \mid \mathbf{f}(x) = \mathbf{f}(y) \text{ for all } x, y \in X^{(1,n-1)} \right\}.$$

The other submodule, which we will denote by U_1, is isomorphic to $S^{(n-1,1)}$. It is the orthogonal complement of U_0, and it consists of those functions whose values sum to zero:

$$U_1 = \left\{ \mathbf{f} \in M^{(1,n-1)} \;\middle|\; \sum_{x \in X^{(1,n-1)}} \mathbf{f}(x) = 0 \right\}.$$

If $\mathbf{f} \in M^{(1,n-1)}$, then we will denote its projection into U_1 by $\widehat{\mathbf{f}}$. Thus $\mathbf{f} = (\mathbf{f} - \widehat{\mathbf{f}}) + \widehat{\mathbf{f}}$ where $\mathbf{f} - \widehat{\mathbf{f}} \in U_0$ and $\widehat{\mathbf{f}} \in U_1$. In fact, if we let $\mathbf{1} \in M^{(1,n-1)}$ denote the function that assigns the value 1 to every tabloid, and let $\mathbf{1}_{\mathbf{f}}$ denote the projection of $\mathbf{f}$ onto $\mathbf{1}$, then $\mathbf{f} - \widehat{\mathbf{f}} = \mathbf{1}_{\mathbf{f}}$.

Next, note that the action of S_n on X^{λ} is transitive. In other words, given two tabloids $x, y \in X^{\lambda}$, there exists at least one permutation $\sigma \in S_n$ such that $\sigma \cdot x = y$. In this case, we have that $\sigma \cdot \mathbf{f}_x = \mathbf{f}_y$. It follows that every module homomorphism defined on M^{λ} is determined by the image of any of the indicator functions $\mathbf{f}_x$. For convenience, when we construct such maps, we will focus on the image of $\mathbf{f}_{x_0}$, where we use x_0 to denote the tabloid in X^{λ} that appears first when the tabloids in X^{λ} are listed lexicographically. More specifically, $x_0 \in X^{\lambda}$ is the tabloid that contains the tableau whose entries, when read from left to right and top to bottom, are the numbers $1, \ldots, n$ in that order.

Finally, suppose λ is a composition of n, and that T is a module homomorphism defined on M^{λ}. The module M^{λ} can be written as the direct sum

$$M^{\lambda} = \ker T \oplus (\ker T)^{\perp}$$

of the kernel of T and its orthogonal complement. We will refer to the submodule $(\ker T)^{\perp}$ as the *effective space* of T, and we will denote it by $E(T)$. Note that the effective space $E(T)$ is isomorphic to (and also determines) the image of T.

3. Voting Theory

For algebraists interested in learning about the mathematics of voting, Donald Saari's papers [**28, 31–33**] and books [**29, 30, 34, 35**] are a fruitful place to start. His work provides a friendly but mathematically sophisticated gateway to the subject, and although his approach is primarily geometric, linear algebra and symmetry are used to great effect.

Some of Saari's key results rely on decomposing vector spaces of voting data into a handful of simple but meaningful subspaces. Despite the geometric flavor of many of his papers, algebraists will easily recognize many of these subspaces as submodules of $M^{(1,\dots,1)}$. This realization has been put to use in papers such as [**6**] and [**7**], and what follows in this section can be partly viewed as an introduction to those two papers.

We begin by considering an election in which there are n candidates labeled $1, \dots, n$. Suppose each voter has been asked to rank the candidates by choosing a tabloid in $X^{(1,\dots,1)}$, where the candidate in the top row of the tabloid is their favorite, the candidate in the second row is their second favorite, and so on.

For example, if there are $n = 3$ candidates, then each voter is being asked to choose one of the following tabloids from $X^{(1,1,1)}$:

$$\begin{array}{|c|}\hline 1\\\hline 2\\\hline 3\\\hline\end{array}\,,\quad \begin{array}{|c|}\hline 1\\\hline 3\\\hline 2\\\hline\end{array}\,,\quad \begin{array}{|c|}\hline 2\\\hline 1\\\hline 3\\\hline\end{array}\,,\quad \begin{array}{|c|}\hline 2\\\hline 3\\\hline 1\\\hline\end{array}\,,\quad \begin{array}{|c|}\hline 3\\\hline 1\\\hline 2\\\hline\end{array}\,,\quad \begin{array}{|c|}\hline 3\\\hline 2\\\hline 1\\\hline\end{array}\,.$$

In this case, a voter who chooses the third tabloid above is saying that she prefers candidate 2 to candidate 1 to candidate 3.

Let $\mathbf{p} \in M^{(1,\dots,1)}$ be the function with the property that $\mathbf{p}(x)$ is the number of voters who chose the tabloid x. The function $\mathbf{p}$ is called a *profile*. Next, let $\mathbf{w} = [w_1, \dots, w_n]^t$ be a column vector in $\mathbb{R}^n$ such that $w_1 \geq \cdots \geq w_n$. The vector $\mathbf{w}$ is called a *weighting vector*, and we will use this vector of "weights" to create a voting procedure as follows.

For each tabloid $x \in X^{(1,\dots,1)}$, if candidate i is in row j of x, then she will be given $\mathbf{p}(x)w_j$ points. We then sum over all of the tabloids in $X^{(1,\dots,1)}$ to find the total number of points received by candidate i. The candidates who receive the most points (there might be ties) are then declared to be winners. We refer to this voting procedure as the *positional voting* procedure associated to $\mathbf{w}$. Observe that the inequality condition above ensures that a voter's "more favored" candidates receive at least as many points from her as her "less favored" candidates.

Note that $\mathbf{w} = [1, 0, \dots, 0]^t$ corresponds to the *plurality* voting procedure, where voters are essentially being asked to vote for only one candidate. When $\mathbf{w} = [1, \dots, 1, 0]^t$, we get the *anti-plurality* voting procedure, where voters are essentially being asked to vote against their least favorite candidate. The *Borda count*, where a candidate receives $(i-1)$ points for each time she is ranked in the ith position by a voter, is given by the vector $\mathbf{w} = [n-1, n-2, \dots, 2, 1, 0]^t$.

Let $\mathbf{w}^{(i)} \in M^{(1,\dots,1)}$ be defined by setting $\mathbf{w}^{(i)}(x) = w_j$ if candidate i is contained in the jth row of the tabloid x. We can use the function $\mathbf{w}^{(i)}$ to find the total number of points candidate i will receive, which is

$$\sum_{x \in X^{(1,\dots,1)}} \mathbf{p}(x)\mathbf{w}^{(i)}(x) = \langle \mathbf{p}, \mathbf{w}^{(i)} \rangle.$$

Furthermore, we can use the tabloids in $X^{(1,n-1)}$ to index the candidates by associating $y \in X^{(1,n-1)}$ with the single candidate in the top row of y. With that in mind, let $\mathbf{c}_i \in M^{(1,n-1)}$ denote the indicator function of the tabloid corresponding to candidate i. We can then create the module homomorphism $T_{\mathbf{w}} : M^{(1,\dots,1)} \to M^{(1,n-1)}$

by setting

$$T_{\mathbf{w}}(\mathbf{f}) = \langle \mathbf{f}, \mathbf{w}^{(1)} \rangle \mathbf{c}_1 + \cdots + \langle \mathbf{f}, \mathbf{w}^{(n)} \rangle \mathbf{c}_n.$$

The coefficient in front of $\mathbf{c}_i$ in $T_{\mathbf{w}}(\mathbf{p})$ is therefore simply the number of points that candidate i received when we apply the positional voting procedure associated with $\mathbf{w}$ to the profile $\mathbf{p}$.

To see the motivation behind the notation defined above, consider encoding the calculation of $T_{\mathbf{w}}(\mathbf{p})$ as a matrix-vector multiplication. For example, let $n = 3$, and suppose $\mathbf{w} = [1, s, 0]^t$ for some s such that $1 \geq s \geq 0$. If the coordinate vector of $\mathbf{p}$ with respect to the usual basis of $M^{(1,1,1)}$ is $[3, 2, 0, 2, 0, 4]^t$, then the coordinate vector of $T_{\mathbf{w}}(\mathbf{p})$ with respect to the usual basis of $M^{(1,2)}$ is given by

$$\begin{bmatrix} 1 & 1 & s & 0 & s & 0 \\ s & 0 & 1 & 1 & 0 & s \\ 0 & s & 0 & s & 1 & 1 \end{bmatrix} \begin{bmatrix} 3 \\ 2 \\ 4 \\ 2 \\ 0 \\ 3 \end{bmatrix} = \begin{bmatrix} 5+4s \\ 6+6s \\ 3+4s \end{bmatrix}.$$

In this case, the functions $\mathbf{w}^{(1)}, \mathbf{w}^{(2)}, \mathbf{w}^{(3)}$ correspond to the three rows of the 3-by-6 matrix, and we are simply dotting each of these rows with the coordinate vector of $\mathbf{p}$ to find the coordinate vector $[5 + 4s, 6 + 6s, 3 + 4s]^t$ of points received by each of the candidates. (Note that $T_{\mathbf{w}}(\mathbf{p}) = (5 + 4s)\mathbf{c}_1 + (6 + 6s)\mathbf{c}_2 + (3 + 4s)\mathbf{c}_3$.)

There is, however, more to see here. Notice that each column of the 3-by-6 matrix can be obtained by permuting the entries of the weighting vector $\mathbf{w} = [1, s, 0]^t$. In fact, if $x_0 \in X^{(1,1,1)}$ is the tabloid with candidate i in row i, and $\sigma \in S_n$, then the column of the matrix corresponding to the tabloid $\sigma \cdot x_0$ corresponds to the function $(1)\mathbf{c}_{\sigma(1)} + (s)\mathbf{c}_{\sigma(2)} + (0)\mathbf{c}_{\sigma(3)}$ in $M^{(1,2)}$.

From this point on we will slightly abuse our notation by identifying the weighting vector $\mathbf{w}$ with the function $w_1\mathbf{c}_1 + \cdots + w_n\mathbf{c}_n \in M^{(1,n-1)}$. If $x_0 \in X^{(1,\ldots,1)}$ is the tabloid with candidate i in row i, and $\widetilde{\mathbf{f}} \in \mathbb{R}S_n$ is defined by setting $\widetilde{\mathbf{f}}(\sigma) = \mathbf{f}(\sigma \cdot x_0)$, we then have that

$$\begin{aligned} T_{\mathbf{w}}(\mathbf{f}) &= \sum_{\sigma \in S_n} \mathbf{f}(\sigma \cdot x_0)(w_1\mathbf{c}_{\sigma(1)} + \cdots + w_n\mathbf{c}_{\sigma(n)}) \\ &= \sum_{\sigma \in S_n} \widetilde{f}(\sigma)(\sigma \cdot \mathbf{w}) \\ &= \widetilde{\mathbf{f}} \cdot \mathbf{w}. \end{aligned}$$

In other words, we may interpret the function $T_{\mathbf{w}}(\mathbf{f})$ as the result of the group algebra element $\widetilde{\mathbf{f}} \in \mathbb{R}S_n$ acting on the function $\mathbf{w} \in M^{(1,n-1)}$. Finding functions $\mathbf{f} \in M^{(1,\ldots,1)}$ so that $T_{\mathbf{w}}(\mathbf{f})$ has certain desirable properties then becomes a question of finding appropriate elements of the group algebra $\mathbb{R}S_n$ to act on $\mathbf{w} \in M^{(1,n-1)}$. This was one of the key insights in [**7**], and it leads to theorems like the following, which is essentially a special case of Theorem 1 in that paper:

THEOREM 1. *Let $n \geq 2$, and suppose $\mathbf{w}_1, \ldots, \mathbf{w}_k \in U_1 \subset M^{(1,n-1)}$ form a linearly independent set of weighting vectors. If $\mathbf{r}_1, \ldots, \mathbf{r}_k \in U_1$, then there exist infinitely many functions $\mathbf{f} \in M^{(1,\ldots,1)}$ such that $T_{\mathbf{w}_i}(\mathbf{f}) = \mathbf{r}_i$ for all i such that $1 \leq i \leq k$.*

To begin to appreciate the connection between Theorem 1 and voting theory, it is first helpful to realize that if $\mathbf{w} = \mathbf{1_w} + \widehat{\mathbf{w}}$, where $\mathbf{1_w} \in U_0$ and $\widehat{\mathbf{w}} \in U_1$, then $T_\mathbf{w}(\mathbf{f}) = \widetilde{\mathbf{f}} \cdot \mathbf{w} = \widetilde{\mathbf{f}} \cdot \mathbf{1_w} + \widetilde{\mathbf{f}} \cdot \widehat{\mathbf{w}}$. Furthermore, the ordinal ranking (i.e., who comes in first, who comes in second, and so on) provided by $T_\mathbf{w}$ is going to depend only on $\widehat{\mathbf{w}}$. After all, $\widetilde{\mathbf{f}} \cdot \mathbf{1_w}$ is contained in U_0, and is therefore a constant function.

Building on this insight, we will say that two weighting vectors $\mathbf{w}, \mathbf{w}' \in M^{(1,n-1)}$ are equivalent, and write $\mathbf{w} \sim \mathbf{w}'$, if and only if there exist $\alpha, \alpha' \in \mathbb{R}$ such that $\alpha > 0$ and $\mathbf{w}' = \alpha\mathbf{w} + \alpha'\mathbf{1}$. It should be clear that in this case, $\mathbf{w}'$ and $\mathbf{w}$ will yield the same ordinal rankings. (Also, note that if $\alpha < 0$ then the ordinal rankings given by $\mathbf{w}'$ would be the reverse of the ordinal rankings given by $\mathbf{w}$.) The following theorem, which follows from Theorem 1 and Theorem 2.3.1 in [**29**], highlights the usefulness of this equivalence relation on weighting vectors when dealing with positional voting procedures:

THEOREM 2. *Let $\mathbf{w}, \mathbf{w}' \in M^{(1,n-1)}$ be weighting vectors. The ordinal rankings of $T_\mathbf{w}(\mathbf{p})$ and $T_{\mathbf{w}'}(\mathbf{p})$ will be the same for all profiles $\mathbf{p} \in M^{(1,\ldots,1)}$ if and only if $\mathbf{w} \sim \mathbf{w}'$.*

In other words, if the weighting vectors $\mathbf{w}$ and $\mathbf{w}'$ are not equivalent, then the ordinal rankings of the outcomes of elections that use $\mathbf{w}$ and $\mathbf{w}'$ might differ, even if the same profile $\mathbf{p} \in M^{(1,\ldots,1)}$ is used in both elections. In fact, by Theorem 1, the results of such elections can be arbitrarily different from one another. This follows from the fact that if $\mathbf{w}$ and $\mathbf{w}'$ are not equivalent, then their projections into U_1 will be linearly independent.

Another way of appreciating the effect that Theorem 1 can have on our understanding of positional voting procedures is to realize that the effective spaces of $T_\mathbf{w}$ and $T_{\mathbf{w}'}$ share very little in common unless $\mathbf{w} \sim \mathbf{w}'$. This is because if $\mathbf{w} \in U_1$ is nonzero, then $E(T_\mathbf{w}) \cong S^{(n-1,1)}$, and thus $E(T_\mathbf{w})$ is an irreducible submodule of $M^{(1,\ldots,1)}$. This allows us to easily prove theorems like the following, which is Theorem 4 in [**7**]:

THEOREM 3. *Let $\mathbf{w}, \mathbf{w}' \in U_1 \subset M^{(1,n-1)}$ be nonzero weighting vectors. Then $E(T_\mathbf{w}) = E(T_{\mathbf{w}'})$ if and only if $\mathbf{w} \sim \mathbf{w}'$. Furthermore, if $E(T_\mathbf{w}) \neq E(T_{\mathbf{w}'})$, then $E(T_\mathbf{w}) \cap E(T_{\mathbf{w}'}) = \{\mathbf{0}\}$.*

In other words, if $\mathbf{w}$ and $\mathbf{w}'$ are not equivalent, then $T_\mathbf{w}$ and $T_{\mathbf{w}'}$ will use very different subspaces of $M^{(1,\ldots,1)}$ from which to pull the information necessary to determine the outcome of an election. The projections of a profile $\mathbf{p} \in M^{(1,\ldots,1)}$ into those subspaces determine the cardinal rankings of the candidates. We should therefore not expect any relationship between the associated ordinal rankings of the candidates.

There are, of course, other voting methods besides positional voting procedures. For example, a *simple ranking scoring function*, or SRSF, takes a profile in $M^{(1,\ldots,1)}$ and uses it to assign points to full rankings instead of individual candidates (see [**5**], and the related paper [**45**] where these are posited as a subset of a much larger class of generalized scoring functions). A winning *ranking* (as opposed to simply a winner, which is usually how one interprets normal positional rules) is a ranking that receives at least as many points as all of the other rankings.

As an example of an SRSF, let $x_0 \in M^{(1,\ldots,1)}$ be the tabloid (i.e., full ranking) that has candidate i in row i. Let $\mathbf{z} \in M^{(1,\ldots,1)}$ be a fixed function, and define

$T_{\mathbf{z}} : M^{(1,\ldots,1)} \to M^{(1,\ldots,1)}$ by setting

$$T_{\mathbf{z}}(\mathbf{f}) = \sum_{\sigma \in S_n} \mathbf{f}(\sigma \cdot x_0)(\sigma \cdot \mathbf{z}) = \widetilde{\mathbf{f}} \cdot \mathbf{z}$$

for all $\mathbf{f} \in M^{(1,\ldots,1)}$. Thus, if $\mathbf{p} \in M^{(1,\ldots,1)}$ is a profile, then a winning ranking in our election would correspond to a tabloid $x \in M^{(1,\ldots,1)}$ with the property that $T_{\mathbf{z}}(\mathbf{p})(x) \geq T_{\mathbf{z}}(\mathbf{p})(y)$ for all $y \in M^{(1,\ldots,1)}$.

How might we find a sensible function $\mathbf{z} \in M^{(1,\ldots,1)}$ to create an SRSF like the one above? One way is to use a metric defined on the tabloids in $X^{(1,\ldots,1)}$. For example, if $d : X^{(1,\ldots,1)} \times X^{(1,\ldots,1)}$ is a metric with maximum distance $d_{\max}$, then we can set $\mathbf{z}(x) = d_{\max} - d(x, x_0)$. A winning ranking with respect to $T_{\mathbf{z}}$ would then correspond to a tabloid that is "closest" to the entire multiset of tabloids chosen by the voters (see [**45**] for a thorough discussion of "proximity rules").

One of the most popular metrics on $X^{(1,\ldots,1)}$ is the *Kendall tau distance*, which measures the number of pairwise disagreements between two rankings. To explain, let $i, j \in \{1, \ldots, n\}$ where $i \neq j$. Let $\mathbf{a}_{ij} \in M^{(1,\ldots,1)}$ be the function defined by setting $\mathbf{a}_{ij}(x) = 1$ whenever candidate i is ranked above j in x (i.e., in the tabloid x, i is in a higher row than j), and setting $\mathbf{a}_{ij}(x) = 0$ otherwise. Kendall's tau distance is then given by

$$d(x, y) = \binom{n}{2} - \sum_{i \neq j} \mathbf{a}_{ij}(x)\mathbf{a}_{ij}(y).$$

The maximum distance between rankings in this case is $\binom{n}{2}$, and therefore we could define the function $\mathbf{z} \in M^{(1,\ldots,1)}$ by setting $\mathbf{z}(x) = \sum_{i \neq j} \mathbf{a}_{ij}(x)\mathbf{a}_{ij}(x_0)$. The resulting voting procedure given by $T_{\mathbf{z}}$ then becomes the well-known and well-studied *Kemeny rule* [**18**]. To distinguish this case, we will denote this particular instance of $T_{\mathbf{z}}$ by $K : M^{(1,\ldots,1)} \to M^{(1,\ldots,1)}$.

Perhaps not surprisingly, we can use the $\mathbf{a}_{ij}$ to apply K to a profile $\mathbf{p} \in M^{(1,\ldots,1)}$. More specifically, with a little bit of effort, and using the fact that the Kendall tau distance is invariant under the action of S_n, one can show that

$$K(\mathbf{p}) = \sum_{i \neq j} \langle \mathbf{p}, \mathbf{a}_{ij} \rangle \mathbf{a}_{ij}$$

(see Proposition 1 in [**41**] for a similar expression). On the other hand, it might not be clear how a voting procedure like the Kemeny rule is related to positional voting procedures. We describe such a relationship next.

First, it turns out that we can turn any positional voting procedure into an SRSF as follows. Let $\mathbf{w} \in M^{(1,n-1)}$ be a weighting vector, and let $\mathbf{b} \in M^{(1,n-1)}$ be the weighting vector for the Borda Count:

$$\mathbf{b} = (n-1)\mathbf{c}_1 + (n-2)\mathbf{c}_2 + \cdots + (1)\mathbf{c}_{n-1} + (0)\mathbf{c}_n.$$

Note that although we will use $\mathbf{b}$ in the construction below, any weighting vector whose weights are strictly decreasing would also work. Next, recall that

$$T_{\mathbf{w}}(\mathbf{f}) = \langle \mathbf{f}, \mathbf{w}^{(1)} \rangle \mathbf{c}_1 + \cdots + \langle \mathbf{f}, \mathbf{w}^{(n)} \rangle \mathbf{c}_n.$$

If we compose $T_{\mathbf{w}}$ with the adjoint $T_{\mathbf{b}}^* : M^{(1,n-1)} \to M^{(1,\ldots,1)}$, then we create the map $T_{\mathbf{b}}^* \circ T_{\mathbf{w}} : M^{(1,\ldots,1)} \to M^{(1,\ldots,1)}$ where

$$(T_{\mathbf{b}}^* \circ T_{\mathbf{w}})(\mathbf{f}) = \langle \mathbf{f}, \mathbf{w}^{(1)} \rangle \mathbf{b}^{(1)} + \cdots + \langle \mathbf{f}, \mathbf{w}^{(n)} \rangle \mathbf{b}^{(n)}.$$

It turns out that if we view $T_{\mathbf{b}}^* \circ T_{\mathbf{w}}$ as an SRSF, then it will return the ranking that agrees with the ranking given by $T_{\mathbf{w}}$ (see Proposition 1 in [**5**]).

For such procedures, individual candidates may then be ranked based on the order they appear in a winning ranking, recovering the winning candidate(s) in the usual interpretation of $T_{\mathbf{w}}$. However, for more general SRSFs (including the Kemeny rule), this may not always be meaningful. For example, suppose there are three candidates A, B, and C, and we are given a profile where two voters prefer ABC, two voters prefer CAB, and only one voter prefers BCA. The Kemeny rule then produces a tie between the rankings ABC and CAB.

Next, let $\mathbf{c}_{ij} \in M^{(1,1,n-2)}$ be the indicator function corresponding to the tabloid that has i in the top row and j in the second row. We will call the map $P : M^{(1,\ldots,1)} \to M^{(1,1,n-2)}$ given by

$$P(\mathbf{f}) = \sum_{i \neq j} \langle \mathbf{f}, \mathbf{a}_{ij} \rangle \mathbf{c}_{ij}$$

the *pairs map* defined on $M^{(1,\ldots,1)}$. Note that if $\mathbf{p} \in M^{(1,\ldots,1)}$ is a profile, then $P(\mathbf{p})$ simply catalogues the number of times each candidate was ranked over another candidate.

For example, when $n = 3$, $M^{(1,\ldots,1)} = M^{(1,1,n-2)}$, and with respect to the usual basis of $M^{(1,1,1)}$, the pairs map P can be encoded as the matrix

$$[P] = \begin{bmatrix} 1 & 1 & 0 & 0 & 1 & 0 \\ 1 & 1 & 1 & 0 & 0 & 0 \\ 0 & 0 & 1 & 1 & 0 & 1 \\ 1 & 0 & 1 & 1 & 0 & 0 \\ 0 & 0 & 0 & 1 & 1 & 1 \\ 0 & 1 & 0 & 0 & 1 & 1 \end{bmatrix}.$$

Note, for example, that the first row of this matrix corresponds to $\mathbf{a}_{12}$, the third row corresponds to $\mathbf{a}_{21}$, and the sum of these two rows corresponds to the constant all-ones function in $M^{(1,1,1)}$.

The pairs map P is not surjective. In particular, the codomain has the decomposition

$$M^{(1,1,n-2)} \cong S^{(n)} \oplus 2S^{(n-1)} \oplus S^{(n-2,2)} \oplus S^{(n-2,1,1)}$$

but the effective space of P is isomorphic to $S^{(n)} \oplus S^{(n-1,1)} \oplus S^{(n-2,1,1)}$ (see p. 682 in [**7**]). Let W_0, W_1, and W_2 denote the corresponding subspaces in $M^{(1,\ldots,1)}$, where $W_0 \cong S^{(n)}$, $W_1 \cong S^{(n-1,1)}$, $W_2 \cong S^{(n-2,1,1)}$, and

$$E(P) = W_0 \oplus W_1 \oplus W_2.$$

Note that if $P^* : M^{(1,1,n-2)} \to M^{(1,\ldots,1)}$ is the adjoint of P, then

$$(P^* \circ P)(\mathbf{f}) = \sum_{i \neq j} \langle \mathbf{f}, \mathbf{a}_{ij} \rangle \mathbf{a}_{ij}.$$

In other words, $K = P^* \circ P$. This should be compared with $T_{\mathbf{b}}^* \circ T_{\mathbf{w}}$ defined above, particularly in the case when $\mathbf{w} = \mathbf{b}$.

If necessary, we can create the matrix encoding of K with respect to the usual basis of $M^{(1,\ldots,1)}$ by simply taking the product of the matrix encodings of P^* and P.

For example, when $n = 3$, the matrix encoding of K is

$$[K] = \begin{bmatrix} 3 & 2 & 2 & 1 & 1 & 0 \\ 2 & 3 & 1 & 0 & 2 & 1 \\ 2 & 1 & 3 & 2 & 0 & 1 \\ 1 & 0 & 2 & 3 & 1 & 2 \\ 1 & 2 & 0 & 1 & 3 & 2 \\ 0 & 1 & 1 & 2 & 2 & 3 \end{bmatrix}$$

which is the product $[P]^t[P]$. Notice the columns of $[K]$ are all permutations of the first column of $[K]$, which corresponds to the function $\mathbf{z} \in M^{(1,1,1)}$ that is based (as was described above) on the Kendall tau distance.

Interestingly, it turns out that the effective space $E(T_\mathbf{b})$ of the Borda Count is $W_0 \oplus W_1$ (see, for example, Section 7 of [**7**]). This is perhaps much more understandable once we realize that

$$\mathbf{b}^{(i)} = \sum_{j:\ i \neq j} \mathbf{a}_{ij}.$$

Furthermore, by Theorem 3, the only nontrivial positional voting procedures that contain W_1 in their effective spaces are those whose weighting vectors are equivalent to $\mathbf{b}$. This begins to explain theorems like the following, which is a combination of parts of Theorem 3 and Theorem 4 in [**36**]:

THEOREM 4. *For $n \geq 3$ candidates, the Borda count always ranks the Kemeny rule top-ranked candidate strictly above the Kemeny rule bottom ranked-candidate. Conversely, the Kemeny rule ranks the Borda count top-ranked candidate strictly above the Borda count bottom-ranked candidate. For any positional voting method other than the Borda count, however, there is no relationship between the Kemeny rule ranking and the positional ranking.*

Using the pairs map P, we can say more about the relationship between the Borda count and the Kemeny rule. More specifically, because $K = P^* \circ P$ is self-adjoint, we know it is orthogonally diagonalizable. In fact, the eigenspaces of K are W_0, W_1, W_2, and $(W_0 \oplus W_1 \oplus W_2)^\perp$ with eigenvalues $\kappa_0 = \frac{n!}{2}\binom{n}{2}$, $\kappa_1 = \frac{(n+1)!}{6}$, $\kappa_2 = \frac{n!}{6}$, and 0, respectively (see Theorem 3 in [**41**]).

If we let $T_i : M^{(1,\ldots,1)} \to W_i$ denote the orthogonal projection onto W_i, we then have that

$$K = \kappa_0 T_0 + \kappa_1 T_1 + \kappa_2 T_2.$$

Furthermore, if $\beta_0 = \frac{(n-1)n!}{2}\binom{n}{2}$, $\beta_1 = \frac{n(n+1)!}{12}$, and $\beta_2 = 0$ then it is possible to show that

$$T_\mathbf{b}^* \circ T_\mathbf{b} = \beta_0 T_0 + \beta_1 T_1 + \beta_2 T_2.$$

The Borda count and Kemeny rule may therefore be viewed as members of the same family of SRSFs whose maps all have the form

$$K_{(\gamma_0,\gamma_1,\gamma_2)} = \gamma_0 T_0 + \gamma_1 T_1 + \gamma_2 T_2$$

where $\gamma_0, \gamma_1, \gamma_2 \in \mathbb{R}$.

When using $K_{(\gamma_0,\gamma_1,\gamma_2)}$, the resulting ordering of the rankings in $X^{(1,\ldots,1)}$ depends only on the ratio γ_2/γ_1. After all, $\gamma_0 T_0$ only contributes scalar multiples of the all-ones function to the outcome. This fact is exploited in [**6**] by setting $\gamma_0 = 0$ and focusing on a one-parameter family of procedures that is interpreted as being "between" the Borda count and Kemeny rule. Such a family allows us, for example, to better understand when properties like being susceptible to the "no-show paradox" can arise in voting (see Proposition 5.16 in [**6**]).

By describing everything up to this point in terms of $M^{(1,\dots,1)}$ and $M^{(1,n-1)}$, we hope it is clear how one might extend all of these ideas to more general settings. For example, suppose $\lambda = (\lambda_1, \dots, \lambda_m)$ is a partition of n. We could ask the voters to rank the candidates by choosing a tabloid in X^{λ}, where the candidates in the top row of their tabloid are their top favorite λ_1 candidates, the candidates in the second row are their next favorite λ_2 candidates, and so on. (See the discussion about *bucket orders* in [**1**].)

When voters are asked to provide rankings by choosing a tabloid from $X^{(1,\dots,1)}$, we say they are giving *full rankings* of the candidates. If they are choosing tabloids from X^{λ} where $\lambda \neq (1,\dots,1)$, then we say they are giving *partial rankings* (of shape λ) of the candidates. We can therefore ask if any of the theorems above can be extended to partial rankings. For some of the theorems, the answer is yes. In fact, the paper by Daugherty et al. [**7**] focused primarily on such theorems for positional voting.

Notice also that the maps $T_{\mathbf{w}}$ and $T_{\mathbf{z}}$ were defined in essentially the same way. In both cases, we chose a vector $\mathbf{v} \in M^{\mu}$ for some partition μ, and then defined $T_{\mathbf{v}} : M^{(1,\dots,1)} \to M^{\mu}$ by setting $T_{\mathbf{v}}(\mathbf{f}) = \widetilde{f} \cdot \mathbf{v}$. We then used $T_{\mathbf{v}}$ and a profile $\mathbf{p} \in M^{(1,\dots,1)}$ to find the "best" tabloid in M^{μ} based on "points" that $T_{\mathbf{v}}(\mathbf{p})$ assigned to the standard basis vectors of M^{μ}. What would happen if we were to extend this construction to situations in which voters choose tabloids in X^{λ} and points are then assigned to tabloids in X^{μ}? Would we learn anything new? For readers interested in exploring this question, Section 3 in [**7**] might be a good place to start. It discusses maps from M^{λ} to $M^{(1,n-1)}$ and what happens to positional voting procedures when voters are asked to provide partial (instead of full) rankings of shape λ.

Another possible direction one might take is to ask what would happen if we were to replace the pairs map in the definition of the Kemeny rule with maps that catalogue information about triples of candidates, or quadruples of candidates, and so on. Doing so would then place the Kemeny rule in another potentially interesting family of voting procedures. We are not aware of any work in this direction, but anyone interested in pursuing such a project would almost certainly benefit from the discussions about inversions in [**10**] and [**26**].

Finally, there are *maximum likelihood estimator* procedures which give the "most likely" ranking of the candidates, assuming that the voters all meant to choose the same ranking but their votes were corrupted by a noise model. Young [**44**] shows that the Borda count is an MLE when the desired outcome is a *single* winner, and that the Kemeny rule is an MLE when the desired outcome is a full ranking. In [**27**], the question of which procedure is an MLE when the desired outcome is a top pair, top triple, and so forth is investigated. Given the connection between SRSFs and MLEs found in [**5**], we suspect that an algebraic approach to MLE procedures has tremendous promise.

4. Game Theory

For algebraists interested in learning about (transferable utility cooperative) game theory, the pioneering work of Kleinberg and Weiss [**19–23**] will quickly make them feel welcome. Consider, for example, the first few lines from the introduction of [**19**]:

> Game theory and algebra become inextricably intertwined once one recognizes that the notion of a permutation of players gives rise to a representation of the symmetric group in the space of automorphisms of $\mathcal{G}$, the vector space of games (in characteristic function form).
>
> In this paper we use this observation to attempt to turn the usual approach to game-theoretic problems on its head by analyzing the space of games as an algebraist might. In this way, we can see if the mathematical structure of $\mathcal{G}$ has anything to say about what constructs are significant from a game-theoretic point of view.

In this section, we focus on the role that tabloids play when navigating Kleinberg and Weiss's work, as well as the recent work of Hernández-Lamoneda, Juárez, and Sánchez-Sánchez [**11**] and Sánchez-Pérez [**38**]. We then take advantage of our use of tabloids to suggest possible extensions of their work.

We begin with some notation and terminology. Let $n \geq 2$, and suppose we have n players. We will label the players with the numbers $1, \ldots, n$, and we will let $N = \{1, \ldots, n\}$ be the set of all of the players.

A *cooperative game* for N is a real-valued function defined on the set of all subsets of N, with the convention that the empty set is always mapped to 0. The set of all such games

$$\mathcal{G} = \{v : 2^N \to \mathbb{R} \mid v(\emptyset) = 0\}$$

models ways in which various subsets ("coalitions") can be allotted a utility ("value") based on the extent to which they are perceived to contribute to the entire set of players ("the grand coalition").

One of the defining challenges in cooperative game theory is to find useful and meaningful *solution concepts*, which are functions defined on $\mathcal{G}$ that are used to determine a "payoff" for each individual player. More specifically, if we index the individual players by tabloids in $X^{(1,n-1)}$ as we did in the last section on voting theory, then we can think of a solution concept as a function

$$\varphi : \mathcal{G} \to M^{(1,n-1)}$$

where if $v \in \mathcal{G}$ and $i \in N$, then the value that φ associates to player i is the coefficient in front of $\mathbf{c}_i$ in $\varphi(v)$. If we denote this value by $\varphi(v)_i$, then

$$\varphi(v) = \varphi(v)_1\mathbf{c}_1 + \cdots + \varphi(v)_n\mathbf{c}_n.$$

It is interesting to note that, algebraically, a solution concept looks similar to a positional voting procedure, in that the payoff that a solution concept assigns to a player is similar to the points that a positional voting procedure assigns to a candidate.

The vector space $\mathcal{G}$ of games is of course much more than just a vector space. It is an $\mathbb{R}S_n$-module, where for all $\sigma \in S_n$ and $v \in \mathcal{G}$,

$$(\sigma \cdot v)(S) = v(\sigma^{-1} \cdot S)$$

for all subsets $S \in 2^N$, and where $\sigma^{-1} \cdot S$ is the set you obtain by applying σ^{-1} to each element of S. Furthermore, the irreducible submodules of $\mathcal{G}$ are straightforward to describe.

First, note that we can write $\mathcal{G} = \mathcal{G}_1 \oplus \mathcal{G}_2 \cdots \oplus \mathcal{G}_{n-1} \oplus \mathcal{G}_n$, where $\mathcal{G}_k$ is the subspace of all games that assign the value 0 to any subset of players that does not have size k. Next, if we index the k-element subsets of players using tabloids in $X^{(k,n-k)}$, then it becomes clear that

$$\mathcal{G}_k \cong M^{(k,n-k)}.$$

When $k \leq n/2$, we know that $M^{(k,n-k)} \cong S^{(n)} \oplus S^{(n-1,1)} \oplus \cdots \oplus S^{(n-k,k)}$. We also know that $M^{(k,n-k)} \cong M^{(n-k,k)}$. Thus, if we let U_j^k denote the submodule of $\mathcal{G}_k$ that is isomorphic to the Specht module $S^{(n-j,j)}$, we then have

$$\begin{aligned}
\mathcal{G}_1 &= U_0^1 \oplus U_1^1 \\
\mathcal{G}_2 &= U_0^2 \oplus U_1^2 \oplus U_2^2 \\
\mathcal{G}_3 &= U_0^3 \oplus U_1^3 \oplus U_2^3 \oplus U_3^3 \\
&\vdots \\
\mathcal{G}_{\lfloor n/2 \rfloor} &= U_0^{\lfloor n/2 \rfloor} \oplus U_1^{\lfloor n/2 \rfloor} \oplus U_2^{\lfloor n/2 \rfloor} \oplus \cdots \oplus U_{\lfloor n/2 \rfloor}^{\lfloor n/2 \rfloor} \\
&\vdots \\
\mathcal{G}_{n-2} &= U_0^{n-2} \oplus U_1^{n-2} \oplus U_2^{n-2} \\
\mathcal{G}_{n-1} &= U_0^{n-1} \oplus U_1^{n-1} \\
\mathcal{G}_n &= U_0^n.
\end{aligned}$$

We may therefore write $\mathcal{G}$ as a direct sum of the irreducible U_j^k, where

$$\mathcal{G} = \left(\bigoplus_{k=1}^{\lfloor n/2 \rfloor} \bigoplus_{j=0}^{k} U_j^k \right) \bigoplus \left(\bigoplus_{k > \lfloor n/2 \rfloor}^{n} \bigoplus_{j=0}^{n-k} U_j^k \right)$$

which is essentially equation (3) in [**19**].

The above decomposition is helpful in cooperative game theory because the consensus since [**40**] is that *linear* and *symmetric* solution concepts are particularly important, and these solution concepts coincide with the set of all $\mathbb{R}S_n$-module homomorphisms from $\mathcal{G}$ to $M^{(1,n-1)}$. Moreover, because the $\mathcal{G}_k$ have such simple decompositions into irreducible submodules, these homomorphisms are easy to describe.

To explain, first recall that $M^{(1,n-1)}$ is the direct sum of U_0 and U_1, where $U_0 \cong S^{(n)}$ and $U_1 \cong S^{(n-1,1)}$. By Schur's lemma, this means that any module homomorphism from $\mathcal{G}_k \cong M^{(k,n-k)}$ to $M^{(1,n-1)}$ must contain U_j^k in its kernel for all $j \geq 2$. We may therefore describe any homomorphism from $\mathcal{G}$ to $M^{(1,n-1)}$ by focusing only on homomorphisms defined on the U_0^k and U_1^k as follows.

Let $T_0^k : U_0^k \to U_0$ and $T_1^k : U_1^k \to U_1$ be any fixed isomorphisms. By a slight abuse of notation, we can view these isomorphisms as homomorphisms from $\mathcal{G}$ to $M^{(1,n-1)}$ by identifying them with $\iota_0 \circ T_0^k \circ p_0^k$ and $\iota_1 \circ T_1^k \circ p_1^k$, respectively, where p_0^k is the projection map onto U_0^k, p_1^k is the projection map onto U_1^k, and ι_0 and ι_1 are the usual inclusion maps into $M^{(1,n-1)}$.

By Schur's lemma, every module homomorphism from $\mathcal{G}$ to $M^{(1,n-1)}$ may be expressed uniquely as a linear combination of the T_0^k and T_1^k. In other words, for every linear symmetric solution concept $\varphi : \mathcal{G} \to M^{(1,n-1)}$, there exist scalars $c_0^1, \dots, c_0^n$ and $c_1^1, \dots, c_1^{n-1}$ such that

$$\varphi = (c_0^1 T_0^1 + \cdots + c_0^n T_0^n) + (c_1^1 T_1^1 + \cdots + c_1^{n-1} T_1^{n-1}).$$

Working with and describing linear symmetric solution concepts therefore becomes a matter of manipulating and communicating the c_0^j and c_1^j. Moreover, there are choices for the T_0^k and T_1^k that make certain properties of solution concepts easy to verify.

For example, consider the following maps that appear in [**22**, **23**]. First, let

$$A(v,k) = \binom{n}{k}^{-1} \sum_{S:\ |S|=k} v(S)$$

denote the average value received by the coalitions of size k, and let

$$\gamma(k) = \binom{n-2}{k-1}.$$

Then define T_0^k by setting $T_0^k(v)_i = k^{-1} A(v,k)$, and define T_1^k by setting

$$T_1^k(v)_i = \gamma(k)^{-1} \sum_{S:\ |S|=k \text{ and } i \in S} [v(S) - A(v,k)].$$

(A somewhat different approach is taken in [**11**], but the results are essentially scaled versions of the maps above.)

As an example of a property that can now be expressed in terms of the c_0^j and c_1^j, consider the property of *efficiency*, where a solution concept φ is said to be *efficient* if

$$\varphi(v)_1 + \cdots + \varphi(v)_n = v(N)$$

for all $v \in \mathcal{G}$. It turns out that a linear symmetric solution concept will be efficient if and only if $c_0^1 = \cdots = c_0^{n-1} = 0$ and $c_0^n = 1$ (see Proposition 2 in [**22**]).

As another example, a solution concept φ is said to be a *marginal value* if there exist scalars $m_1, \dots, m_n$ such that

$$\varphi(v)_i = \sum_{S:\ i \in S} m_{|S|}(v(S) - v(S - i))$$

for all $v \in \mathcal{G}$. In this case, it turns out we can recover φ by setting

$$c_0^k = k\left[m_k \binom{n-1}{k-1} - m_{k+1}\binom{n-1}{k}\right]$$

and

$$c_1^k = \gamma(k)\,[m_k + m_{k+1}]$$

where we define $m_{n+1} = 0$ (see Proposition 1 in [**23**]).

The above approach to describing linear symmetric solution concepts becomes even more powerful when we ask about solution concepts that have more than one such property. For example, using the insights above, we may easily verify that there must be precisely one linear symmetric solution concept that is also an efficient marginal value. It is the well-studied *Shapley value* [**40**], which is obtained by setting $c_0^1 = \cdots = c_0^{n-1} = 0$, $c_0^n = 1$, and $c_1^1 = \cdots = c_1^{n-1} = \frac{1}{n-1}$ (see Section 3 of [**23**]; see also Section 4.3 of [**11**]).

Similarly, [**11**] uses this type of analysis to prove a simple and intuitive criterion for a solution concept to be a "self-dual" marginal value. First, they define the *duality operator* $* : \mathcal{G} \to \mathcal{G}$ by setting $(*v)(S) = v(N) - v(N - S)$ for all $v \in \mathcal{G}$. A linear symmetric solution φ is then said to be *self-dual* if $\varphi(*v) = \varphi(v)$ for all $v \in \mathcal{G}$. It turns out that a linear symmetric marginal value will be self-dual if and only if the m_k defined above have the property that $m_j = m_{n-j-1}$ for all $j < n$ (see Proposition 6 of [**11**]).

By viewing a linear symmetric solution concept as a module homomorphism $\varphi : \mathcal{G} \to M^{(1,n-1)}$, it is now easy to see how one might replace $M^{(1,n-1)}$ by $M^{(k,n-k)}$ to get linear symmetric solution concepts that assign payoffs to k-element subsets of players instead of individual players. Such an idea was explored in [**19**], but there is room for significant additional investigation, as the conjectures in that paper seem to imply. From a representation-theoretic point of view, we could easily begin by including in the discussion above module homomorphisms defined on the U_j^k when $j \geq 2$.

In some sense, this would begin to address the question posed in some of the papers referenced above as to whether the common kernel of all linear symmetric solution concepts has *useful* structure. After all, this common kernel is the sum of all of the U_j^k where $j \geq 2$, and each of these submodules would begin to contribute to generalized solution concepts that involved payoffs for pairs, triples, and so on. Finding meaningful properties of the associated generalized solution concepts could be challenging, but doing so might ultimately prove to be illuminating, especially if they were to enhance our understanding of cherished properties (e.g., being a marginal value) that solution concepts might possess.

The approach taken in [**11**] is extended in [**38**] to so-called *games in partition function form*. In this setting, the value of a coalition S depends not just on the coalition, but on how the players *not* in the coalition are themselves partitioned into coalitions. In this way, an "embedded coalition" is a pair (S, Q) such that Q is a (set) partition of N and $S \in Q$. The dimension of the resulting game space is much larger, and its decomposition into irreducible submodules has only been done for $n = 3$ and $n = 4$, which are (unsurprisingly) mostly U_0 and U_1 for those n. Can a decomposition of the general case be described? If so, can it be used to describe properties (e.g., efficiency) of the associated generalized solution concepts? (See, for example, Corollary 4 in [**38**].)

Finally, it is also possible to decompose *non-cooperative games* [**16, 17**]. In the simplest version of this setting, players have strategies they individually decide upon, yielding varying payoffs. As is shown in [**17**], we can decompose such games into "fully competitive" zero-sum components and "cooperative" components in which all of the players receive the same payoff. The main theorem in [**16**] takes a different approach, arguing that the zero-sum component is not necessarily identical to the strategic component. As an example of their results, they show that the space of 2×2 games may be decomposed into a four-dimensional subspace that uniquely determines the Nash outcome, a two-dimensional subspace consisting solely of payoffs for each player which does not affect any strategic analysis, and a two-dimensional subspace governing other behavioral features. It would be useful to better understand the role that representation theory plays here. Some first steps in this direction are being taken by Jessie [**15**].

5. Conclusion

As we have seen, the permutation representations arising from the action of the symmetric group on tabloids provide a unifying framework for understanding and extending foundational ideas in both voting theory and game theory. There is, however, much more work that could be done, and we encourage interested readers to consider how they might use these and other ideas to contribute to our understanding of voting theory, game theory, and other mathematical behavioral sciences.

References

[1] Michael Ackerman, Sul-Young Choi, Peter Coughlin, Eric Gottlieb, and Japheth Wood, *Elections with partially ordered preferences*, Public Choice **157** (2013), no. 1-2, 145–168 (English).

[2] Anna E. Bargagliotti and Michael E. Orrison, *Linear rank tests of uniformity: understanding inconsistent outcomes and the construction of new tests*, J. Nonparametr. Stat. **24** (2012), no. 2, 481–495, DOI 10.1080/10485252.2011.649282. MR2921148

[3] Tullio Ceccherini-Silberstein, Fabio Scarabotti, and Filippo Tolli, *Harmonic analysis on finite groups*, Cambridge Studies in Advanced Mathematics, vol. 108, Cambridge University Press, Cambridge, 2008. Representation theory, Gelfand pairs and Markov chains. MR2389056

[4] Michael Clausen and Ulrich Baum, *Fast Fourier transforms*, Bibliographisches Institut, Mannheim, 1993. MR1270670

[5] Vincent Conitzer, Matthew Rognlie, and Lirong Xia, *Preference functions that score rankings and maximum likelihood estimation*, Proceedings of the 21st International Jont Conference on Artifical Intelligence (San Francisco, CA, USA), IJCAI'09, Morgan Kaufmann Publishers Inc., 2009, pp. 109–115.

[6] Karl-Dieter Crisman, *The Borda count, the Kemeny rule, and the permutahedron*, The mathematics of decisions, elections, and games, Contemp. Math., vol. 624, Amer. Math. Soc., Providence, RI, 2014, pp. 101–134, DOI 10.1090/conm/624/12475. MR3308273

[7] Zajj Daugherty, Alexander K. Eustis, Gregory Minton, and Michael E. Orrison, *Voting, the symmetric group, and representation theory*, Amer. Math. Monthly **116** (2009), no. 8, 667–687, DOI 10.4169/193009709X460796. MR2572103

[8] Persi Diaconis, *Group representations in probability and statistics*, Institute of Mathematical Statistics Lecture Notes—Monograph Series, 11, Institute of Mathematical Statistics, Hayward, CA, 1988. MR964069

[9] Persi Diaconis and Nicholas Eriksson, *Markov bases for noncommutative Fourier analysis of ranked data*, J. Symbolic Comput. **41** (2006), no. 2, 182–195, DOI 10.1016/j.jsc.2005.04.009. MR2197154

[10] J. P. Grossman and Gregory Minton, *Inversions in ranking data*, Discrete Math. **309** (2009), no. 20, 6149–6151, DOI 10.1016/j.disc.2009.04.030. MR2552651

[11] L. Hernández-Lamoneda, R. Juárez, and F. Sánchez-Sánchez, *Dissection of solutions in cooperative game theory using representation techniques*, Internat. J. Game Theory **35** (2007), no. 3, 395–426, DOI 10.1007/s00182-006-0036-3. MR2304546

[12] Jonathan Huang, Carlos Guestrin, and Leonidas Guibas, *Fourier theoretic probabilistic inference over permutations*, J. Mach. Learn. Res. **10** (2009), 997–1070. MR2520800

[13] Gordon James and Adalbert Kerber, *The representation theory of the symmetric group*, Encyclopedia of Mathematics and its Applications, vol. 16, Addison-Wesley Publishing Co., Reading, Mass., 1981. With a foreword by P. M. Cohn; With an introduction by Gilbert de B. Robinson. MR644144

[14] Gordon James and Martin Liebeck, *Representations and characters of groups*, 2nd ed., Cambridge University Press, New York, 2001. MR1864147

[15] Daniel T. Jessie, *Games and symmetry*, (preprint).

[16] Daniel T. Jessie and Donald G. Saari, *Cooperation in n-player repeated games*, The mathematics of decisions, elections, and games, Contemp. Math., vol. 624, Amer. Math. Soc., Providence, RI, 2014, pp. 189–206, DOI 10.1090/conm/624/12469. MR3308278

[17] Adam Kalai and Ehud Kalai, *Cooperation in strategic games revisited*, The Quarterly Journal of Economics **128** (2013), no. 2, 917–966.

[18] John G. Kemeny, *Mathematics without numbers*, Daedalus **88** (1959), 571–591.

[19] Norman L. Kleinberg and Jeffrey H. Weiss, *Algebraic structure of games*, Math. Social Sci. **9** (1985), no. 1, 35–44, DOI 10.1016/0165-4896(85)90005-8. MR795555

[20] Norman L. Kleinberg and Jeffrey H. Weiss, *Equivalent n-person games and the null space of the Shapley value*, Math. Oper. Res. **10** (1985), no. 2, 233–243, DOI 10.1287/moor.10.2.233. MR793881

[21] Norman L. Kleinberg and Jeffrey H. Weiss, *The orthogonal decomposition of games and an averaging formula for the Shapley value*, Math. Oper. Res. **11** (1986), no. 1, 117–124, DOI 10.1287/moor.11.1.117. MR830112

[22] Norman L. Kleinberg and Jeffrey H. Weiss, *Weak values, the core, and new axioms for the Shapley value*, Math. Social Sci. **12** (1986), no. 1, 21–30, DOI 10.1016/0165-4896(86)90045-4. MR857902

[23] Norman L. Kleinberg and Jeffrey H. Weiss, *On membership and marginal values*, Internat. J. Game Theory **42** (2013), no. 2, 357–373, DOI 10.1007/s00182-013-0367-9. MR3053899

[24] Risi Kondor, *A Fourier space algorithm for solving quadratic assignment problems*, Proceedings of the Twenty-First Annual ACM-SIAM Symposium on Discrete Algorithms, SIAM, Philadelphia, PA, 2010, pp. 1017–1028. MR2809723

[25] Risi Kondor, *Non-commutative harmonic analysis in multi-object tracking*, Bayesian time series models, Cambridge Univ. Press, Cambridge, 2011, pp. 277–294. MR2894243

[26] John I. Marden, *Analyzing and modeling rank data*, Monographs on Statistics and Applied Probability, vol. 64, Chapman & Hall, London, 1995. MR1346107

[27] Ariel D. Procaccia, Sashank J. Reddi, and Nisarg Shah, *A maximum likelihood approach for selecting sets of alternatives*, Proceedings of the UAI 2012 Conference on Uncertainty in Artificial Intelligence, 2012.

[28] Donald G. Saari, *The ultimate of chaos resulting from weighted voting systems*, Adv. in Appl. Math. **5** (1984), no. 3, 286–308, DOI 10.1016/0196-8858(84)90011-3. MR755382

[29] Donald G. Saari, *Geometry of voting*, Studies in Economic Theory, vol. 3, Springer-Verlag, Berlin, 1994. MR1297124

[30] Donald G. Saari, *Basic geometry of voting*, Springer-Verlag, Berlin, 1995. MR1410265

[31] Donald G. Saari, *Explaining all three-alternative voting outcomes*, J. Econom. Theory **87** (1999), no. 2, 313–355, DOI 10.1006/jeth.1999.2541. MR1713824

[32] Donald G. Saari, *Mathematical structure of voting paradoxes. I. Pairwise votes*, Econom. Theory **15** (2000), no. 1, 1–53, DOI 10.1007/s001990050001. MR1731508

[33] Donald G. Saari, *Mathematical structure of voting paradoxes. II. Positional voting*, Econom. Theory **15** (2000), no. 1, 55–102, DOI 10.1007/s001990050001. MR1731509

[34] Donald G. Saari, *Chaotic elections!*, American Mathematical Society, Providence, RI, 2001. A mathematician looks at voting. MR1822218

[35] Donald G. Saari, *Decisions and elections*, Cambridge University Press, Cambridge, 2001. Explaining the unexpected. MR2006479

[36] Donald G. Saari and Vincent R. Merlin, *A geometric examination of Kemeny's rule*, Soc. Choice Welf. **17** (2000), no. 3, 403–438, DOI 10.1007/s003550050171. MR1762588

[37] Bruce E. Sagan, *The symmetric group*, 2nd ed., Graduate Texts in Mathematics, vol. 203, Springer-Verlag, New York, 2001. Representations, combinatorial algorithms, and symmetric functions. MR1824028

[38] Joss Sánchez-Pérez, *Application of the representations of symmetric groups to characterize solutions of games in partition function form*, Oper. Res. Decis. **24** (2014), no. 2, 97–122. MR3243693

[39] Jean-Pierre Serre, *Linear representations of finite groups*, Springer-Verlag, New York-Heidelberg, 1977. Translated from the second French edition by Leonard L. Scott; Graduate Texts in Mathematics, Vol. 42. MR0450380

[40] L. S. Shapley, *A value for n-person games*, Contributions to the theory of games, vol. 2, Annals of Mathematics Studies, no. 28, Princeton University Press, Princeton, N. J., 1953, pp. 307–317. MR0053477

[41] Eric Sibony, *Borda count approximation of Kemeny's rule and pairwise voting inconsistencies*, Proceedings of the NIPS 2014 Workshop on Analysis of Rank Data, 2014.

[42] Alan D. Taylor, *Social choice and the mathematics of manipulation*, Outlooks, Cambridge University Press, Cambridge; Mathematical Association of America, Washington, DC, 2005. MR2157533

[43] Audrey Terras, *Fourier analysis on finite groups and applications*, London Mathematical Society Student Texts, vol. 43, Cambridge University Press, Cambridge, 1999. MR1695775
[44] Peyton Young, *Condorcet's theory of voting*, The American Political Science Review **82** (1988), no. 4, 1231–1244.
[45] William S. Zwicker, *Consistency without neutrality in voting rules: When is a vote an average?*, Math. Comput. Modelling **48** (2008), no. 9-10, 1357–1373, DOI 10.1016/j.mcm.2008.05.031. MR2459706

Department of Mathematics and Computer Science, Gordon College, 255 Grapevine Road, Wenham, Massachusetts 01984
E-mail address: karl.crisman@gordon.edu

Department of Mathematics, Harvey Mudd College, 301 Platt Boulevard, Claremont, California 91711
E-mail address: orrison@hmc.edu

Contemporary Mathematics
Volume **685**, 2017
http://dx.doi.org/10.1090/conm/685/13750

Modeling the distribution of distance data in Euclidean space

Ruth Davidson, Joseph Rusinko, Zoe Vernon, and Jing Xi

ABSTRACT. Phylogenetic inference-the derivation of a hypothesis for the common evolutionary history of a group of species- is an active area of research at the intersection of biology, computer science, mathematics, and statistics. One assumes the data contains a phylogenetic signal that will be recovered with varying accuracy due to the quality of the method used, and the quality of the data.

The input for distance-based inference methods is an element of a Euclidean space with coordinates indexed by the pairs of organisms. For several algorithms there exists a subdivision of this space into polyhedral cones such that inputs in the same cone return the same tree topology. The geometry of these cones has been used to analyze the inference algorithms. In this chapter, we model how input data points drawn from DNA sequences are distributed throughout Euclidean space in relation to the space of tree metrics, which in turn can also be described as a collection of polyhedral cones.

1. Introduction to phylogenetics

A *phylogeny* is a mathematical model of the common evolutionary history of a group of *taxa* X, where X may be a set of genes, species, or multiple individuals sampled within a population. Phylogenies are commonly represented by a *phylogenetic tree* T which is a connected, acyclic graph in which the degree-one vertices, or *leaves* of T, are labeled with the elements of X. The tree T is a representation of the evolutionary history of the set X [**33**].

In many phylogenetic reconstruction pipelines one creates a nucleotide or protein sequence alignment, uses the alignment to estimate pairwise dissimilarities between taxa, and then uses the set of dissimilarities to infer a phylogenetic tree. One class of methods used in the third step of such a pipeline are distance-based algorithms: examples include Neighbor Joining (NJ) [**32**], Unweighted Pair Group Method with Arithmetic Mean (UPGMA) [**35, 36**] (Algorithm 3.1), and FastME [**14**]. These algorithms seek an edge-weighted tree such that the pairwise distances between taxa on the tree reflect the dissimilarities computed from the data. While not as popular as maximum-likelihood methods, distance-based algorithms remain relevant for reasons beyond their speed relative to other methods. For example,

2010 *Mathematics Subject Classification.* Primary 05B25, 05C05, 46N30, 62-07, 62P10, 65C60, 92B10, 97K80.

they serve as components of other popular methods, such as the species tree inference method NJst [**27**] and the DNA sequence alignment method MUSCLE [**16**].

It is possible to analyze the theoretical properties of distance-based algorithms by considering which sets of input data return particular output trees. The set of input data can be viewed as a subset of Euclidean space. It was shown geometrically that NJ solves its optimization problem well in low dimension in [**17**], and in [**11, 12**], geometry provided new insights into biases in the structure of trees reconstructed from phylogenetic data [**2**]. In this chapter we use computational tools in a geometric setting to observe that pairwise dissimilarity data computed from real sequence data will not be uniformly distributed across Euclidean space, but instead will be clustered around an object corresponding to pairwise distances between taxa induced by phylogenetic trees. We derive models for the distribution of this data and discuss possible applications for these models and tools.

2. The space of additive distance matrices embedded in $\mathbb{R}^{\binom{n}{2}}$

2.1. Dissimilarity maps used in distance-based phylogenetic inference as points in Euclidean space.

DEFINITION 2.1. For a fixed set X of n taxa, a *dissimilarity map* is an $n \times n$ symmetric matrix δ where $\delta(u, v)$ is a measure of evolutionary difference between taxa u and v.

There are many popular ways to estimate such a measure that assume models of sequence evolution such as Jukes-Cantor (JC) [**24**], Kimura two-parameter (K2P) [**25**], or the general time-reversible model (GTR) [**22**]. For example, the JC model of DNA sequence evolution corresponds to a measure of dissimilarity between two DNA sequences of the same length representing taxa u and v:

$$\delta(u, v) = -\frac{3}{4}\log\left(1 - \frac{4}{3}\left(\frac{h}{l}\right)\right) \tag{2.1}$$

where h is the number of places that the sequences differ and l is the shared length of the two sequences. Equation 2.1 is known as the *Jukes-Cantor Correction* because it returns a measure of dissimilarity between taxa based on the expected number of mutations given by the JC model of sequence evolution instead of the observed number of differences in the sequences.

Dissimilarity maps are the inputs to distance-based algorithms. By our definition, $\delta(u, v) = \delta(v, u)$ for all pairs $\{u, v\} \subset X$. Furthermore, $\delta(u, u) = 0$ because δ is a measure of evolutionary distance. So each dissimilarity map δ is also a vector in the Euclidean space $\mathbb{R}^{\binom{n}{2}}$ with coordinates labeled by all distinct pairs of taxa. Finally, we can assume δ is in the positive orthant, which we denote $\mathbb{R}_{\geq 0}^{\binom{n}{2}}$; as we see in Equation 2.1, popular dissimilarity measures such as JC, K2P, and GTR return positive numbers as estimates of evolutionary difference.

The output of a distance-based algorithm is an additive distance matrix (sometimes called a tree metric, though this term has more than one usage in the literature). An *additive distance matrix*, which we will denote as $d(u, v)$ or simply d, can be realized by a phylogenetic tree T with edge weights that are real numbers, where $d(u, v)$ for two distinct elements u and v of X is the sum of the weights of the edges in the unique path between u and v in T. *Ultrametric* additive distance matrices can be realized by a tree with a root vertex r that is not labeled with a element of X, and satisfy the property that for every two elements u and v of X, $d(u,r) = d(v,r)$. Figure 1 shows an ultrametric on three taxa.

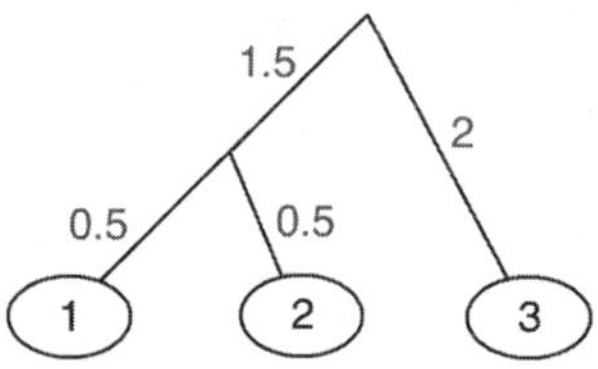

FIGURE 1. An ultrametric on three taxa

2.2. Polyhedral descriptions of the outputs of distance-based methods as points in Euclidean space. Recall that a *polyhedral cone* is the nonnegative span of a set of vectors in $\mathbb{R}^d$ for $d \geq 1$, commonly referred to as the $\mathcal{V}$-representation. Since polyhedral cones are polyhedra, they also have an $\mathcal{H}$-representation as the closed (in the traditional sense of the usual topology on $\mathbb{R}^d$) intersection of a finite number of half-spaces defined by hyperplanes. A *face* of a polyhedral cone is the intersection of the cone with a hyperplane that defines a half-space in $\mathbb{R}^d$ that entirely contains the cone. See [**42**] for a comprehensive introduction to polyhedral theory.

DEFINITION 2.2. A *fan* is a family $\mathcal{F}$ of polyhedral cones in $\mathbb{R}^d$ such that:

(1) if $P \in \mathcal{F}$ then every nonempty face of P is in $\mathcal{F}$, and
(2) if $P_1, P_2 \in \mathcal{F}$ then $P_1 \cap P_2$ is a face of both P_1 and P_2.

Informally, polyhedral fans are special collections of cones that are easy to work with because they "play well together." The set of all additive distance matrices on n taxa, denoted $\mathcal{T}_n$, and the set of ultrametric distance matrices on n taxa, denoted $\mathcal{ET}_n$, are each classified as both a polyhedral fan and a tropical variety. These spaces are studied in [**3**] and [**37**], respectively. We follow the notation of [**12**] to describe these spaces and note that $\mathcal{ET}_n \subset \mathcal{T}_n \subset \mathbb{R}^{\binom{n}{2}}$ for all $n \geq 3$. In other words, the spaces of additive matrices $\mathcal{ET}_n$ and $\mathcal{T}_n$ are *embedded* in the Euclidean space of dissimilarity maps in a very natural way.

In this chapter we model how three-taxon samples of distance data drawn from dissimilarity maps on n taxa are distributed in relation to $\mathcal{ET}_3$. We model this data in relation to $\mathcal{ET}_3$ rather than $\mathcal{T}_3$ because (1) most of the restrictions of the samples to three taxa already lie in $\mathcal{T}_3$ and (2) the condition of ultrametricity is of particular interest to biologists as it indicates that the molecular clock assumption is valid on the sample, which means that evolution occurred at a constant rate over time throughout the entire phylogeny.

We write the restriction of a dissimilarity map to a three-taxon sample δ as a point $(x, y, z) = (\delta(u,v), \delta(u,w), \delta(v,w))$ in $\mathbb{R}^{\binom{3}{2}} = \mathbb{R}^3$. We wish to understand geometric properties of the distribution of points δ in relation to $\mathcal{ET}_3$.

The space $\mathcal{ET}_3$ (see Figure 2) is a two-dimensional polyhedral fan consisting of three two-dimensional cones

$$\mathrm{Span}_{\geq 0}\{(1,1,1),(0,1,1)\},\ \mathrm{Span}_{\geq 0}\{(1,1,1),(1,0,1)\},$$

and

$$\text{Span}_{\geq 0}\{(1,1,1),(1,1,0)\},$$

whose pairwise intersections consist solely of the one-dimensional cone $\text{Span}_{\geq 0}\{(1,1,1)\}$. We refer to the two-dimensional (top-dimensional) cones in the fan $\mathcal{ET}_3$ as the *wings*, where

$$\text{Wing } 1 = \text{Span}_{\geq 0}\{(1,1,1),(0,1,1)\},$$

$$\text{Wing } 2 = \text{Span}_{\geq 0}\{(1,1,1),(1,0,1)\},$$

and

$$\text{Wing } 3 = \text{Span}_{\geq 0}\{(1,1,1),(1,1,0)\}.$$

We refer to the one-dimensional cone in $\mathcal{ET}_3$ $\text{Span}_{\geq 0}\{(1,1,1)\}$ as the *spindle*.

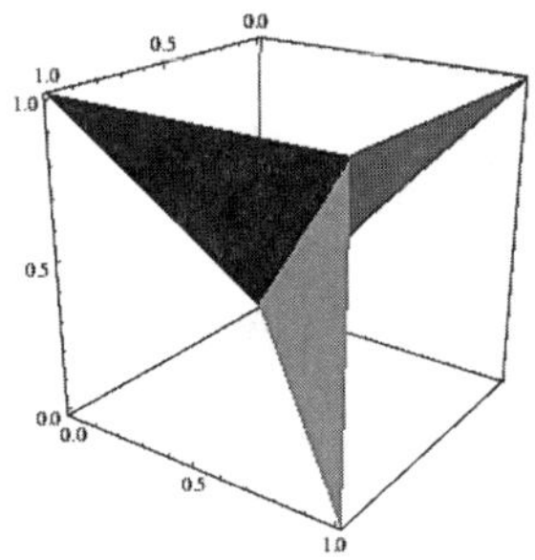

FIGURE 2. Ultrametric outputs $\mathcal{ET}_3$

We introduce a new set of coordinates which describe a dissimilarity map in relation to the geometry of $\mathcal{ET}_3$. Let δ be a dissimilarity map. The relationship of δ to $\mathcal{ET}_3$ can be described by (1) W, the nearest wing to δ, (2) D_1, the signed distance from δ to the nearest wing, (3) D_2, the distance from the point on the nearest wing to the spindle, and (4) D_3, the distance from the point on the spindle to the origin $(0,0,0)$. Then δ can be written as a sum of three pairwise-orthogonal vectors with lengths D_1, D_2, and D_3 representing a path from δ to the origin that begins with a choice of the nearest wing. We call the vector $(W,\ D_1,\ D_2,\ D_3)$ the *path trace* of δ, where $W = \text{Wing } i,\ i \in \{1,2,3\}$. The path trace is an equivalent representation of δ as δ can be uniquely determined by the path trace.

When analyzing a distance-based algorithm, one should emphasize dissimilarity maps likely to arise as outputs of a distance-approximation method based on a popular statistical model of sequence evolution such as JC, K2P, and GTR, which were introduced in Section 1. In the following section we will review distance-based methods and see that they also induce polyhedral structures; studying these structures has shed light on the behavior of the algorithms themselves.

3. Geometry of distance-based methods

3.1. Optimization problems associated to distance-based methods.

PROBLEM 3.1. The Least-Squares Phylogeny problem (LSP) asks, for a given dissimilarity map $\delta \in \mathbb{R}^{\binom{n}{2}}$, what is the additive distance matrix d that minimizes the ordinary Euclidean distance given by the formula

$$\sqrt{\sum_{x,y\in X} (\delta(x,y) - d(x,y))^2}.$$

The distance-based methods Unweighted Pair Group Method with Arithmetic Mean (UPGMA) [**35, 36**] and Neighbor-Joining (NJ) [**32**] are approximations to LSP [**19, 32**]. However, an alternative interpretation of NJ [**15, 21**] is that NJ performs a heuristic search, guided by a linear transformation of δ known as the *Q-criterion* at each agglomeration step, that minimizes a tree-length estimate due to Yves Pauplin [**29**] known as the "Balanced Minimum Evolution" (BME) criterion.

Day showed that both Problem 3.1 and finding the tree minimizing the BME criterion are NP-hard [**13**]. Thus polynomial-time, heuristic, distance-based algorithms such as BIONJ [**20**], Weighbor [**5**], and FastME [**14**] remain essential for solving Problem 3.1 as well as for finding a tree minimizing the BME criterion. These algorithms take dissimilarity maps as inputs, may outperform NJ in terms of topological accuracy under certain conditions, and exhibit superior immunity to reconstruction pathologies well-known to biologists such as long-branch attraction [**4**].

3.2. Polyhedral decompositions of $\mathbb{R}^{\binom{n}{2}}$ induced by distance-based algorithms. Given a distance-based algorithm, there exists a closed-form description of the input space for a phylogenetic reconstruction method due to the natural subdivision of said space induced by the decision criteria of the method. This subdivision consists of regions containing dissimilarity maps that return additive distance matrices realized by the same combinatorial type of tree upon application of the method. For example, the distance-based algorithms NJ and UPGMA induce subdivisions of the input space into families of polyhedral cones defined by the sets of linear inequalities defining the decision steps in these algorithms. To illustrate, we review why there is a polyhedral description of the UPGMA algorithm below.

The inequalities coming from the decision steps in UPGMA (Algorithm 3.1) are linear combinations of the original input coordinates, and define a family of polyhedral cones that completely partition the input space. This representation of a polyhedral cone in terms of a set of inequalities is known as an *H-representation.* The *V-representation* of these cones-i.e. a description in terms of their extreme rays-is given in [**11**]. Knowledge of both the H- and V-representations of the cones into which UPGMA divides the input space gives a complete discrete geometric description of the input space, allowing for computations and further conjectures about how this discrete geometric description might affect the performance and behaviors of the algorithm in practice.

While UPGMA outputs ultrametrics, and so maps $\mathbb{R}_{\geq 0}^{\binom{n}{2}}$ to cones in $\mathcal{ET}_n$ indexed by combinatorial rooted trees, Neighbor Joining maps $\mathbb{R}_{\geq 0}^{\binom{n}{2}}$ to cones in $\mathcal{T}_n$ indexed by unrooted tree shapes. As is the case with UPGMA, NJ partitions the input space into a family of polyhedral cones with an H-representation arising naturally from the decision criterion in Step (2) of Algorithm 3.1 which are linear inequalities in terms of the coordinates of the input space. Though the problem has been studied in [**18**], a complete description of the V-representation, or extreme rays, for the cones in the polyhedral partition of the input space induced by NJ is still unknown.

Algorithm 3.1 UPGMA

- Input: a dissimilarity map $\delta \in \mathbb{R}_{\geq 0}^{n(n-1)/2}$ on $[n]$.
- Output: an ultrametric $d \in \mathbb{R}_{\geq 0}^{n(n-1)/2}$.
  (1) Initialize an unordered set partition $\pi_n = 1|2|\cdots|n$, and set $\delta^n = \delta$.
  (2) For $i = n-1, \ldots, 1$ do
    - From partition $\pi_{i+1} = \lambda_1^{i+1}|\cdots|\lambda_{i+1}^{i+1}$ and distance vector $\delta^{i+1} \in \mathbb{R}_{\geq 0}^{(i+1)i/2}$ choose j, k be so that $\delta^{i+1}(\lambda_j^{i+1}, \lambda_k^{i+1})$ is minimized.
    - Set π_i to be the partition obtained from π_{i+1} by merging λ_j^{i+1} and λ_k^{i+1} and leaving all other parts the same. Let $\lambda_i^i = \lambda_j^{i+1} \cup \lambda_k^{i+1}$.
    - Create new distance $\delta^i \in \mathbb{R}_{\geq 0}^{i(i-1)/2}$ by $\delta^i(\lambda, \lambda') = \delta^{i+1}(\lambda, \lambda')$ if λ, λ' are both parts of π_{i+1} and

$$\delta^i(\lambda, \lambda_i^i) = \frac{|\lambda_j^{i+1}|}{|\lambda_i^i|}\delta^{i+1}(\lambda, \lambda_j^{i+1}) + \frac{|\lambda_k^{i+1}|}{|\lambda_i^i|}\delta^{i+1}(\lambda, \lambda_k^{i+1})$$

      otherwise.
    - For each $u \in \lambda_j^{i+1}$ and $v \in \lambda_k^{i+1}$, set $d(u,v) = \delta^{i+1}(\lambda_j^{i+1}, \lambda_k^{i+1})$.
  (3) Return: A combinatorial rooted tree T with edge weights $w : E(T) \to \mathbb{R}_{\geq 0}$ and the ultrametric $d_{T,w}$ realized by T.

3.3. Using polyhedral geometry to analyze distance-based algorithms. The intersection of the polyhedral decomposition of the input space cones with the piece of the hypersphere $\mathbb{S}_{\binom{n}{2}-1}$ in the section of a Euclidean space with all non-negative coordinates, which we denoted as $HS_{\geq 0}$ in Section 2, provides a finite measure of how frequently the algorithm returns a particular topology sometimes called the *spherical volume* of the cone. Estimation of spherical volumes of polyhedral cones has been used to study the behavior of distance-based phylogenetic methods on small numbers of taxa. For example, in [**17**] estimated spherical volumes were used to assess the agreement of regions in the input space for NJ and the Balanced Minimum Evolution (BME) criterion, which as mentioned above is one of the optimization problems NJ seeks to solve. Also, in [**11**], estimated spherical volumes of cones in the input space for UPGMA indicated that UPGMA may be biased against unbalanced tree topologies, and in [**12**] the notion of spherical volume of small neighborhoods of the input space was further used to investigate the behavior of NJ and UPGMA. These findings assume that observed dissimilarity maps are uniformly distributed when projected on the sphere.

3.4. Limitation to the uniform distribution model. A key biological motivation for properly modeling the distribution of distance data is the so-called *rogue taxa phenomenon* in which the inclusion of a taxon in the estimation of a phylogeny results in reduced accuracy. See [**1**] for an example of a recent method designed to deal with this problem. In [**10**] polyhedral geometry was used to investigate the impact of including an additional taxon on the topology of a BME tree. In a simulation the authors showed that by including distances to an extra taxon and therefore lifting the problem to a higher-dimensional space, one could completely transform the BME topology for the set of taxa corresponding to the original distance matrix in a large number of cases, and that the effect worsened as the number of taxa grew.

But in [**41**] the effect of adding one additional taxon to a small tree was studied for a viral data set as a biological analog of the simulation study done in [**10**], which indicates that this effect may have been overstated due to the model of distances used in the simulation in [**10**]. Additional study of this effect using biological data may lead to more speculation about the true frequency of this effect. Motivated by the failure of the uniform distribution to capture the observed biological features we introduce two families of distribution functions in Section 4 that may better model the dissimilarity maps computed from DNA sequence data.

4. Modeling dissimilarity maps using statistical distributions and geometry

4.1. Two model families for dissimilarity maps. We investigate two families of models of dissimilarity maps based on the underlying structure of $\mathcal{ET}_n$. To construct our first model we use biological data as input, and fit a distribution function that describes the individual coordinates of the path trace associated to the dissimilarity map. We refer to this viewpoint as Model Family (1). In the second model of distance data we assume the data has been generated under the Yule-Harding model with noise accounting for the non-ultrametric tree-like features in the data. Precisely, we use a normal random variable to account for the regular Euclidean distance of an input point from the space $\mathcal{ET}_n$. We then compute the induced distribution of the associated coordinates of the path trace. We refer to this as Model Family (2).

4.2. Using biological data to develop the two model families. To ensure the comparison of these two models reflects biological data we fit the models to data drawn from the biological literature. TreeBASE [**23**] is a public, open-access website maintained by the Phyloinformatics Research Foundation, Inc. Data publicly available from TreeBASE is restricted to data associated to publications that have been submitted for peer review. TreeBASE includes gene, species, and population phylogenies as well as the data used to infer these phylogenies. We used data matrices representing sequence alignments of nucleotide data, inferred distances from these data matrices using the open-source software package MEGA 5 [**40**], and then extracted triples of distances from the large distance matrices to create large samples of dissimilarity maps in three dimensions.

In this chapter we highlight the findings for two data sets we obtained from TreeBASE. In [**28**] the authors are testing support for the classification of Boletineae as a sub-order as well as the families Boletaceae and Paxillaceae, and they find support for these using a phylogenetic analysis of three different genes. The paper [**38**] uses 9 genes to study the phylogenetic relationships of 29 species all within the same family: 22 of the taxa are in the *Daucus* genus and 7 are from related genera. All 29 species belong to the Umbelliferae family.

5. Developing Model Family (1)

It is equally likely that a data point δ is closest to any of the three Wings, so $p_W(\text{Wing } i) = \Pr(W = \text{Wing } i) = 1/3$, $i \in \{1, 2, 3\}$. We model D_1 using a

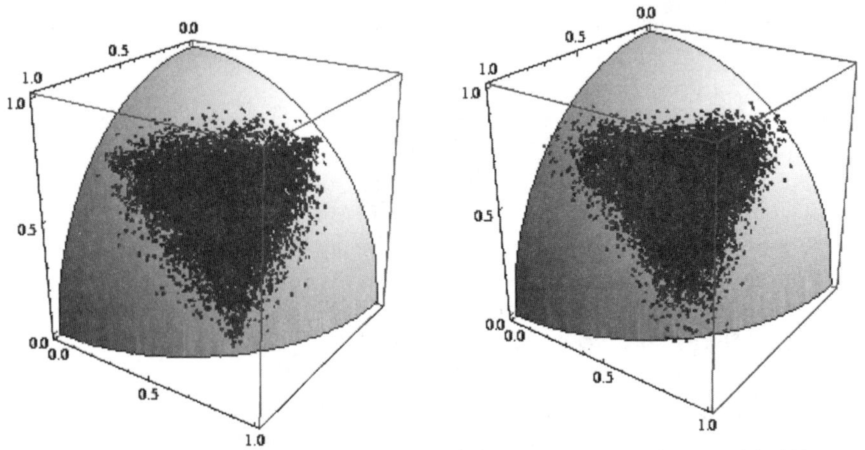

(A) Data points from [**28**]

(B) Simulated data (Model (1)-(A) with parameters estimated from [**28**]

FIGURE 3. A comparison of real versus simulated dissimilarity maps. Data points projected onto the unit sphere.

truncated normal distribution, and D_2 and D_3 are modeled in Model (1)-(A) using truncated generalized extreme value distributions (EVDs). For a comprehensive introduction to extreme value theory, see [**9**]. For Model (1)-(A), we designate the probability density function f_{D_3} for D_3 to be the truncated EVD with parameters α_s and β_s on domain $(0, +\infty)$ (the truncation forbids negative values):

$$f_{D_3}(d_3|\alpha_s, \beta_s) = \frac{e^{\frac{\alpha_s - d_3}{\beta_s} - e^{\frac{\alpha_s - d_3}{\beta_s}}}}{\beta_s c_{\alpha_s,\beta_s}} I_{\{x>0\}},$$

where c_{α_s,β_s} is the normalizing constant and I is the indicator function.

Next, $f_{D_2|D_3}(d_2|d_3, \theta)$ follows the resulting distribution obtained by assuming that $r_1 = \frac{d_2}{d_3}$ follows an EVD with α and β on domain $(0, 1]$, where the truncation is determined by the geometric setting. Note that, conditional on $D_3 = d_3$, $f_{D_2|D_3}$ also follows a truncated EVD. We consider r_1 first and let $d_2 = r_1 d_3$ so that the estimation of α and β does not depend on the actual value of d_3. Similarly we define $r_2 = \frac{d_1}{\min(d_3 + d_2/\sqrt{2},\ d_2/2)}$ and model r_2 by the truncated normal distribution with mean 0 and variance σ^2 on domain $[-\frac{1}{\sqrt{3}}, \frac{1}{\sqrt{3}}]$. Then $d_1 = r_2 \min(d_3 + d_2/\sqrt{2},\ d_2/2)$ will also follow a truncated normal distribution with independently estimated variance σ^2.

The EVD distributions in Model (1)-(A) are replaced with Gamma distributions in Model (1)-(B). We chose the distributions for Model (1)-(A) and Model (1)-(B) for our model by fitting the path trace to histograms taken from triples of distances estimated from biological data using MEGA [**40**].

6. Model Family (2)

To describe Model Family (2), we first describe the Yule-Harding pure birth process (Algorithm 6.1) which generates random rooted, binary phylogenetic trees with leaf set $[n]$. This is also commonly known as the Yule Process but we do

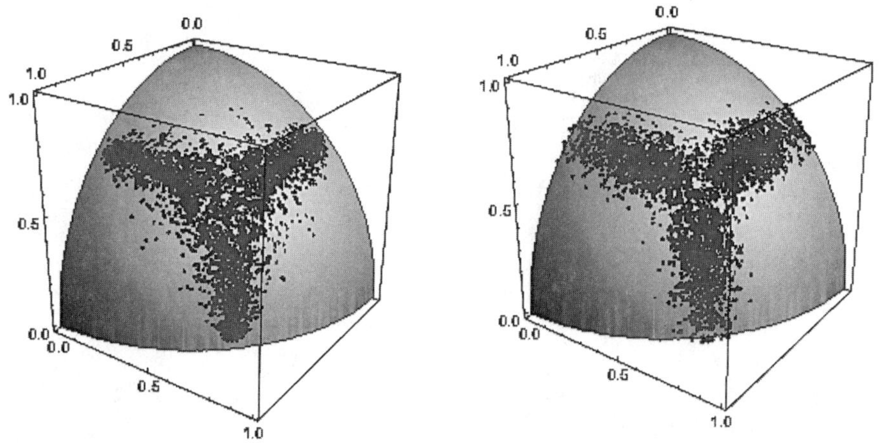

(A) Data points from [**38**]

(B) Simulated data (Model (1)-(A) with parameters estimated from [**38**]

FIGURE 4. A comparison of real versus simulated dissimilarity maps. Data points projected onto the unit sphere.

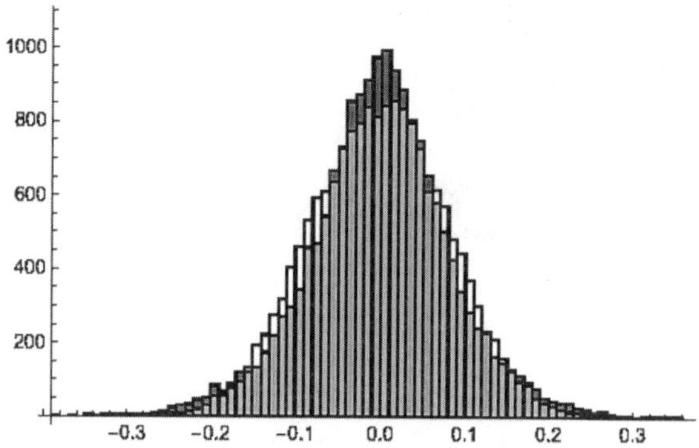

FIGURE 5. Histograms of Real and Simulated Data from Model (1)-(A) $D_1|D_2, D_3$ from [**28**]

emphasize that extinction is often incorporated into the model, in which case it is called a "birth-death" process. We ignore the "death" part of the model for simplicity. Recall that a *pendant edge* is an edge in a phylogenetic tree incident to a leaf.

Given a tree T with n leaves generated under the Yule-Harding pure birth model we investigate the distributions for D_2 and D_3 for a subtree obtained by restricting T to three randomly selected taxa. We define D_1 in the same way as for Model Family (1). We determine in this section that D_2 and D_3 should follow Gamma distributions. Throughout this section we will refer to the *graph* distance in a tree T, which is the number of edges in the unique path between a pair of taxa in T.

As mentioned above, dissimilarity maps estimated from DNA sequences is almost never ultrametric, but we use ultrametric trees - i.e. the geometric object $\mathcal{ET}_3$

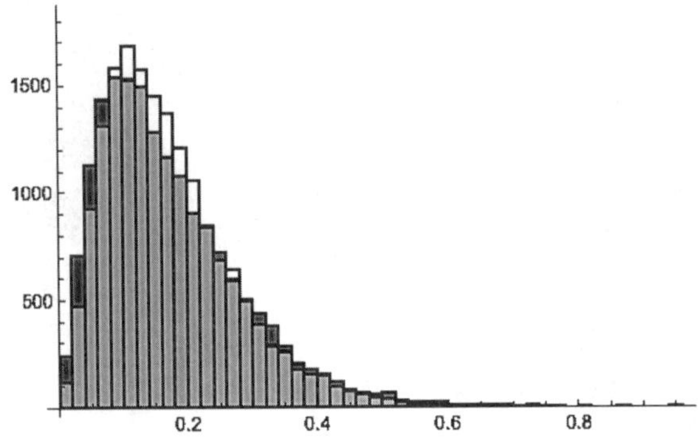

FIGURE 6. Histograms of Real and Simulated Data from Model (1)-(A) $D_2|D_3$ from [**28**]

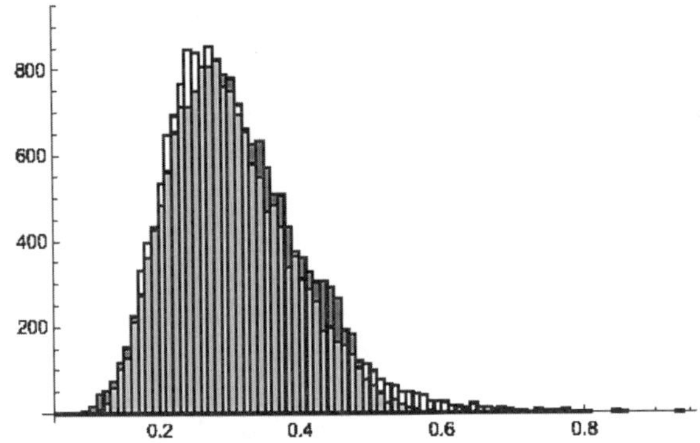

FIGURE 7. Histograms of Real and Simulated Data from Model (1)-(A) D_3 from [**28**]

- as we did for Model Family (1) as a reference for constructing Model Family (2). We again consider the perpendicular projection onto $\mathcal{ET}_3$ of a dissimilarity map δ, so that $\mathcal{ET}_3$ is a baseline object for constructing our model. We consider three cases, which are represented by three labeled tree topologies pictured in Figure 8. We may again, as in Model Family (1) assume that it is equally likely that a data point δ is closest to one of the three wings.

The three topologies are illustrated in Figure 8 with branch lengths labeled, where a and b separate the two most closely related taxa, c separates the single internal node of the tree and the root, and d goes from the root to the third taxa.

$$\begin{aligned}\text{Wing}\,1 &= (b+c+d, \tfrac{2a+b+c+d}{2}, \tfrac{2a+b+c+d}{2})\\ \text{Wing}\,2 &= (\tfrac{a+2b+c+d}{2}, a+c+d, \tfrac{a+2b+c+d}{2})\\ \text{Wing}\,3 &= (\tfrac{a+b+2c+2d}{2}, \tfrac{a+b+2c+2d}{2}, a+b)\end{aligned}$$

Under the assumption that T is ultrametric we can substitute $A = a = b$, $B = c = d - a$. Since these trees are subtrees of an n-taxon tree it is important to note that A and B are likely the sum of multiple branch lengths in the larger tree T. We can use the projections and corresponding ultrametric coordinates to give A and B as

Algorithm 6.1 Yule Process

- Input: Leaf set $[n] = \{1, \ldots, n\}$, $n \geq 2$.
- Output: a rooted, binary phylogenetic $[n]$-tree.
- Initialize: Randomly select two leaves x and y from $[n]$ with uniform probability. Identify these two leaves as the leaf set of a rooted binary tree T_1. Set $S_1 = [n] \setminus \{x, y\}$ and $L_1 = \{x, y\}$.
- While $S_i \neq \emptyset$:
  - Randomly select an element x_i of S_i with uniform probability.
  - Randomly select a pendant edge $e_i = (u_i, y_i)$ of T_i with probability determined by the uniform distribution on the set of pendant edges of T_i. Here u_i is a binary internal vertex and $y_i \in L_i$.
  - Subdivide e_i by adding a new vertex v_i.
  - Update T_{i+1} as the tree with leaf set $L_{i+1} = L_i \cup \{x_i\}$, $V(T_{i+1}) = V(T_i) \cup \{v_i, x_i\}$ and $E(T_{i+1}) = E(T_i) \setminus \{e_i\} \cup \{(u_i, v_i), (v_i, y_i), (v_i, x_i)\}$. Update $S_{i+1} = S_i \setminus \{x_i\}$.
- Return T_{n-2+1}, a rooted, binary phylogenetic tree with leaf set $[n]$.

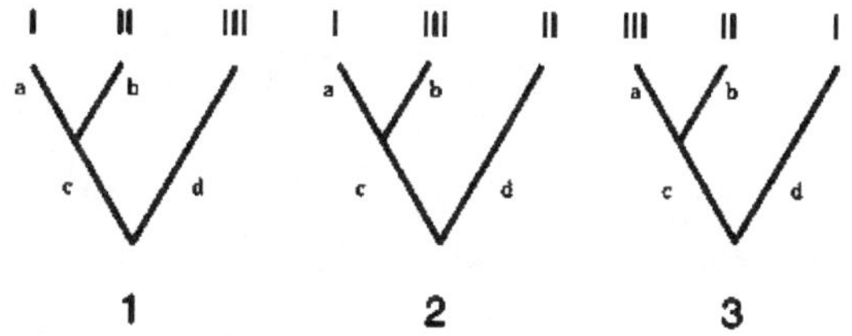

FIGURE 8. Three taxa tree topologies with edge lengths.

functions of the input data.

$$\begin{array}{lll} \text{Wing 1}: & A = \frac{b+c+d}{2} & B = \frac{2a-b-c-d}{4} \\ \text{Wing 2}: & A = \frac{a+c+d}{2} & B = \frac{2b-a-c-d}{4} \\ \text{Wing 3}: & A = \frac{a+b}{2} & B = \frac{2c+2d-a-b}{4} \end{array}$$

6.1. Spindle-origin distance.

PROPOSITION 6.1. Under the Yule-Harding model, we may describe the distribution of the set of unscaled distances of the projection from wing to spindle, D_2, as follows: $D_2|\alpha_{D_2}$ has a Gamma distribution with parameters $\frac{2}{\sqrt{3}}\lambda$ and α_{D_2}, where λ is the birth rate and α_{D_2} has expected value $E[\alpha_{D_2}] = 3(\mu_n - 2(1 - \frac{\mu_n}{n-1}))$ where $\mu_n = \sum_{j=2}^{n} \frac{1}{j} \approx ln(n) + \gamma + \frac{1}{2n} - \frac{1}{12n^2}$, and $\gamma \approx 0.5772156649$ denotes the Euler-Mascheroni constant.

PROOF. The sum of n independent exponential random variables with parameter λ is a Gamma distribution with parameters $\alpha = n$ and λ [**26**]. We use this to show $D_2|\alpha_{D_2}$ follows a Gamma distribution, and then determine the expected value of α_{D_2} based on the graph distances between the three taxa.

Suppose we fix a three-taxon subtree T_0. We compute D_2 in terms of the branch lengths of T_0 by finding the Euclidean norm of the orthogonal projection from the point on the wing to the spindle. We find that the D_2 is given by

$$\frac{2}{\sqrt{3}}D,$$

where $D = 2A+2(A+B) = a+b+c+d$. We note that D is the sum of the branch lengths of T_0. Let the number of branches included in D be α_{D_2}. Under the Yule-Harding model branch lengths follow an exponential distribution with parameter λ. Thus, D is the scalar multiple of the sum of exponential branch lengths, which implies D has distribution $\Gamma(\lambda, \alpha_{D_2})$ and D_2 follows a Gamma distribution under the Yule-Harding pure birth model. Notice that here α_{D_2} is a constant given the fixed subtree.

Then we see that $D = a+b+c+d = \frac{1}{2}(d(\mathrm{I},\mathrm{II}) + d(\mathrm{I},\mathrm{III}) + d(\mathrm{II},\mathrm{III}))$, which implies $E[D] = \frac{1}{2}(E[d(\mathrm{I},\mathrm{II})] + E[d(\mathrm{I},\mathrm{III})] + E[d(\mathrm{II},\mathrm{III})])$. Since the subtree is randomly chosen, we have $E[D] = \frac{3}{2}E[d(\mathrm{I},\mathrm{II})]$. Therefore $E[D] = E[E[D|\alpha_{D_2}]] = E[\alpha_{D_2}/\lambda] = E[\alpha_{D_2}]/\lambda = \frac{3}{2}E[d_{ij}]/\lambda$, i.e. $E[\alpha_{D_2}] = \frac{3}{2}E[d_{ij}]$, where $E[d_{ij}]$ is the expected graph distance between a random subset of two taxa in an Yule tree with n. Steel and McKenzie [**39**] showed

$$E[d_{ij}] = 2\left(\mu_n - 2\left(1 - \frac{\mu_n}{n-1}\right)\right).$$

Therefore

$$E[\alpha_{D_2}] = 3\left(\mu_n - 2\left(1 - \frac{\mu_n}{n-1}\right)\right).$$

Additionally we note that the scalar multiple kX of a random variable X that follows a Gamma distribution with parameters α_{D_2} and λ is also a Gamma distribution with parameters α_{D_2} and $k\lambda$. Thus we expect the spindle-to-origin distance to be $\Gamma\left(\alpha_{D_2}, \frac{2}{\sqrt{3}}\lambda\right)$. □

6.2. Wing-spindle distance.

PROPOSITION 6.2. *Under the Yule-Harding model, the distribution of the set of distances of the projection from wing to spindle can be described as follows: $D_3|\alpha_{D_3}$ has a Gamma distribution with parameters $2\sqrt{\frac{2}{3}}\lambda$ and α_{D_3}, where λ is the birth rate.*

We use similar argument to show $D_3|\alpha D_3$ also follows a Gamma distribution. The Euclidean norm of the projection from the spindle to the origin is $2\sqrt{\frac{2}{3}}B$. For a fixed subtree under the Yule-Harding pure birth model, B is the sum of exponential random variables and therefore follows a Gamma distribution.

In order to determine the expected value for the shape parameter we need to know the expected graph distance in T between the unique internal node and the root in T_0. Steel and McKenzie [**39**] gave the expected value for the distance from the unique internal node of two taxa to the root in T_0. It is possible that a refinement of this work could lead to an explicit formula for the expected value of the shape parameter in terms of the number of taxa. But without such an explicit formula it is simpler to estimate the parameter directly from the data.

7. Data analysis

7.1. Method for scoring the deviation of a model distribution. Given a distribution function for pairwise distance data, one might wish to measure how well it conforms with the data it was designed to model.

In this section we present a method for scoring a model distribution's deviation from a data sample on $HS_{\geq 0}$. We use this method to compare how well our model families, as well as the uniform distribution, model a test data set. We use a numerical test statistic motivated by the assumption that for large samples, the proportion of the sample in any region of $HS_{\geq 0}$ should be the same as the integral of the proposed density function over that region.

Let S be a sample of N triples of pairwise distance data which has been orthogonally projected onto $HS_{\geq 0}$. Let P be a random sample of triples of distance data drawn from a distribution function γ and then projected onto the unit sphere. We define a test statistic to determine how different S and P are.

We select k randomly generated points v_i on the $HS_{\geq 0}$. Each such point divides $HS_{\geq 0}$, into three spherical triangles

$$T_1 = \{v_i, (1,0,0), (0,1,0)\},$$

$$T_2 = \{v_i, (1,0,0), (0,0,1)\}, \quad \text{and} \quad T_3 = \{v_i, (0,1,0), (0,0,1)\}.$$

For each $1 \leq i \leq k$ we define the triples $\hat{S}$ (resp $\hat{P}$) $= (s_1, s_2, s_3)$ where $s_j = \frac{|S \cap T_j|}{|S|}$. Then we can define the test statistic:

$$\rho(N,k) = \frac{\sum_{i=1}^{k} \sqrt{(s_1 - p_1)^2 + (s_2 - p_2)^2 + (s_3 - p_3)^2}}{k}. \tag{7.1}$$

For large values of N and k, ρ should approach zero.

Note that if a density function f was known for γ, then the p_i could be computed by integrating f along T_1, T_2, and T_3. If f is the density function for the uniform distribution, then as $n \to \infty$, $\hat{P}$ will approach

$$\left(\frac{Area(T_1)}{\frac{\pi}{6}}, \frac{Area(T_2)}{\frac{\pi}{6}}, \frac{Area(T_3)}{\frac{\pi}{6}} \right).$$

7.2. Fitting data sets from Treebase.org to Model Families (1) and (2). We used custom Mathematica software available on our supplementary materials website to test the ability of our different model families to fit biological data sets. The results for ten data sets are shown in Table 1. Rows are indexed by the ten data sets and the entries in the tables are the score based on our geometric test statistic explained in Section 7.1.

We explain the labels for the columns: Models (1)-(A) and (1)-(B) are from Model Family (1) where (1)-(A) indicates the choice of the extreme value distribution and (1)-(B) indicates the choice of the Gamma distribution. Uniform is the model where input points are uniformly distributed in the input space. Lower scores are best, as explained in Section 7.1, and the best score for each data set is in bold text. Rows are indexed by the data matrix name on Treebase.org: M18755 is the data matrix we used from [**28**], and we were no longer able to find the study [**38**] in Treebase.org by the time of this publication, so we have made the distances we calculated from that data matrix available in our online supplementary materials. The row corresponding to that data matrix is labeled "Daucus."

In Table 1, for each data set, we used $k = 10000$ sample points for the area subdivisions in our scoring system. For small data matrices, we used all possible triples of input points $(\delta(u,v), \delta(u,w), \delta(v,w)) = (x, y, z)$, i.e. $N = \binom{n}{3}$ where n was the number of taxa in the matrix. For data matrices with 50 or more taxa, we used approximately $N = 20,000$ data points, where N varied slightly because our method for random sampling of points $(\delta(u,v), \delta(u,w), \delta(v,w))$ produced duplicates.

While the results in Table 1 are interesting, and definitely show that for these datasets, the uniform distribution perform poorly, analysis of many more data sets would be necessary to establish a clear trend. It also appears that Model (1)-(B) which blends the geometric features of Model Family (1), with the choice of Gamma distribution motivated by Model Family (2), may provide the best overall accuracy.

TABLE 1. Scores $\rho(N, k)$ from each model for data sets with Jukes-Cantor Distances from Treebase.org. Model of best fit is in bold.

Data Matrix	Model (1)-(A)	Model (1)-(B)	Model 2	Uniform
M1807	0.0208444	0.0276282	**0.01446**	0.0892374
M1789	**0.00944452**	0.00961268	0.0154542	0.26513
M2566	0.0624382	**0.04826421**	0.0736874	0.060538
M25733	0.0140685	**0.0131326**	0.0334055	0.264626
M18755	0.00826286	**0.00819624**	0.0134024	0.293712
M25665	0.0141961	**0.0141181**	0.0276884	0.237157
Daucus	0.00879329	**0.00672433**	0.0148475	0.294249
M536	0.0251412	**0.019827**	0.0215093	0.130195
M806	**0.0282066**	0.0331788	0.0359551	0.153135
M1169	**0.0115309**	0.0122847	0.0177641	0.328366

8. Motivations: choice of distributions in Model Families (1) and (2)

The variance of the normal distribution parametrizes D_1 in both model families (1) and (2). A value of $d_1 = 0$ means that a data point is in $\mathcal{ET}_3$, and so is an ultrametric additive distance matrix. Therefore larger d_1 should correspond to higher variations in the rates of evolution between the three taxa for the data point; i.e. deviation from the molecular clock. The normal distribution is thus a natural choice for both model families.

We first chose the EVD to model D_2 and D_3 because histograms of the corresponding steps in path traces of the real data points closely resembled the probability density function for the EVD. Figures 5, 6, and 7 show the histograms of data points of the type $(\delta(u,v), \delta(u,w), \delta(v,w))$ plotted together with histograms of data points generated using Model (1)-(A). However, since both the Gamma and EVD are transformations of the exponential distribution, [**26**], and because of our results from Section 6, we felt it was necessary to include Model (1)-(B).

Our model families seem to be capable of capturing rather subtle features of the biological data beyond how far the data deviate from the molecular clock. We observed that in some datasets, such as in [**38**], data points are less distributed in the center (i.e. around the spindle). This phenomenon is visible in Figures 4a and

4b. This means that the distribution of D_2 has lower density when D_2 is close to 0 and near the boundary. We comment on this feature for this data set because the density functions of the EVD and Gamma distributions can accommodate this feature of the data with appropriate parameter choices in both Model Families (1)-(A) and (1)-(B), while the density function for the exponential distribution on branch lengths used in Model Family (2) cannot.

9. Non-ultrametric trees and application to larger data sets

The model families described below emphasize the geometry of ultrametric trees ($\mathcal{ET}_3$). However, a similar analysis could be conducted relative to the more general tree space $\mathcal{T}_n$. With only three taxa, every dissimilarity map that satisfies the triangle inequality is already realizable as a pairwise distance on a tree [**34**]. Therefore, it is more interesting to consider the relationship between maps and the more general tree space in the context of $\mathcal{T}_4$.

The geometry of $\mathcal{T}_4$ can be described as the union of three *superwings*

$$W_{14} = \{P \in \mathbb{R}^6 | x_12 + x_34 = x_{13} + x_{24} \leq x_{14} + x_{23}\},$$

$$W_{13} = \{P \in \mathbb{R}^6 | x_12 + x_34 = x_{14} + x_{23} \leq x_{13} + x_{24}\},$$

and

$$W_{12} = \{P \in \mathbb{R}^6 | x_13 + x_24 = x_{14} + x_{23} \leq x_{12} + x_{34}\}.$$

While this model begins with an emphasis on general tree space, we shift our emphasis to the condition of ultrametricity through a projection to $\mathcal{ET}_3$ at a later stage.

In the earlier sections of this chapter we assumed our biological data arose from the addition of Gaussian noise to phylogenetic trees identified as points in Euclidean space. We continue with this paradigm and model the distance of a dissimilarity map to the nearest superwing using a normal distribution. Next we project the image of the point on the superwing to the *superspindle* or set of coordinates with $x_{12} + x_{34} = x_{13} + x_{24} = x_{14} + x_{23}$. This distance could be modeled by a Gamma distribution.

As the sum of branch lengths under the Yule-Harding model we can assume $x_{12} + x_{34}$ follow a Gamma distribution. By knowing this sum we can reconstruct all coordinates using only x_{12}, x_{13}, and x_{23}. These triples of coordinates lie in $\mathbb{R}^3$ and correspond to triples of pairwise distances drawn from a tree on n taxa and can thus be modeled using the findings in Section 4.

Moreover, when there are more than four taxa the geometry of tree space is simply the intersection of superwings over all sets of four taxa. This follows directly from Buneman's four-point condition (or three point condition in the ultrametric setting) [**6**]. While it is possible to build more complicated distributions relating dissimilarity maps to $\mathcal{T}_n$ or $\mathcal{ET}_n$ such models would require a larger set of parameters, and may necessarily contain more information than the minimal cases. However, the following question is paramount for turning the theoretical work of this chapter into a practical tool for computational biologists.

QUESTION 9.1. How does a distribution of dissimilarity maps on subsets of 3 (or 4) of n taxa induce a distribution of dissimilarity maps on n taxa?

10. Applications and future work

The models in this chapter provide new tools for interpreting analytic results about distance-based phylogenetic methods using geometry. Analyses of trends in the shapes of trees using notions such as balance statistics has led to interesting insights and questions about model assumptions in the past [**2**]. The tools presented in Sections 5 and 6 establish a framework for evaluating suitability of model assumptions for phylogenetic data. Recently, the TreeBASE website (accessed June 15, 2016) stated: "as of April 2014, TreeBASE contained data for 4,076 publications written by 8,777 different authors" [**23**]. A systematic analysis of more alignment matrices using the software available in our supplementary materials might give a clearer picture of trends in biological data that relate the model families presented here, providing empirical support for choosing one model family as more suitable for further development and study.

Our use of the geometric object $\mathcal{ET}_3$ as the backbone of our models serves many purposes. For example, constructing Model Families (1) and (2) around $\mathcal{ET}_n$ provides a unifying framework to study which Model Family is better suited to modeling biological data. The Yule-Harding pure birth process does produce ultrametric trees, which are points in $\mathcal{ET}_n$. So, comparing the Model Families (1) and (2) in this geometric setting may provide insight into the implications of assuming the Yule-Harding model in data simulation.

Furthermore, the Mathematica software in our supplementary materials can be used directly to estimate the deviation of a biological data set from the molecular clock, via the amount of variance in the normal distribution parameter estimated as D_1. We have already called attention to the fact that biological data is rarely ultrametric, but in the biology and computer science communities, especially in the field of phylogenomics, the assumption of a molecular clock is still necessary to provide theoretical guarantees in many instances. For example, in [**30**] it is shown that without the assumption of a molecular clock, no theoretical guarantees bounding gene tree estimation error can be made.

Also, some phylogenomic methods that bypass the need for gene tree estimation rely on the assumption of a molecular clock for their theoretical guarantees. For example, the recently developed method SVDquartets [**7**] relies on the identifiability result in [**8**] for theoretical guarantees, and requires a molecular clock. So, we argue that establishing the deviation of a data set from the molecular clock using our software would be useful in determining whether or not it is reasonable to apply a method that is only guaranteed to work well on data that follows a molecular clock to that specific data set.

The results shown in this chapter are restricted to distances computed under the Jukes-Cantor model of sequence evolution. We performed the same experiments with the K2P model, but the results on the data sets we investigated were so similar for each data set that we omitted these findings. However, future work should include an analysis of both distances computed from amino acid data as well as more general statistical models of sequence evolution such as GTR [**22**] and GTR+Γ [**31**].

10.1. Future research questions.

QUESTION 10.1. Is there a correlation between the number of missing taxa in a tree and the parameters of the Gamma distribution in Model Family (2)? How does this shed light on how closely a set of sampled species may be related in the Tree of Life?

QUESTION 10.2. What is the likelihood of the rogue taxa phenomenon using Model Family (2), which more naturally extends to $n > 3$ than Model Family (1)?

QUESTION 10.3. Can one provide an inductive and computationally tractable model for the distribution of distance data for pairwise dissimilarity vectors on n taxa for Model Family (1)? If so, how will this model compare to Model Family (2)

QUESTION 10.4. For sequence data generated under the coalescent model, is there a correlation between the estimated Gamma distribution parameters across the different gene trees? Could the variation among these parameters inform the design of summary methods?

11. Supplementary material

All supplementary materials, including the distance matrixes computed from the alignment data matrices downloaded from Treebase.org, and Mathematica software for generating simulated distance data, can be found at goo.gl/08PUC5.

Acknowledgements

R.D. was partially supported by the National Science Foundation (DMS 0954865). J.R. was partially supported by grants from the National Center for Research Resources (5 P20 RR016461) and the National Institute of General Medical Sciences (8 P20 GM103499) from the National Institutes of Health. Z.V. was partially supported by the National Institute of General Medical Sciences (8 P20 GM103499) from the National Institutes of Health. J. X. was partially supported by the David and Lucille Packard Foundation. We also thank James Degnan, Megan Owen, Mike Steel, Seth Sullivant, Caroline Uhler, Tandy Warnow, and Ruriko Yoshida for helpful discussions over the course of this project.

References

[1] A. J. Aberer, D. Krompass, and A. Stamatakis, *Pruning rogue taxa improves phylogenetic accuracy: an efficient algorithm and webservice*, Syst. Biol. **62** (2013), no. 1, 162–166.
[2] David J. Aldous, *Stochastic models and descriptive statistics for phylogenetic trees, from Yule to today*, Statist. Sci. **16** (2001), no. 1, 23–34, DOI 10.1214/ss/998929474. MR1838600
[3] Federico Ardila and Caroline J. Klivans, *The Bergman complex of a matroid and phylogenetic trees*, J. Combin. Theory Ser. B **96** (2006), no. 1, 38–49, DOI 10.1016/j.jctb.2005.06.004. MR2185977
[4] J. Bergsten, *A review of long-branch attraction*, Cladistics **32** (2005), 163–193.
[5] W. M. Bruno, N. D. Socci, and A. L. Halpern, *Weighted Neighbor-Joining: a likelihood-based approach to distance- based phylogeny reconstruction*, Mol. Biol. Evol. **17** (2000), 189–197.
[6] Peter Buneman, *A note on the metric properties of trees*, J. Combinatorial Theory Ser. B **17** (1974), 48–50. MR0363963
[7] J. Chifman and L. Kubatko, *Quartet inference from SNP data under the coalescent model*, Bioinform. **30** (2014), no. 23, 3317–3324.
[8] Julia Chifman and Laura Kubatko, *Identifiability of the unrooted species tree topology under the coalescent model with time-reversible substitution processes, site-specific rate variation, and invariable sites*, J. Theoret. Biol. **374** (2015), 35–47, DOI 10.1016/j.jtbi.2015.03.006. MR3341829
[9] Stuart Coles, *An introduction to statistical modeling of extreme values*, Springer Series in Statistics, Springer-Verlag London, Ltd., London, 2001. MR1932132

[10] María Angélica Cueto and Frederick A. Matsen, *Polyhedral geometry of phylogenetic rogue taxa*, Bull. Math. Biol. **73** (2011), no. 6, 1202–1226, DOI 10.1007/s11538-010-9556-x. MR2799936

[11] Ruth Davidson and Seth Sullivant, *Polyhedral combinatorics of UPGMA cones*, Adv. in Appl. Math. **50** (2013), no. 2, 327–338, DOI 10.1016/j.aam.2012.10.002. MR3003350

[12] R. Davidson and S. Sullivant, *Distance-based phylogenetic methods around a polytomy*, IEEE/ACM Trans. Comput. Biol. Bioinf. **11** (2014), no. 1, 325–335.

[13] William H. E. Day, *Computational complexity of inferring phylogenies from dissimilarity matrices*, Bull. Math. Biol. **49** (1987), no. 4, 461–467, DOI 10.1016/S0092-8240(87)80007-1. MR908160

[14] R. Desper and D. Gascuel, *Fast and accurate phylogeny reconstruction algorithms based on the minimum-evolution principle*, J. Comput. Biol. **9** (2002), no. 5, 687–705.

[15] R. Desper and O. Gascuel, *Theoretical foundation of the balanced minimum evolution method of phylogenetic inference and its relationship to weighted least-squares tree fitting*, Mol. Biol. Evol. **21** (2004), no. 3, 587–598.

[16] R. C. Edgar, *MUSCLE: multiple sequence alignment with high accuracy and high throughput*, Nucl. Acids Res. **32** (2004), no. 5, 1792–1797.

[17] K. Eickmeyer, P. Huggins, L. Pachter, and R. Yoshida, *On the optimality of the Neighbor-Joining algorithm*, Alg. Mol. Biol. **3** (2008), doi:10.1186/1748–7188–3–5.

[18] K. Eickmeyer and R. Yoshida, *Partitioning the sample space on five taxa for the Neighbor-Joining algorithm*, http://arxiv.org/abs/math/0703081.

[19] C. Fahey, S. Hosten, N. Krieger, and L. Timpe, *Least squares methods for equidistant tree reconstruction*, (2008), http://arxiv.org/abs/0808.3979.

[20] O. Gascuel, *BIONJ: an improved version of the NJ algorithm based on a simple model of sequence data*, Mol. Biol. Evol. **14** (1997), 685–695.

[21] O. Gascuel and M. Steel, *Neighbor-Joining revealed*, Mol. Biol. Evol. **23** (2006), no. 11, 1997–2000.

[22] X. Gu, Y.X. Fu, and W. H. Li, *Maximum likelihood estimation of the heterogeneity of substitution rate among nucleotide sites*, Mol. Biol. Evol. **12** (1995), 546–557.

[23] The Phyloinformatics Research Foundation Incorporated, *TreeBASE web*, 2010, pp. http://treebase.org/treebase–web/home.html.

[24] T. H. Jukes and C. R. Cantor, *Evolution of protein molecules*, Mamm. Protein Metab. **3** (1969), 21–132.

[25] M. Kimura, *A simple method for estimating evolutionary rates of base substitutions through comparative studies of nucleotide sequences*, J. Mol. Evol. **16** (1980), 111–120.

[26] Lawrence M. Leemis and Jacquelyn T. McQueston, *Univariate distribution relationships*, Amer. Statist. **62** (2008), no. 1, 45–53, DOI 10.1198/000313008X270448. MR2416896

[27] L. Liu and L. Yu, *Estimating species trees from unrooted gene trees*, Syst. Biol. **60** (2011), 661–667.

[28] M. E. Nuhn, M. Binder, A. F. S. Taylor, R. E. Halling, and D. S. Hibbett, *Phylogenetic overview of the Boletineae*, Fungal Biol. **117** (2013), 479–511.

[29] Y. Pauplin, *Direct calculation of a tree length using a distance matrix*, J. Mol. Evol. **51** (2000), 41–47.

[30] S. Roch and T. Warnow, *On the robustness to gene tree estimation error (or lack thereof) of coalescent-based species tree methods*, Syst. Biol. (2015), doi: 10.1093/sysbio/syv016.

[31] J. S. Rogers, *Maximum likelihood estimation of phylogenetic trees is consistent when substitution rates vary according to the invariable sites plus Gamma distribution*, Syst. Biol. **50** (2001), 713–722.

[32] N. Saitou and M. Nei, *The Neighbor-Joining method: a new method for reconstructing phylogenetic trees*, Molecular Biology and Evolution **4** (1987), 406–425.

[33] Charles Semple and Mike Steel, *Phylogenetics*, Oxford Lecture Series in Mathematics and its Applications, vol. 24, Oxford University Press, Oxford, 2003. MR2060009

[34] Charles Semple and Mike Steel, *Phylogenetics*, Oxford Lecture Series in Mathematics and its Applications, vol. 24, Oxford University Press, Oxford, 2003. MR2060009

[35] R. R. Sokal and C. Michener, *A statistical method for evaluating systematic relationships*, Univ. Kansas Sci. Bull. **38** (1958), 1409–1438.

[36] Peter H. A. Sneath and Robert R. Sokal, *Numerical taxonomy*, W. H. Freeman and Co., San Francisco, Calif., 1973. The principles and practice of numerical classification; A Series of Books in Biology. MR0456594

[37] David Speyer and Bernd Sturmfels, *The tropical Grassmannian*, Adv. Geom. **4** (2004), no. 3, 389–411, DOI 10.1515/advg.2004.023. MR2071813

[38] D. Spooner, P. Rojas, M. Bonierbale, L.A. Mueller, M. Srivastav, D. Senalik, and P. Simon, *Molecular phylogeny of Daucus*, Syst. Bot. **38** (2013), 850–857.

[39] Mike Steel and Andy McKenzie, *Properties of phylogenetic trees generated by Yule-type speciation models*, Math. Biosci. **170** (2001), no. 1, 91–112, DOI 10.1016/S0025-5564(00)00061-4. MR1818785

[40] K. Tamura, D. Peterson, N. Peterson, G. Stecher, M. Nei, and S. Kumar, *MEGA5: Molecular evolutionary genetics analysis using maximum likelihood, evolutionary distance, and maximum parsimony methods*, Mol. Biol. Evol. **28** (2011), 2731–2739.

[41] K.M. Westover, J.P. Rusinko, J. Hoin, and M. Neal, *Rogue taxa phenomenon: a biological companion to simulation analysis*, Mol. Phylogenetics Evol. **1** (2013), 1–3.

[42] Günter M. Ziegler, *Lectures on polytopes*, Graduate Texts in Mathematics, vol. 152, Springer-Verlag, New York, 1995. MR1311028

Department of Mathematics, University of Illinois Urbana-Champaign
E-mail address: `redavid2@illinois.edu`

Department of Mathematics and Computer Science, Hobart and William Smith Colleges
E-mail address: `rusinko@hws.edu`

Department of Mathematics, Washington University in St. Louis
E-mail address: `zoe.vernon@wustl.edu`

Department of Mathematics, North Carolina State University
E-mail address: `jxi2@ncsu.edu`

Contemporary Mathematics
Volume **685**, 2017
http://dx.doi.org/10.1090/conm/685/13754

Geometric combinatorics and computational molecular biology: Branching polytopes for RNA sequences

Elizabeth Drellich, Andrew Gainer-Dewar, Heather A. Harrington, Qijun He, Christine Heitsch, and Svetlana Poznanović

Abstract. Questions in computational molecular biology generate various discrete optimization problems, such as DNA sequence alignment and RNA secondary structure prediction. However, the optimal solutions are fundamentally dependent on the parameters used in the objective functions. The goal of a parametric analysis is to elucidate such dependencies, especially as they pertain to the accuracy and robustness of the optimal solutions. Techniques from geometric combinatorics, including polytopes and their normal fans, have been used previously to give parametric analyses of simple models for DNA sequence alignment and RNA branching configurations. Here, we present a new computational framework, and proof-of-principle results, which give the first complete parametric analysis of the branching portion of the nearest neighbor thermodynamic model for secondary structure prediction for real RNA sequences.

1. Motivation

Over the past four decades, improvements in biotechnology have greatly accelerated the amount of biological sequence data available. Yet, a fundamental challenge in computational molecular biology remains to reliably infer functional information from the linear encoding of DNA, RNA, and protein molecules.

As articulated in the central dogma of molecular biology, genetic information is stored in DNA from which it is transcribed into messenger RNA, and then translated into proteins by ribosomal and transfer RNA. However, as always in biology, a wealth of complexity lurks below the surface of this basic principle. Historically, most interest was focused on DNA sequences (as the cellular genome) and protein structures (as the cellular machinery). Since the early 2000's, though, attention has increasingly turned to RNA as many more critical functions have been revealed, including gene splicing, editing, and regulation.

2010 *Mathematics Subject Classification.* Prmary 92D10.

Key words and phrases. Parametric analysis, secondary structure, polytope.

The authors' collaboration was funded by the Mathematics Research Communities program of the AMS.

The fifth author was supported in part by a Career Award at the Scientific Interface from the Burroughs Wellcome Fund.

The sixth author was also supported in part by NSF grant DMS-1312817. HAH gratefully acknowledges the support of EPSRC Fellowship EP/K041096/1.

Like DNA, RNA is a sequence of nucleic acids, abbreviated `A`, `C`, `G`, and `U` (instead of `T`), which form the familiar Watson-Crick pairings. Unlike the canonical double-stranded DNA helix, most RNA molecules are naturally single-stranded and the intra-sequence base pairings are an integral component of the three-dimensional structure. This is in contrast to the more subtle amino acid interactions which govern the formation of protein structures. However, knowing the structure of a protein or RNA molecule is critical to understanding and manipulating its cellular functions.

The structure of an RNA molecule is understood hierarchically, from the linear biochemical chain through the planar set of canonical base pairings to the 3D molecule, whose structure includes more complicated tertiary interactions like pseudoknots and base triples. Given the prevalence of RNA sequence data and the difficulties, both experimental and computational, of determining 3D structures, the majority of attention has focused on the set of base pairs, known as the *secondary structure*. In particular, one seeks to answer: *What are the native base pairs for a given RNA sequence?* As explained below, it is possible to efficiently compute an optimal secondary structure under the current Nearest Neighbor Thermodynamic Model (NNTM). However, when these minimum free energy (MFE) predictions are compared to structures derived from information-theoretic means, the current gold-standard, the average accuracy for longer ribosomal RNA sequences is only 40%. Hence, it is critical to understand which aspects of RNA base pairing are not captured well by the NNTM.

One known weakness of the current model is the energy function which governs the branching of an RNA secondary structure. For computational reasons, the entropic cost is modeled as an affine function with three parameters. A very natural question to ask is: *How does the optimal secondary structure depend on the branching loop parameters?*

Such a question is an example of a parametric analysis, which illuminates the dependencies of the solution on the underlying optimization parameters. As we illustrate here, methods from geometric combinatorics, specifically polytopes and their normal fans, can be used to answer this question, and others including accuracy and stability. The outline of this paper is as follows. Section 2 contains the basic RNA terminology and a brief overview of the geometric tools needed for parametric analysis. In Section 3 we first give background on the NNTM, specifically the way the multibranch loops are scored. Then we introduce branching polytopes and explain how to compute them using the software `pmfe` which we have made available online. Finally, we present a few biological questions and discuss mathematical problems related to the branching polytopes they raise. The paper ends with a summary in Section 4.

2. Background

2.1. Computational molecular biology. Nucleotide sequences can be modeled as strings over a finite alphabet. The questions related to DNA and RNA are amenable to analysis using discrete mathematical techniques from algebra, combinatorics, and topology. We refer the reader to the following references for mathematical approaches for studying computational molecular biology [**2, 8, 12, 24, 25, 36**]. Of particular interest here are techniques from geometric combinatorics, which can inform parametric analysis of models arising from optimization questions, such as DNA sequence alignment and RNA secondary structure prediction.

Previous parametric analyses in molecular biology have primarily focused on DNA. Given sequences of DNA, an important problem is to find regions that have been evolutionary conserved. The first step requires *aligning* the sequences by adding spaces to the sequences until they are the same length. The *DNA sequence alignment* problem is to find the best alignment between sequences given some scoring function [**35**]. The optimal solution can be determined via dynamic programming optimization which scores various features and provides global alignment. The optimization fundamentally depends on parameters, and the notion of 'best matching' sequences may depend on the value of these parameters. Techniques from geometric combinatorics can be used [**5**, **35**] to analyze systematically how sequence alignment depends on these parameters. The mathematical background will be presented in Section 2.2.

The problem we focus on here is prediction of the secondary structure of a single given RNA sequence. Unlike DNA, which typically forms a fully base-paired double helix, RNA typically exists as a single strand whose bases can pair with one another to form diverse and complex secondary structures. The secondary structure of a given RNA molecule serves as a scaffold for the 3D structure [**32**]. As shown in Figure 1, an RNA secondary structure consists of stacked base pairs (*helices*) and single-stranded regions (*loops*). The *degree* of a loop is the number of helices emerging from it (or, equivalently, the number of base pairings in the loop). *Hairpin loops* have degree 1, *internal loops* have degree 2, and *multibranch loops* have degree greater than 2; each RNA molecule also has a (possibly empty) *exterior loop*, which includes the unpaired bases not contained in other loops. Figure 1 shows a secondary structure with all of these substructures labeled.

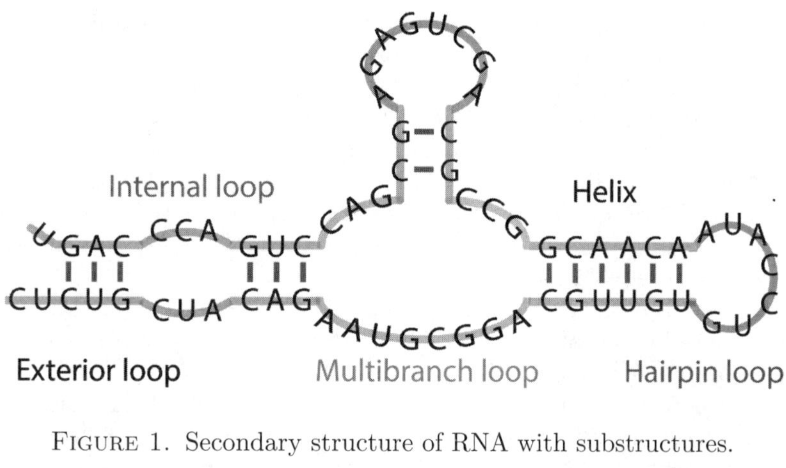

FIGURE 1. Secondary structure of RNA with substructures.

The base pairings in the secondary structure are difficult to determine experimentally, necessitating mathematical models to predict the structure [**40**]. Indeed secondary structure prediction has been an active area of research for several decades (see survey in [**19**]). One approach for predicting secondary structure is calculating the *free thermodynamic energy* of the RNA molecule according to the standard NNTM [**17**, **33**]. In this model, bases that bond are generally energetically favorable and *stabilize* the RNA with a negative free energy, whereas unpaired bases are energetically disfavorable and *destabilize* the RNA with a positive free energy. Each substructure is assigned an energy value, and the total energy of the secondary structure is determined by summing over its substructures [**17**, **33**].

If we further assume that secondary structure has no crossing base pairs, we could then generate secondary structure of an RNA sequence recursively over its subsequences. This allows us to compute the minimum energy structure efficiently using a dynamic programming algorithm. This approach is widely studied, and several software packages are available to perform the appropriate computations; widely-used implementations include `RNAfold` from the Vienna RNA package [**13**], `mfold` [**41**], and `RNAstructure` [**26**]. However, the optimal structure found using the thermodynamic model is often not the correct one. The prediction depends on the thousands of parameter values in the objective function, affecting the substructure prediction.

Current versions of the scoring model [**17**, **33**] have over eight thousand parameters, but we focus our attention on three related to multibranch loops such as that at the center of Figure 1. (We will discuss some details of the model and its parameters in Section 3.1.) We perform parametric analysis using geometric combinatorics, which has previously been applied to a simplified tree model of RNA base pairing that focuses on branching configurations [**14**], and extend this to study the branching of real RNA sequences under the TURNER99 NNTM parameters. In the next section we introduce the required combinatorial and geometric notions to carry out this parametric analysis.

2.2. Polytopes and geometric combinatorics. The optimization of the NNTM objective function, with an emphasis on branching loop calculations, can be presented as a simple linear programming (LP) problem, and thus well-established mathematical tools can be used to tackle the optimization problem parametrically. To analyze optimization problems with large sets of feasible solutions, such as all possible secondary structures that could form from a particular sequence of RNA, a (relatively) concise description of the possibilities is required. This can be achieved by constructing the convex polytope associated with the LP problem at hand.

The theory of convex polytopes and normal fans is well-developed, so we will not reproduce it here. Rather we present just the definitions needed for a rigorous description of the results in Section 3. More information on these applications of polytopes is given in [**4**, **24**, **31**]; readers wishing to gain a thorough understanding of the theory can see [**3**, **11**, **39**].

The theory of linear programming starts from the following question: Given a set of feasible solutions X in $\mathbb{R}^n_{\geq 0}$ and a vector $\mathbf{A} \in \mathbb{R}^n$, for which $\mathbf{x} \in X$ is the *objective function* $\mathbf{Ax}$ minimized?

For a fixed $\mathbf{A}$, this is a straightforward question; however as $\mathbf{A}$ varies, different vectors in the set X will minimize $\mathbf{Ax}$. If X is a large or infinite set it may be simpler to deal with a geometric region containing X, than with the original set of feasible solutions itself. A *convex polytope* is a bounded region in $\mathbb{R}^n$ which can be described by a finite number of linear inequalities $\mathbf{a}_i\mathbf{x} \leq c_i$. A hyperplane is a *supporting hyperplane* for a polytope P if it intersects P non-trivially but no interior point of P is on the hyperplane. A k-*face* of P is a k-dimensional intersection of P with a supporting hyperplane. A 0-face is a *vertex* of P, a 1-face is an *edge*, and so on, with the $n-1$ dimensional faces called *facets*. The (unique) n-face is defined to be the entire polytope.

The *convex hull* of a set X is the smallest convex set containing X. If X is a finite set, its convex hull P is the set

$$P = \left\{ \sum_{\mathbf{x} \in X} b_{\mathbf{x}} \mathbf{x} : \text{ each } b_{\mathbf{x}} \geq 0 \text{ and } \sum_{\mathbf{x} \in X} b_{\mathbf{x}} = 1 \right\}. \tag{1}$$

Equivalently, Eq. 1 can be used as a definition of a convex polytope—that is, a convex polytope is the convex hull of a finite set. It is a fundamental theorem in the field that these two definitions of convex polytopes are equivalent. Figure 2 shows the convex hull of a finite set in $\mathbb{R}^2$.

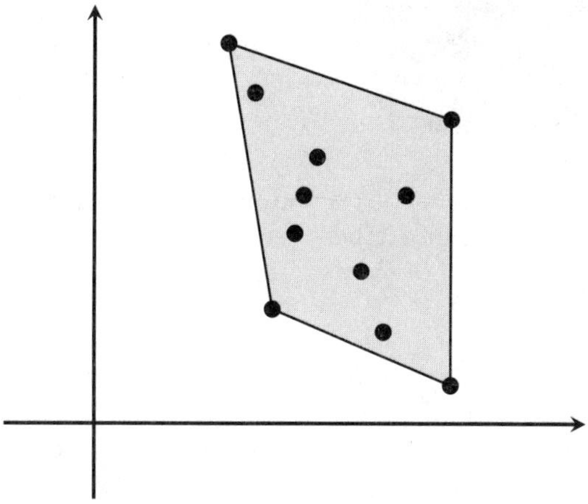

FIGURE 2. The convex hull of the set X: a polytope with four 0-faces, four 1-faces, and one 2-face.

The boundary of a polytope is the union of its facets and contains all faces except the n-face. If the polytope P is the convex hull of a set X of feasible solutions, then the boundary of P contains all optimal solutions (i.e. all vectors that minimize $\mathbf{A}\mathbf{x}$ for non-zero $\mathbf{A}$). Thus, studying the polytope rather than the original set X simplifies the optimization problem.

In practice, however, computing the convex hull of the feasible set X is typically a hard problem. We usually know too little about the set of feasible solutions to construct the convex hull directly. In computational biology, it is usually the case that the feasible set is defined implicitly by some intrinsic rules of the linear programming problem. The real challenge, then, is to extract information about the feasible set from these intrinsic rules. In this chapter, we consider two different methods for constructing the convex hull: polytope propagation and incremental convex hull. We will discuss these methods further in Section 3.2.

The dual object to the polytope is its *normal fan*. The normal fan is a collection of regions of $\mathbb{R}^n$ called *cones*. If F is a face of the polytope, the cone corresponding to F, denoted C_F, is the set of all vectors $\mathbf{B}$ such that any point $\mathbf{y} \in F$ minimizes the product $\mathbf{B}\mathbf{x}$—that is,

$$C_F = \{\mathbf{B} \in \mathbb{R}^n : \mathbf{B}\mathbf{y} \leq \mathbf{B}\mathbf{x} \text{ for all } \mathbf{x} \in P, \mathbf{y} \in F\}.$$

The normal fan is represented visually, as in Figure 3, by drawing the vectors $\mathbf{B} \in C_F$ with their heads at the face F.

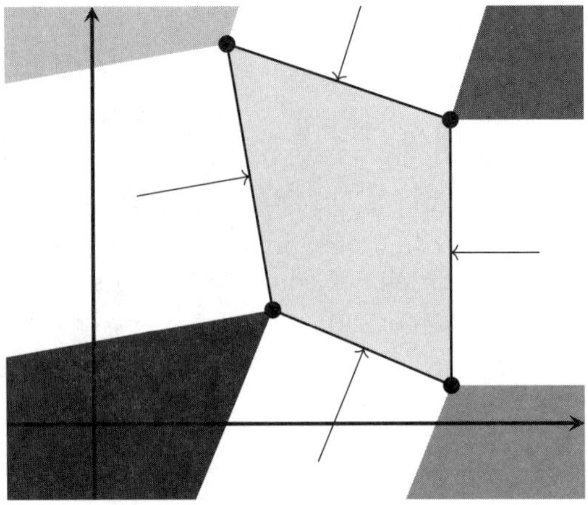

FIGURE 3. The cones corresponding to the vertices of polytope P, drawn "pointing into" their vertices. Also shown: the normal vectors to each 1-face of the polytope.

The normal fan of a polytope of dimension d is a partition of the $\mathbb{R}^d$; every d-vector lies in some cone, and the cones are preserved by positive scalar multiplication. The cones from Figure 3, for example, if drawn so that their heads are the origin, clearly fill the entire space $\mathbb{R}^2$, as shown in Figure 4.

Once we have constructed the convex hull of feasible solutions for a particular LP problem, the normal fan provides ways to discuss the stability of a parameter vector $\mathbf{A}$. If $\mathbf{A}$ is near the boundary of a cone of the normal fan, then a small change in $\mathbf{A}$ will result in a different point of the polytope optimizing the function. However if $\mathbf{A}$ is far from any boundaries of its cone, then small changes will not change which point of the polytope optimizes $\mathbf{Ax}$.

In Section 3 we use the NNTM free energy calculation as our objective function and the space of all feasible secondary structures over a given RNA sequence as the point set. Treating this objective function as a linear functional and applyling the machinery of linear programming produces what we term the *branching polytope*, which we use to analyze the parameters in the model related to multibranch loops.

3. RNA branching polytope

3.1. NNTM. In Sections 1 and 2.1, we introduced the discrete optimization paradigm for RNA secondary structure prediction. Before we consider applications of geometric combinatorics for parametric analysis of this method, we review it in greater detail.

Discrete optimization methods for RNA secondary structure prediction derive their objective function from the NNTM. If we make the standard simplifying assumption to forbid pseudoknots, we can formulate the free-energy minimization problem recursively; dynamic programming can then be used to find the minimum energy and a traceback algorithm can determine a corresponding structure.

The specific objective function used for NNTM analysis has evolved substantially over the years (cf. [**22, 34, 42**]), with a significant increase in the number

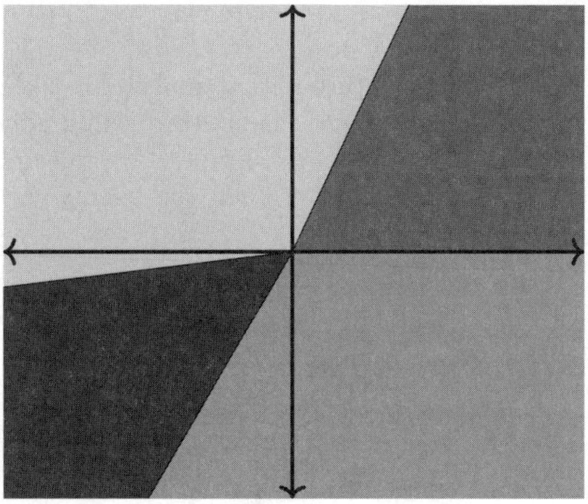

FIGURE 4. The normal fan of the polytope fills $\mathbb{R}^2$.

of parameters. The TURNER99 version considered here, obtained from the Nearest Neighbor Database [**33**], has over eight thousand parameters representing the energy contributions of various small substructures—some (~ 300) measured directly through experiments, others (~ 7000 inferred indirectly from experimental data, and finally a handful through machine learning techniques to tune the model. Among those inferred through machine learning are three connected to multibranch loops, such as the one at the center of Figure 1.

In particular, the TURNER99 version of the model we study assigns to a given multibranch loop the energy

$$\Delta G_{\text{initiation}} = a + b \cdot (\#\text{ unpaired nucleotides}) + c \cdot (\#\text{ branching helices}) \tag{2}$$

where a, b, and c are three of the learned parameters. The free energy changes associated with multibranch loops have in fact been studied experimentally [**6**, **18**]. However, these results do not map neatly onto the linear function from Eq. 2, which is a computational simplification—a logarithmic dependence on loop size would be more biophysically accurate, but the dynamic programming approach requires a recursive decomposition that would not be possible with such a function. Recent work has led to the development of new approaches to loop energy estimation [**1**, **38**], which rank structures more accurately but do not translate neatly into the discrete optimization setting.

Rather than focus on specific values of parameters, we investigate the behavior of the model as a function of the parameters, hoping to gain some insight into (for example) how sensitive the model is to each one and whether certain families of sequences exhibit increased or reduced sensitivity. If we could compute the collection of all secondary structures T for a given sequence S, we could attack this question directly. Unfortunately, this collection is far too large to compute explicitly—the total number of secondary structures on n bases is known to grow exponentially with n [**28**], and specific sequences of biologically-relevant lengths may have 10^{50} or more possible structures—so we need to trim the problem down in scope.

To do this, we mathematically reformulate the NNTM as a simple linear functional focusing on the branching loop energy calculations. For a given RNA sequence, let T be a secondary structure with x multibranch loops, y unpaired nucleotides in those multibranch loops, and z helices branching from those multibranch loops. Let $a_{99} = 3.4$, $b_{99} = 0$, and $c_{99} = 0.4$ be the values of the parameters a, b, and c in the TURNER99 assignment. Then the energy ΔG associated to the secondary structure T is given by

$$\Delta G_T = a_{99}x + b_{99}y + c_{99}z + w \tag{3}$$

for a "leftover" energy value w which can be computed using the model. We call the vector $\langle x, y, z, w\rangle$ the (branching) *energy signature* of the secondary structure T. Crucially, this energy signature does not actually depend on the TURNER99 values of a, b, and c; the numbers x, y, and z are simply the integer counts of certain substructures, while the leftover energy w represents the non-multibranch-loop components of the energy calculation including the energy of the helices that meet the multibranch loops. Treating all other energy components as a single variable w allows us to focus on the multibranch parameters exclusively while also linearizing the optimization problem.

Let a, b, c, and d be four arbitrary rationals. For a given secondary structure T on a given RNA sequence, we can compute an associated parameterized energy:

$$\Delta G_T(a, b, c, d) = ax + by + cz + dw \tag{4}$$

for the energy signature $\langle x, y, z, w\rangle$ of T.

Given a particular parameter vector $\langle a, b, c, d\rangle$ and RNA sequence S, we can find the secondary structure T which minimizes Eq. 4 using existing software; the details of this computation are discussed in Section 3.3. However, to understand how the model behaves for *all possible* values of the parameters a, b, and c, we will need to bring the powerful machinery of linear programming and polyhedral geometry to bear on the problem.

3.2. Polyhedral methods and convex hulls. Although the collection of secondary structures for a given sequence is enormous, it is nevertheless finite, and its corresponding energy signatures occupy a bounded region of $\mathbb{R}^4$. We thus shift our attention to this region—specifically, to the *convex hull* of the collection of signature vectors for the sequence which we refer to as the *branching polytope*. Unlike a classical LP problem, in which the feasible region is explicitly defined by inequalities, the collection of all branching signatures is *implicitly* defined by the NNTM and the RNA sequence. As a result, we have little information about the branching polytope. On the other hand, we have a dynamic programming algorithm that can find an energy-minimizing RNA configuration for *any particular* set of parameters. Hence it is natural to try to construct the convex hull of the signature vectors using this dynamic programming algorithm.

The first approach we consider is polytope propagation [**23**]. Polytope propagation has been used in parametric analysis of various problems, such as DNA sequence alignment and hidden Markov model for gene annotation [**24**]. This algorithm computes the convex hull by recursive convex hull and Minkowski sum computations on unions of polytopes. The recursions are related to the dynamic programming algorithm for calculating MFE structures. Namely, the dynamic programming algorithms we use involve decomposing a problem into a sum of smaller problems, each of which is a minimization problem involving only addition. The minimization operations can be translated into taking convex hulls of unions of polytopes, while the sums can be translated into polytope Minkowski sums, transporting the algorithm directly into the geometric domain.

However, while the translation from dynamic programming to polytope algebra is mathematically very elegant, polytope propagation might not be the most efficient way to compute the RNA branching polytope. The intermediate polytope computations required in this approach (convex hulls and Minkowski sum) are expensive for complicated polytopes, and these costs quickly add up. In fact, empirical results show that polytope propagation is often outperformed by incremental convex hull approaches, especially for high-dimensional models [**5**]. In addition, applying the polytope propagation approach would require re-implementing the optimization algorithm from scratch without taking advantage of the existing fast, peer-reviewed software for NNTM prediction.

As a result, we instead use a variant of the *Beneath-and-Beyond* approach. The core idea of Beneath-and-Beyond was introduced by Grünbaum in [**11**] as a method to find the facets of the convex hull of a given point set. Huggins [**15**] developed his `iB4e` algorithm, which applies this idea to specialized LP problems such as ours. The basic idea of `iB4e` method is to build the convex hull incrementally—that is, to add one vertex at a time to an already constructed convex hull, by systematically solving LP problems to either expand the hull or confirm that some part of it matches the final hull. The objective vectors are chosen so that after each iteration, either a new vertex of the convex hull is found or a facet of the convex hull is confirmed. This ensures that the method requires running the LP solver for no more than $O(V+F)$ objective vectors, if the convex hull has V vertices and F facets. The algorithm `iB4e` applies to abstract convex-hull-finding problems in any dimension, so we describe it here in full generality.

Before the main loop of the algorithm can be applied, we must perform an initialization step which finds a collection of points whose affine hull is the same as the affine hull of the branching polytope we aim to construct. (Several approaches to this step are possible; for example in $\mathbb{R}^n$ one might run the LP solver with the $2n$ basis vectors $\langle \pm 1, 0, \ldots, 0\rangle, \langle 0, \pm 1, \ldots, 0\rangle, \ldots, \langle 0, \ldots, 0, \pm 1\rangle$.) We then compute the convex hull of these points and label all facets of the convex hull as 'tentative,' as illustrated in Figure 5a.

We then begin the main loop of the algorithm. At each step, we pick a tentative facet, use its outer normal vector as objective function, and apply the LP solver with this objective. If the signature of the optimal RNA structure for these parameters is not outside the known hull, as illustrated in Figure 5b, that facet becomes 'confirmed' and the process is restarted with another tentative facet. However, if the resulting signature is outside the known hull, as illustrated in Figure 5c, we add it to the vertex set, compute the new convex hull, and label the newly added facets as tentative.

The process is repeated until all facets are confirmed. In the case of a particular RNA sequence, the set of all possible secondary structures is finite and thus set of signatures is also finite. The process must terminate and its end result is the convex hull of the set of *all* solutions to the LP solver. This is the branching polytope, and once we have computed it we can easily compute its normal fan using standard methods and proceed to study the model.

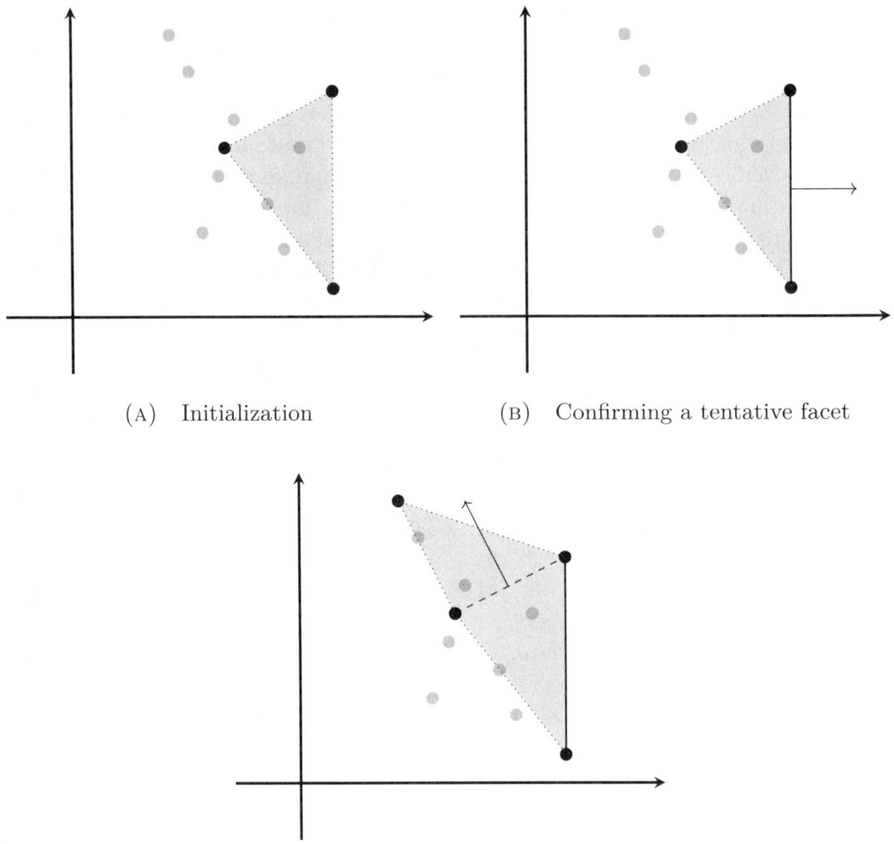

(A) Initialization (B) Confirming a tentative facet

(C) Extending a tentative facet

FIGURE 5. The Beneath-and-Beyond algorithm for polytope construction

3.3. Computing the branching polytope with pmfe. It is conceptually natural to apply the Beneath-and-Beyond algorithm to compute the branching polytope of a particular RNA sequence by treating the parametrized NNTM minimization problem as a linear program. However, coupling the two algorithms to perform the computation requires some care. We introduce the software `pmfe`, available publicly at `https://github.com/AMS-MRC-disc-math-bio/pmfe`. Instructions to build and install the software are provided in the file `README.md` of that repository. Alternatively, a ready-to-use version of the software is available in a Docker container, posted publicly on the Docker hub as `agdphd/pmfe`.

The `pmfe` package has three components (see Figure 6): `iB4e`, `findmfe`, and `scorer`.

- `iB4e` is an implementation of Huggins' algorithm as a header-only template library in C++, taking advantage of the CGAL computational geometry library [**30**] to handle the geometric computations.
- `findmfe` takes a parameter vector $\mathbf{P} = \langle a, b, c, d \rangle$ and an RNA sequence S as input and returns as output a secondary structure on S which has

minimal energy with respect to **P**. From the available software packages which implement NNTM optimization, we chose to base `findmfe` on `GTfold` [**20**] because it is parallelized and thus able to take advantage of modern multi-core desktop computers. Some modifications were required to adapt it to our use. For computational efficiency, `GTfold` performs all arithmetic with fixed precision of two decimal places; we converted it to use arbitrary-precision rational arithmetic based on the GMP library [**10**], which increases running time by an order of magnitude but ensures that the results are exact even when parameters are not multiples of 1/100. We also added an interface to modify the multibranch parameters $\langle a, b, c\rangle$ and the dummy scaling parameter d programmatically, allowing the Beneath-and-Beyond algorithm to treat `GTfold` as a function of these four variables and call it repeatedly.

- `scorer` takes an RNA sequence S and a secondary structure T and returns the signature vector $\langle x, y, z, w\rangle$ of T according to Eq. 3. (In particular, it computes the value of w by computing the energy decomposition of T under the TURNER99 parameters.)

We combine these three components with the control flow illustrated in Figure 6 to create the `pmfe` software package. Given an RNA sequence S, the `iB4e` module repeatedly generates parameter vectors according to Huggins' algorithm. Each of these vectors is sent to the `findmfe` module as a query, which generates a secondary structure with minimal energy. Each such secondary structure is sent to the `scorer` module to find its signature vector, which is then sent back to `iB4e` as the response to the query. After some number of iterations of this loop, the convex hull of these signature points s is found to be equal to the branching polytope, which is returned to the user. (For convenience, we in fact provide the user with a file containing both the signatures s and their associated structures T in condensed dot-bracket notation.)

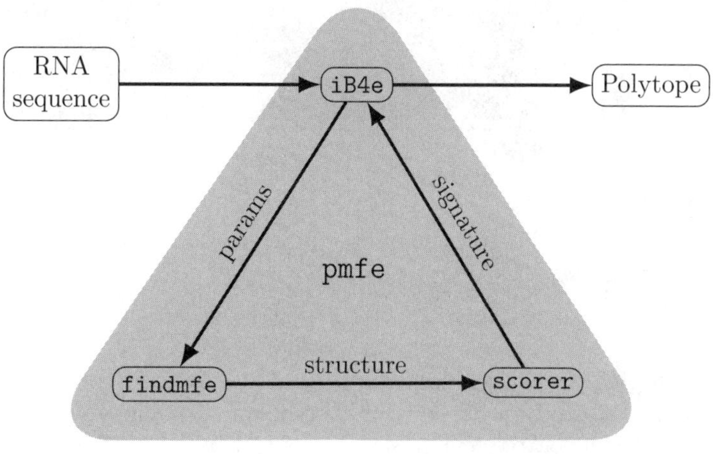

FIGURE 6. Control flow of the `pmfe` software

Since the software `pmfe` which we introduce is complex and extensively modifies the peer-reviewed codebase of `GTfold`, we provide a testing framework to ensure correctness. Tests are available to ensure that `pmfe` returns the same structures and scores as `GTfold`, using both the classical TURNER99 parameters [**17**] and various modified parameter sets for a variety of natural and synthetic RNA sequences. Details for running these tests are available in the `pmfe` repository.

We also provide a module `rna_poly` in the SageMath computer algebra system [**29**] which can be used to study the branching polytopes produced by `pmfe`. This can be used to generate a variety of visualizations of the normal fan of the branching polytope. In particular, it provides a simple interface for taking the $d = 1$ slice of the fan, giving a polyhedral partition of $\mathbb{R}^3$ whose regions are collections of parameters $\langle a, b, c\rangle$ which yield the same secondary structure for the sequence under study. The results are illustrated in Figure 7, which shows the $b = 0, d = 1$ slice of the normal fan of the branching polytope for a *H.sapiens* tRNA sequence. The TURNER99 parameter values $a_{99} = 3.4, c_{99} = 0.4$ are marked with a circle just northeast of the origin.

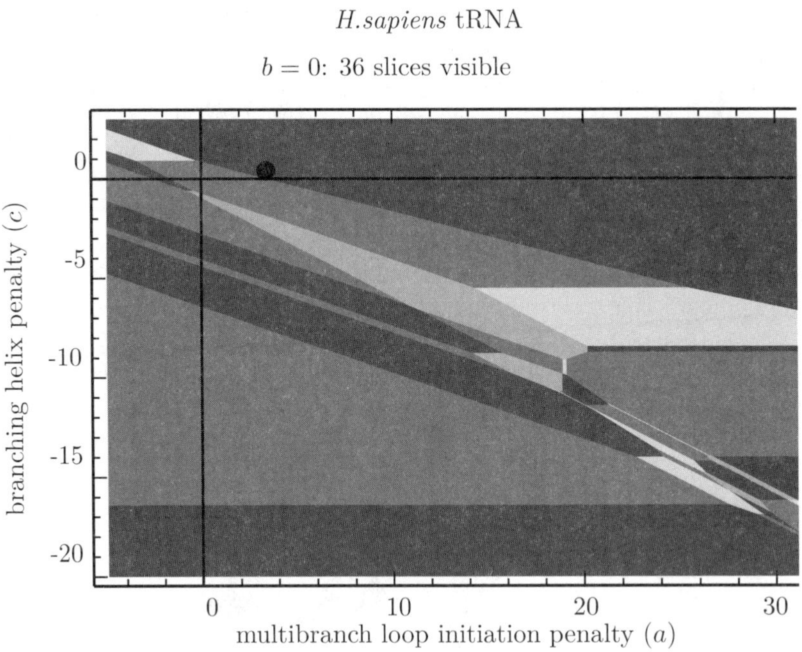

FIGURE 7. The intersection of the parameter space decomposition for a *H.sapiens* tRNA sequence with $b = 0$.

3.4. Biological questions leading to mathematical problems. The RNA branching polytope and the associated polyhedral decomposition of the three-dimensional multibranch parameter space obtained by intersecting its normal fan with $d = 1$ encapsulate the dependency of the optimization on the parameters a, b, c. Various questions related to accuracy and stability can be addressed through the analysis of the polytope and the polyhedral decomposition. However, answering such questions in a manner that is biologically relevant requires consideration of certain subtleties and this leads to interesting mathematical problems. Here we discuss a few biological questions that could be addressed through the parametric analysis and related mathematical problems that arise.

Question 1: *How robust is the optimization for a given RNA sequence?*

In the strictest sense, the prediction is sensitive if a variation of the parameters within an allowed margin of error produces different optimal secondary structures. This occurs if the distance of the TURNER99 parameters (a_{99}, b_{99}, c_{99}) from the boundary of the polyhedron that contains this point is smaller than the allowed margin of error. For example, Figure 7 shows the position of (a_{99}, b_{99}, c_{99}) within the polyhedral decomposition of a *H.sapiens* tRNA (only the $b = 0$ slice is depicted, which is the baseline value in TURNER99). Its distance to the boundary is 0.13699 and, for instance, optimization using $a = 3.39130, b = -0.14786, c = 0.37391$ yields structures with a different signature.

However, this by itself does not mean that the optimization is very sensitive; analysis of the structure space is required so that one can quantify the sensitivity. Namely, two structures with different signatures may still have a lot of structural similarities (for example, very similar long helices and branching pattern), in which case one would not necessarily say that the prediction is very sensitive. Moreover, since the structure-to-signature map is not one-to-one, when quantifying the structural differences in the optimal structures one actually needs to compare two sets of structures, each one corresponding to a given signature. In short, in the assessment of robustness we face the following problem.

Problem 1: *Define a good representative of the set of structures that correspond to the same signature.*

The problem of representing a whole set of structures by a single one has appeared before, for example in the context of compact representation of a sample of structures. However, the methods developed for this (e.g. sfold clustering [**7**] and profiling [**27**]) usually assume structural similarities of the elements in the set. Here, though, we are presented with a different set-up: the structures mapping to the same signature do not a priori have any common motifs. Therefore, choosing one consensus structure as the other methods do may not yield a reasonable representative of the whole set.

Question 2: *How much can the prediction of secondary structure be improved for a given sequence?*

For sequences for which the native structure has been obtained experimentally or via comparative sequence analysis, one could assess the accuracy of the NNTM by comparing it to the MFE structure. In fact, since the structures corresponding to all signatures on the boundary of the branching polytope can in practice be obtained fairly efficiently [**37**], one can precisely determine branching parameters that would yield a signature corresponding to a structure that is closest to the native one. However, here we again face the problem that besides the most accurate one, that signature corresponds to a whole set of MFE structures that might be very different from the native structure. In such a case, the almost accurate structure might be unrecognizable. Therefore, in assessing accuracy, the problem of finding a good representative structure for a given signature is still very relevant. Furthermore, even if we can solve this problem, the most accurate prediction obtained this way may require a triple of parameters (a, b, c) that is very different from (a_{99}, b_{99}, c_{99}), which may not be very desirable. Namely, even though (a_{99}, b_{99}, c_{99}) were not obtained experimentally, they still might have some biological relevance since the initiation point for the genetic algorithm used was suggested by stabilities determined by optical melting for an RNA multibranch loop with three branching helices [**17**]. This suggests the following natural problem.

Problem 2: *Maximize the improvement in accuracy while minimizing parameter change.*

One mathematical subproblem here is to formulate an objective function which appropriately measures the difference between two RNA secondary structures. While there are simple structure metrics that can be very efficiently computed (e.g. the base pair metric based on symmetric set difference), other metrics may be more appropriate to quantify the structural differences [**21**].

Another important subproblem is formulating an appropriate metric to measure changes in parameter values. Standard metrics such as the p-metrics can be applied, since the parameter vectors are taken from $\mathbb{R}^4$, but it is not clear which is most appropriate. Furthermore, careful consideration should be given to how each component of the parameter vector is weighted: a unit change in each coordinate may have significantly different impacts on the secondary structure. For example, since a given multibranch loop in a structure has several unpaired nucleotides, a small change (e.g. by 0.1) in b in general has bigger effects on the prediction than the same change in a.

Question 3: *Is there a choice of branching parameters that would improve the prediction accuracy for a family of sequences?*

Homologous sequences perform the same function in different species and they fold into essentially the same structure. However, their MFE prediction accuracies can vary significantly. For instance, within the tRNA family, the accuracy calculated as the F-measure (the harmonic mean of the MFE sensitivity and positive predictive value against true positive base pairs) ranges from the minimal 0 to the maximal 1 [**27**]. Hence, it is natural to try to find a set of parameters which yields the highest accuracy for a family of sequences. For this, one needs to address the question: *Are there any patterns in the dependency of the prediction on the parameters across a family of homologous sequences?* Mathematically, one could phrase this as a problem of finding similarities between polytopes in $\mathbb{R}^4$ or between polyhedral decompositions of $\mathbb{R}^3$. For example, Figure 8 shows the $b = 0$ slice of the polyhedral decompositions for a tRNA sequence of *L.delbrueckii*. By visually comparing this figure to Figure 7 one can notice similarities in the directions of the splitting hyperplanes. The challenge of course is to quantify such similarities. So, we face the following problem.

Problem 3: *Find a way to compare branching polytopes.*

Comparison of polytopes is a problem that has appeared before, for instance in relation to the traveling salesman problem in optimization [**9**]. Even in optimization problems when one works with relaxation polytopes that are, loosely speaking, "close", comparing relaxations is not well understood and different comparison methods have been suggested (e.g. [**16**]). In our case, an additional challenge is the fact that the polytopes to be compared do not a priori satisfy any mathematically tractable assumptions for closeness.

4. Conclusions

The affine energy function which governs the branching of an RNA secondary structure under the NNTM optimization is a known weakness of the thermodynamic

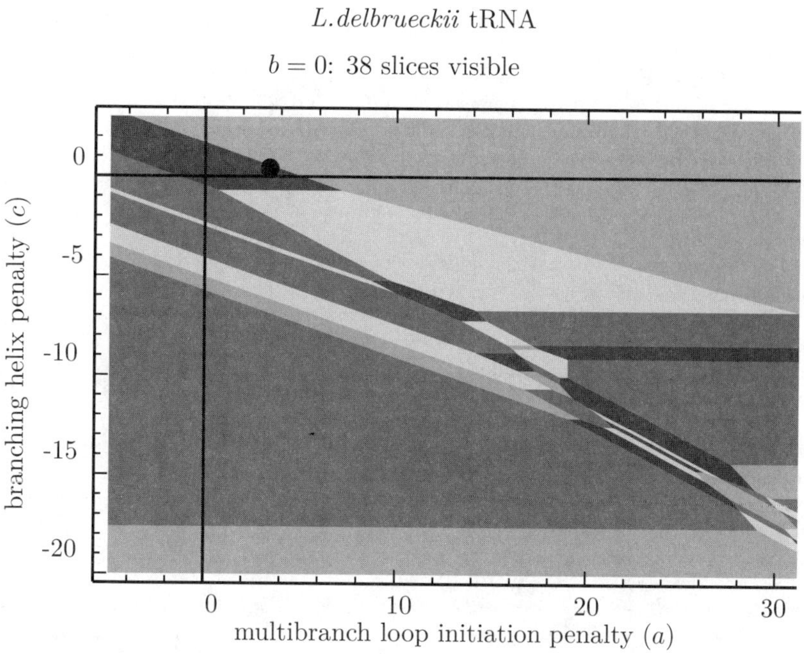

FIGURE 8. The $b = 0$ slice of the parameter space decomposition for a *L.delbrueckii* tRNA sequence.

model, and the methods from geometric combinatorics outlined here offer significant potential to assess its accuracy and stability through a parametric analysis. While simplified models are certainly more tractable, the new `pmfe` computational framework makes possible the analysis of branching polytopes for real RNA sequences for the first time. As illustrated by the qualitative similarities between Figures 7 and 8, this approach is capturing some interesting characteristics which are preserved between two different sequences. Moreover, as discussed, moving beyond the qualitative comparison of polytope slices to quantify their similarities and differences presents some interesting mathematical challenges. Consequently, it remains to be seen whether the observed patterns in these proof-of-principle results are resulting from the structure of the thermodynamic optimization and/or the coding of structural motifs in the base pairing of these RNA sequences. Resolving this dichotomy will reveal much about the interplay between mathematics and biology in the prediction of RNA secondary structures.

References

[1] Daniel P Aalberts and Nagarajan Nandagopal, *A two-length-scale polymer theory for RNA loop free energies and helix stacking*, RNA **16** (2010), no. 7, 1350–1355.

[2] Andrew D Bates and Anthony Maxwell, *DNA topology*, Oxford University Press, Oxford, UK, 2005.

[3] Dimitris Bertsimas and John N Tsitsiklis, *Introduction to linear optimization*, Athena Scientific Series in Optimization and Neural Computation, Athena Scientific, 1997.

[4] Jesús De Loera, Raymond Hemmecke, and Matthias Köeppe, *Algebraic and geometric ideas in the theory of discrete optimization*, MOS-SIAM Series on Optimization, Society for Industrial and Applied Mathematics, Philadelphia, PA, USA, 2012.

[5] Colin N Dewey, Peter M Huggins, Kevin Woods, Bernd Sturmfels, and Lior Pachter, *Parametric alignment of Drosophila genomes*, PLoS Comput Biol **2** (2006), no. 6, e73.

[6] Joshua M Diamond, Douglas H Turner, and David H Mathews, *Thermodynamics of three-way multibranch loops in RNA*, Biochemistry **40** (2001), no. 23, 6971–6981.

[7] Ye Ding, Chi Yu Chan, and Charles E Lawrence, *RNA secondary structure prediction by centroids in a Boltzmann weighted ensemble*, RNA **11** (2005), no. 8, 1157–1166.

[8] Richard Durbin, Sean R Eddy, Anders Krogh, and Graeme Mitchison, *Biological sequence analysis: Probabilistic models of proteins and nucleic acids*, Cambridge University Press, 1998.

[9] Michel X Goemans, *Worst-case comparison of valid inequalities for the TSP*, Mathematical Programming **69** (1995), no. 1-3, 335–349.

[10] Torbjörn Granlund and the GMP development team, *GNU MP: The GNU Multiple Precision Arithmetic Library*, 6.0.0a, 2014. `http://gmplib.org/`.

[11] Branko Grünbaum, *Convex polytopes*, Graduate Texts in Mathematics, vol. 221, Springer-Verlag, New York, 2003.

[12] Dan Gusfield, *Algorithms on strings, trees and sequences: computer science and computational biology*, Cambridge University Press, 1997.

[13] Ivo L Hofacker, Walter Fontana, Peter F Stadler, L Sebastian Bonhoeffer, Manfred Tacker, and Peter Schuster, *Fast folding and comparison of RNA secondary structures*, Monatshefte für Chemie/Chemical Monthly **125** (1994), no. 2, 167–188.

[14] Valerie Hower and Christine E Heitsch, *Parametric analysis of RNA branching configurations*, Bulletin of Mathematical Biology **73** (2011), no. 4, 754–776.

[15] Peter Huggins, *iB4e: A software framework for parametrizing specialized LP problems*, Mathematical Software–ICMS 2006, 2006, pp. 245–247.

[16] Jon Lee and Walter D Morris, *Geometric comparison of combinatorial polytopes*, Discrete Applied Mathematics **55** (1994), no. 2, 163–182.

[17] David H Mathews, Jeffrey Sabina, Michael Zuker, and Douglas H Turner, *Expanded sequence dependence of thermodynamic parameters improves prediction of RNA secondary structure*, Journal of Molecular Biology **288** (1999), no. 5, 911–940.

[18] David H Mathews and Douglas H Turner, *Experimentally derived nearest-neighbor parameters for the stability of RNA three- and four-way multibranch loops*, Biochemistry **41** (2002), no. 3, 869–80.

[19] David H Mathews and Douglas H Turner, *Prediction of RNA secondary structure by free energy minimization*, Current Opinion in Structural Biology **16** (2006), no. 3, 270–278.

[20] Amrita Mathuriya, David A Bader, Christine E Heitsch, and Stephen C Harvey, *GTfold: a scalable multicore code for RNA secondary structure prediction*, Proceedings of the 2009 ACM Symposium on Applied Computing, 2009, pp. 981–988.

[21] Vincent Moulton, Michael Zuker, Michael Steel, Robin Pointon, and David Penny, *Metrics on RNA secondary structures*, Journal of Computational Biology **7** (2000), no. 1-2, 277–292.

[22] Ruth Nussinov, George Pieczenik, Jerrold R Griggs, and Daniel J Kleitman, *Algorithms for loop matchings*, SIAM Journal on Applied mathematics **35** (1978), no. 1, 68–82.

[23] Lior Pachter and Bernd Sturmfels, *Parametric inference for biological sequence analysis*, Proceedings of the National Academy of Sciences of the United States of America **101** (2004), no. 46, 16138–16143.

[24] Lior Pachter and Bernd Sturmfels, *Algebraic statistics for computational biology*, Vol. 13, Cambridge University Press, 2005.

[25] Pavel Pevzner, *Computational molecular biology: an algorithmic approach*, MIT Press, 2000.

[26] Jessica S Reuter and David H Mathews, *RNAstructure: software for RNA secondary structure prediction and analysis*, BMC Bioinformatics **11** (2010), no. 1, 129.

[27] Emily Rogers and Christine E Heitsch, *Profiling small RNA reveals multimodal substructural signals in a Boltzmann ensemble*, Nucleic Acids Research (2014), gku959v1.

[28] PR Stein and MS Waterman, *On some new sequences generalizing the Catalan and Motzkin numbers*, Discrete Mathematics **26** (1979), no. 3, 261–272.

[29] W. A. Stein et al., *Sage Mathematics Software (Version 6.7)*, The Sage Development Team, 2015. `http://www.sagemath.org`.

[30] The CGAL Project, *CGAL user and reference manual*, 4.6, CGAL Editorial Board, 2015.

[31] Rekha R Thomas, *Lectures in geometric combinatorics*, Vol. 33, American Mathematical Society, 2006.

[32] Ignacio Tinoco and Carlos Bustamante, *How RNA folds*, Journal of Molecular Biology **293** (1999), no. 2, 271–281.

[33] Douglas H Turner and David H Mathews, *NNDB: the nearest neighbor parameter database for predicting stability of nucleic acid secondary structure*, Nucleic Acids Research (2009), gkp892.

[34] Michael Waterman, *Secondary structure of single-stranded nucleic acids*, Studies on foundations and combinatorics, 1978, pp. 167–212.

[35] Michael S Waterman, *Parametric and ensemble sequence alignment algorithms*, Bulletin of Mathematical Biology **56** (1994), no. 4, 743–767.

[36] Michael S Waterman, *Introduction to computational biology: Maps, sequences and genomes*, CRC Press, 1995.

[37] Stefan Wuchty, Walter Fontana, Ivo L Hofacker, Peter Schuster, et al., *Complete suboptimal folding of RNA and the stability of secondary structures*, Biopolymers **49** (1999), no. 2, 145–165.

[38] Jian Zhang, Ming Lin, Rong Chen, Wei Wang, and Jie Liang, *Discrete state model and accurate estimation of loop entropy of RNA secondary structures*, The Journal of Chemical Physics **128** (2008), no. 12, 125107.

[39] Günter M Ziegler, *Lectures on polytopes*, Graduate Texts in Mathematics, Springer-Verlag, 1995.

[40] Michael Zuker, *RNA folding prediction: The continued need for interaction between biologists and mathematicians*, Some mathematical questions in biology: DNA sequence analysis, 1986, pp. 87–124.

[41] Michael Zuker, *On finding all suboptimal foldings of an RNA molecule*, Science **244** (1989), no. 4900, 48–52.

[42] Michael Zuker and Patrick Stiegler, *Optimal computer folding of large RNA sequences using thermodynamics and auxiliary information*, Nucleic Acids Research **9** (1981), no. 1, 133–148.

Department of Mathematics, University of North Texas, Denton, Texas 76203
E-mail address: `elizabeth.drellich@unt.edu`

Department of Mathematics and Computer Science, Hobart and William Smith Colleges, Geneva, New York 14456
Current address: Center for Quantitative Medicine, UConn Health, Farmington, Connecticut 06051
E-mail address: `andrew.gainer.dewar@gmail.com`

Mathematical Institute, University of Oxford, Andrew Wiles Building, Radcliffe Observatory Quarter, Woodstock Road, Oxford OX2 6GG United Kingdom
E-mail address: `harrington@maths.ox.ac.uk`

Department of Mathematical Sciences, Clemson University, Clemson, South Carolina 29634
E-mail address: `qhe@clemson.edu`

School of Mathematics, Georgia Institute of Technology, Atlanta, Georgia 30332
E-mail address: `heitsch@math.gatech.edu`

Department of Mathematical Sciences, Clemson University, Clemson, South Carolina 29634
E-mail address: `spoznan@clemson.edu`

Contemporary Mathematics
Volume **685**, 2017
http://dx.doi.org/10.1090/conm/685/13751

Polyhedral approaches to learning Bayesian networks

David Haws, James Cussens, and Milan Studený

ABSTRACT. Learning Bayesian network structure is the NP-hard task of finding a directed acyclic graph that best fits real data. Two integer vector encodings exist – family variable and characteristic imset – which model the solution space of BN structure. Each encoding yields a polytope, the family-variable and characteristic imset polytopes respectively. It has been shown that learning BN structure using a decomposable and score equivalent scoring criteria (such as BIC) is equivalent to optimizing a linear function over either the family-variable or characteristic imset polytope. This monograph is primarily intended for readers already familiar with BN but not familiar with polyhedral approaches to learning BN. Thus, this monograph focuses on the family-variable and characteristic imset polytopes, their known faces and facets, and more importantly, deep connections between their faces and facets. Specifically that many of the faces of the family variable polytope are superfluous when learning BN structure. Sufficient background on Bayesian networks, graphs, and polytopes are provided. The currently known faces and facets of each polytope are described. Deep connections between many of the faces and facets of family-variable and characteristic polytope are then summarized from recent results. Lastly, a brief history and background on practical approaches to learning BN structure using integer linear programming over both polytopes is provided.

1. Introduction

Bayesian networks (BNs) are graphical models used extensively in probabilistic reasoning [**44**], machine learning, and statistics [**32**]. They are modeled by *directed acyclic graphs* (DAGs) where nodes of the graph correspond to random variables and the structure of the graph determines conditional independences of the underlying probabilistic model. Learning *Bayesian network structure* is the task of finding the structure that best fits statistically according to a given database [**40**]. That is, given a database of observations, one attempts to find the optimal structure (graph). BN structure learning is both immensely useful while at the same time very difficult and known the be NP-hard [**7**].

Learning BN structure has evolved into roughly three approaches – with heavy cross pollination: conditional independence test-based [**22, 49, 65**], Markov chain Monte Carlo [**24, 33, 37, 38**], and optimization/score based [**13, 19, 27, 30**]. The GES algorithm by Chickering and Meek is the most popular and widely used

2010 *Mathematics Subject Classification.* Primary 52B12, 90C27, 68Q32.

Research of Milan Studený has been supported from the project GAČR n. 1320012S.

heuristic optimization approach [**10**]. The classic GES algorithm however is not guaranteed to find the global optimal solution, and thus researchers developed alternative approaches, for example a dynamic programming methodology [**48**]. A branch-and-bound (not in the integer programming sense) approach has also been developed which has the benefit that stopping at any time returns a best known solution [**21**].

However, as a trade-off, dynamic programming approaches only work on smaller models due to exponential demands on memory. Smaller models here means fewer number of random variables. As for score based approaches, learning Bayesian network structure entails maximizing a *scoring* criteria which is a real function of the BN structure (graph) and of a database [**40**]. That is, given a database of observations, one attempts to find the optimal structure (graph) according to a scoring criteria.

Two DAGs may in fact define the same statistical model, and are thus called *Markov equivalent*. The scoring criteria therefore should be *score equivalent*, i.e. score Markov equivalent graphs equally. It is also critical that the scoring criteria be *decomposable*, that is, a sum of scores of local properties of the graph [**9**]. The scoring criteria will be described in more detail, but it suffice to say they are reasonable, and familiar. For example the well-known BIC criteria [**46**] is a Markov equivalent and decomposable scoring criteria.

Polyhedral approaches are the central focus here and are a score based optimization approach which use integer or linear programming to learn BN structure. They too aim to find exact solutions, however typically at the price of only working on smaller models. Polyhedral approaches to learning BN structure rely on integer vector representations of either DAGs or Markov equivalence classes. Another closely related zero-one vector encoding of conditional independence is given by [**42**]. In practice, approaches have even blended the representations. The key insight is that score equivalent and decomposable scoring criteria (e.g. BIC) are in fact an affine function on these representations. Thus, the main idea of the polyhedral approach is to encode DAGs or Markov equivalence classes and then maximize the affine scoring criteria. This is equivalent to solving an integer linear program over the integer points in a polytope defined by the convex hull of the integer vector representatives of all DAGs or all Markov equivalence classes. Thinking geometrically, the encoding defines a convex body while the scoring criteria along with the database of observations defines a direction. The goal is to find the integer point in the convex body furthest along the direction defined by the scoring criteria and database.

Two integer vector encodings dominate polyhedral approaches to learning BN: *characteristic imsets* (cimsets) and *family variables*. The former (characteristic imsets) encode Markov equivalence classes, while the latter (family variables) encode DAGs directly. The precursor to characteristic imsets, was first introduced in [**52**] in the form of standard imsets. Therein Studený first proposed to use such vector representations in a polyhedral approach to learn BN. The actual introduction of characteristic imsets was given in [**26, 60**] where a one-to-one correspondence with standard imsets was provided. The later encoding using family variables first appeared in [**14, 16, 28**] where they were used specifically for polyhedral approaches to learning BN structure. Both approaches have seen success, although recent results reveal that family variable approaches seem to have better practical success. Of critical importance here is that strong theoretical connections have recently been shown between these two important polytopes [**17**]. Specifically it has been shown through these theoretical connections that many of the faces of the family variable polytope are superfluous when learning BN structure.

The focus of this monograph is the overview of polyhedral approaches to learning BN structure and the deep connections between the various polytopes and methodologies. Specifically the connection between the polytope defined by cim-sets and the polytope defined by family variables. Note that [**59**] covered some of this material, focusing primarily on the characteristic imset polytope and some techniques for solving IP programs for special graphs. This older chapter also lacked many of the currently known facets of both the family variable and characteristic imset polytope. Additionally, herein other related polytopes will be described as well as linear relaxations used in practical approaches to learning BN structure. Sufficient background on BN, characteristic imsets, polytopes, and other miscellaneous required ideas will be provided. Scoring functions as well as primary motivations for why performing BN structure learning is useful are not covered here as others have explained these issues in detail, for example see [**29**].

2. Background

This section is devoted to the required background material, such as Bayesian networks, graphs, and polyhedral theory. The notation and form of the following borrows heavily from the authors previous works [**14, 16, 17, 52, 54, 55, 57–60**].

2.1. Graphs. Graphs herein have N as the set of nodes and are *hybrid* in the sense that two types of edges between (distinct) nodes $i, j \in N$ are allowed, namely directed edges, called *arrows*, and denoted like $i \to j$ or $j \leftarrow i$, and undirected edges, called *lines*, and denoted like $i — j$ or $j — i$. Multiple edges are not allowed between two nodes. If there is an edge between nodes i and j, we say they are *adjacent*. A graph is *undirected* if it only has lines; it is *directed* if it only has arrows.

A *cycle* of length $m \geq 3$ in H is a sequence $\rho : i_0, i_1, \ldots, i_m = i_0$, where $i_1, \ldots, i_m$ are distinct nodes, and, for each $r = 0, \ldots, m-1$, (i_r, i_{r+1}) is an edge in H. The cycle ρ is *chordless*, or minimal, if there is no other edge in H between nodes in $\{i_1, \ldots, i_m\}$, where $i_m = i_0$, besides those which form the cycle ρ. An undirected graph is *chordal* if it has no chordless cycle of length $m \geq 4$.

The cycle ρ is *directed* if $i_r \to i_{r+1}$ in H for each $r = 0, \ldots, m-1$. A directed graph is *acyclic* if it has no directed cycle. The set of directed acyclic graphs on nodes N will be denoted by $\mathsf{DAGS}\,(N)$. An equivalent definition of an acyclic directed graph G is that there exists an ordering $b_1, \ldots, b_{|N|}$ of all nodes of G which is consistent with the direction of arrows: $b_i \to b_j$ in G implies $i < j$. The set of *parents* of a node $i \in N$ in a (directed) graph G is the set $pa_G(i) := \{j \in N : j \to i \text{ in } G\}$.

The cycle ρ is *semi-directed* if $i_0 \to i_1$ in H and, for each $r = 1, \ldots, m-1$ one has either $i_r \to i_{r+1}$ in H or $i_r — i_{r+1}$ in H. We say that a hybrid graph H is *3-acyclic* if it has no semi-directed cycle of length 3.

A set of nodes $C \subseteq N$ in a hybrid graph H is *connected* if, for every pair $i, j \in C$, there exists an undirected path between i and j, that is, a sequence of (distinct) nodes $i = i_1, \ldots, i_s = j$, $s \geq 1$ such that $i_r — i_{r+1}$ in H for $r = 1, \ldots, s-1$.

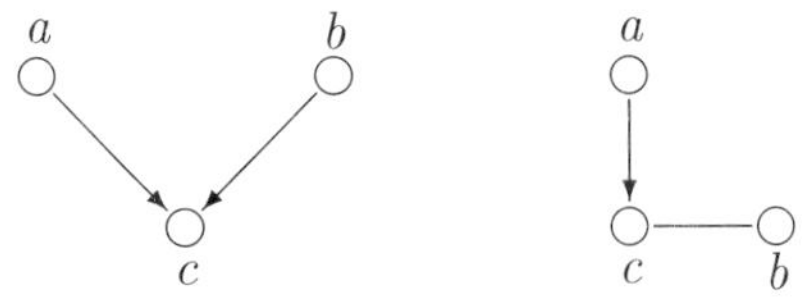

FIGURE 1. The immorality $a \to c \leftarrow b$ and the flag $a \to c — b$.

The maximal connected sets (with respect to inclusion) are called the *connected components* of H.

A hybrid graph H is a *chain graph* if (all) its components can be ordered into a sequence $(C_1, \ldots, C_m)$, $m \geq 1$, called a *chain*, such that if $i \to j$ in H, then $i \in C_r$ and $j \in C_s$ with $r < s$. An equivalent definition of a chain graph is that it is a hybrid graph which has no semi-directed cycle, or alternatively, which has no semi-directed chordless cycle; see [**51**, Lemma 2.1]. Thus, chain graphs are, in fact, acyclic hybrid graphs, they involve acyclic directed graphs. Also, clearly, every chain graph is a 3-acyclic hybrid graph.

An *immorality* in H is an induced subgraph (over three nodes) of the form $i \to k \leftarrow j$, that is, i and j are not adjacent in H. A *flag* in a hybrid graph H is an induced subgraph (over three nodes) of the form $i \to j — k$, which means that i and k are not adjacent. See Figure 1.

2.2. Bayesian Networks. Let N be a finite non-empty set of *variables* where the trivial case is avoided by assuming $n := |N| \geq 2$. In the statistical context the elements of N correspond to random variables under consideration where in the graphical context they correspond to nodes.

Each variable $i \in N$ is assigned a finite (individual) *sample space* X_i (of possible values). To avoid technical problems assume $|\mathsf{X}_i| \geq 2$ for each $i \in N$. The *joint sample space* is the Cartesian product $\mathsf{X}_N : \prod_{i \in N} \mathsf{X}_i$. For $A \subseteq N$, the *projection* of an element $x = [x_i]_{i \in N}$ of X_N to A is the configuration (list) of values $x_A = [x_i]_{i \in A}$. The symbol X_A will denote the respective *projection space*. Thus, for $A = \emptyset$, the projection space $\mathsf{X}_\emptyset$ is a singleton, consisting of the empty list; for $\emptyset \neq A$ one has $\mathsf{X}_A = \prod_{i \in A} \mathsf{X}_i$.

A *Bayesian network* (BN) is a pair (G, P), where G is an acyclic directed graph having N as the set of nodes and P a *Markovian* probability distribution (with respect to G) on the *joint sample space* X_N. This means that P satisfies conditional independence restrictions determined by G; see [**32**, § 3.2.2] for details. The *BN structure* defined by an acyclic directed graph G is the class of Markovian probability distributions with respect to G on (fixed) $\mathsf{X}_N = \prod_{i \in N} \mathsf{X}_i$.

2.3. Markov Equivalence. It is important to note that different graphs with the same set of nodes can be *Markov equivalent*, which means they define the same BN structure. The classic graphical characterization of equivalent graphs by Verma and Pearl [**64**] states that two graphs are Markov equivalent if and only if they have the same adjacencies and immoralities. Every Markov equivalence class can be represented by the graph below.

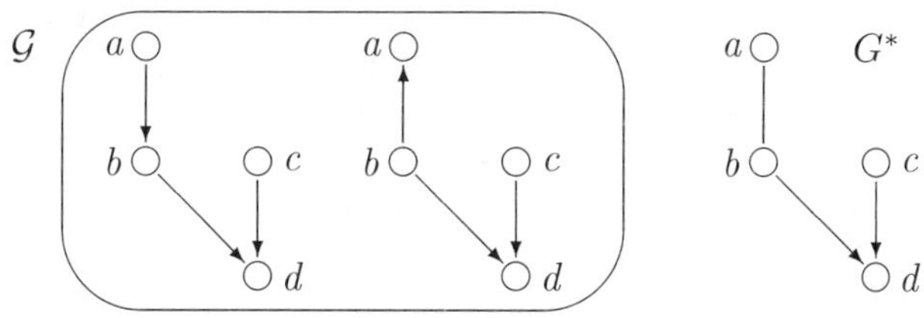

FIGURE 2. An equivalence class $\mathcal{G}$ of acyclic directed graphs and the essential graph G^* of $\mathcal{G}$.

DEFINITION 2.1. Let $\mathcal{G}$ be a Markov equivalence class of acyclic directed graphs over N. The *essential graph* G^* of $\mathcal{G}$ is a hybrid graph over N such that:

- $i \to j$ in G^* if $i \to j$ in every G from $\mathcal{G}$,
- $i - j$ in G^* if $\exists G_1, G_2 \in \mathcal{G}$ with $i \to j$ in G_1 and $i \leftarrow j$ in G_2.

See Figure 2 for example. The symbol $G \sim H$ for $G, H \in \mathsf{DAGS}(N)$ will mean the graphs G and H are Markov equivalent. The well-known transformational characterization of Markov equivalence [**8**, Theorem 2] says $G \sim H$ if and only if there exists a sequence $G = G_1, \ldots, G_m = H$, $m \geq 1$ in $\mathsf{DAGS}(N)$ such that, for $i = 1, \ldots, m-1$, the graph G_{i+1} is obtained from G_i by *covered arc reversal*: G_i has an arrow $a \to b$ with $pa_{G_i}(b) = \{a\} \cup pa_{G_i}(a)$ and G_{i+1} is obtained from G_i by replacing $a \to b$ in G_i by $b \to a$ in G_{i+1}; the remaining arrows are unchanged.

2.4. Polytopes, LP, & IP. For full details and background on polyhedral theory see [**5, 45, 66, 67**]. The *convex hull* of $\{\mathbf{a}_1, \ldots, \mathbf{a}_m\} \subset \mathbb{R}^n$ is

$$\mathrm{conv}(\mathbf{a}_1, \ldots, \mathbf{a}_m) := \left\{ \mathbf{x} \in \mathbb{R}^n : \mathbf{x} = \sum_{i=1}^{m} \lambda_i \mathbf{a}_i,\ \sum_{i=1}^{m} \lambda_i = 1,\ \lambda_i \geq 0 \right\}.$$

A *vertex* or *extreme point* of a polytope P is any point $z \in P$ that can not be written as a linear combination of two non-equal points in P. The convex hull of finitely many points defines a *polytope* P, called a *vertex representation.* The dimension $dim(P)$ of a polytope $P \subsetneq \mathbb{R}^n$ is the dimension of the smallest affine subspace containing P. Or equivalently one minus the number of affinely independent vertices of P. By convention a singleton has dimension 0 and the empty set has dimension -1. One says $F \subseteq P$ is a *face* of the polytope P if there exists a vector $\mathbf{c}$ such that $F = \arg\max_{\mathbf{x} \in P} \mathbf{c} \cdot \mathbf{x}$. For convenience, the empty set $\emptyset$ is the *empty face* of P. Every face F of P is also a polytope. The dimension of a face is one less than the number of affinely independent vectors in the face. If the dimension of P is d, a face is a *facet* if it is of dimension $d-1$. The faces of a polytope defined a poset given by inclusion of faces, where the empty set is the minimal face and the polytope is the maximal face. Thus, a facet can be defined as a sub-maximal face with respect to inclusion. A point $\mathbf{x} \in P$ is a *vertex* if and only if it can not be written as a convex combination of points from $P \backslash \{\mathbf{x}\}$. Vertices are the 0-dimensional faces of P.

Additionally a polytope can be defined by the intersection of half-spaces: for $\{\mathbf{a}_1, \ldots, \mathbf{a}_l\} \subseteq \mathbb{R}^n$ and $\{b_1, \ldots, b_l\} \subseteq \mathbb{R}$

$$P := \{\mathbf{x} : \mathbf{a}_i \cdot \mathbf{x} \leq b_i,\ 1 \leq i \leq l\} \subsetneq \mathbb{R}^n,$$

or in matrix notation simply $P = \{\mathbf{x} \in \mathbb{R}^n : A\mathbf{x} \leq \mathbf{b}\}$, where A is given by the rows $\mathbf{a}_i$ and $\mathbf{b}$ is composed of b_i. Such a polytope is said to be in *hyper-plane* represented form. Note, here we only considered bounded polyhedra, i.e. polytopes not containing infinite rays. A vector $\mathbf{u}$ and real-value b is a *valid inequality* for a polytope P if $\langle \mathbf{x}, \mathbf{u} \rangle \leq b$ for all $x \in P$. Geometrically, this means the polytope P lies entirely in the half-space given by $\mathbf{u}$ and b. A valid inequality always defines a face F of P (possibly the empty face): $F := \{\mathbf{x} \in P : \langle \mathbf{x}, \mathbf{u} \rangle \leq b\}$.

It is possible to transform from vertex to hyper-plane representations and vice-versa, but at immense computational cost, and in many cases it is not feasible. As it will be demonstrated below, one can trivially define a polytope in its vertex representation, but the difficult challenge remains then to provably find its hyper-plane representation.

Given a polytope P and an *objective* $\mathbf{c} \in \mathbb{R}^n$, *linear programming* (LP) is understood as finding $\mathbf{x}^* \in \arg\max\{\mathbf{x} \cdot \mathbf{c} : \mathbf{x} \in P\}$. Linear programming is known to be polynomial time solvable as well as fast in practice mainly due to the Simplex method as well as a long history of study and improvements.

Integer programming (IP) is the task of finding $\mathbf{z}^* \in \arg\max\{\mathbf{z} \cdot \mathbf{c} : \mathbf{z} \in P,\ \mathbf{z} \in \mathbb{Z}^n\}$. Here the distinction is that one must find an optimal integer point in the polytope P. Integer programming is vastly more difficult than LP, and is known to be NP-hard. If it happens that P is *integral*, i.e. $\mathrm{conv}\,\{x \in P : x \in \mathbb{Z}^n\} = P$, then IP over P can be done using LP. However, this is rarely the case and in practice, one often relies on branch-and-bound techniques coupled with cutting planes [**3, 31**]. In branch-and-bound, the solution space is split conditioned on one variable, and bounds are estimated using LP sub-problems along with cutting planes. Entire branches can be eliminated during branch-and-bound, thus narrowing in on an integer solution. Cutting planes is the technique of starting from an LP relaxation (see below) of the IP problem, then progressively adding additional linear constraints to the problem to 'cut' off non-integer solutions in the hopes of narrowing down to an integer solution. Looking for cutting planes is still worthwhile even if one has no hope of eventually ending up with an integer solution, since adding cutting planes improves the dual bound which can speed up branch-and-bound due to additional pruning of branches.

In order to solve an IP over a polytope in vertex representation, one does not necessarily need the hyper-plane representation. A *linear relaxation* of P is a polytope $P' \supseteq P$ such that $P \cap \mathbb{Z}^n = P' \cap \mathbb{Z}^n$. That is, P' contains P and they share the exact same integer points. Thus, solving an IP over P' is equivalent to solving an IP over P.

Both LP and IP require a polytope to have a hyper-plane representation. Many LP and IP solvers exist, some open source and some commercial. The most dominant and long standing solver is ILOG CPLEX [**1**] which performs both LP and IP. Another popular solver is SCIP [**2**].

A polytope P is *rational* if its vertices are rational which is equivalent to the hyper-plane representation of P having rational form: the coefficients of all inequalities and right hand sides are rational. The *cone* of $\{\mathbf{a}_1, \ldots, \mathbf{a}_m\} \subset \mathbb{R}^n$ is

$$\mathrm{cone}(\mathbf{a}_1, \ldots, \mathbf{a}_m) := \left\{ \mathbf{x} \in \mathbb{R}^n : \mathbf{x} = \sum_{i=1}^{m} \lambda_i \mathbf{a}_i,\ \lambda_i \geq 0 \right\}.$$

3. BN Representations & BN Polytopes

Multiple integer vector representations of BN are defined here along with their accompanying vector-represented polytope.

3.1. Family Variable. Let G be a directed acyclic graph, i.e. $G \in \mathsf{DAGS}\,(N)$. A node a together with its parent set $\eta_G(a)$ will be called a *family*. Note that any directed graph is determined by its $n := |N|$ families, i.e. one needs only to define the parents of each node to fully describe any graph. The index set of family-variable vectors will be

$$\Upsilon := \{\, (a\,|\,B) \,:\, a \in N \wedge \emptyset \neq B \subseteq N \setminus \{a\}\,\}\,.$$

Note that families with empty parent sets are not included.

The symbol η_G, called the *DAG-code* of G, will be used to denote the family-variable vector encoding of a directed graph G:

$$\eta_G(a\,|\,B) = \begin{cases} 1 & \text{if } B = pa_G(a), \\ 0 & \text{otherwise,} \end{cases} \qquad \text{for } (a\,|\,B) \in \Upsilon.$$

Given two vectors $v, w \in \mathbb{R}^{\Gamma}$, where Γ is a non-empty finite index set, say $\Gamma = \Upsilon$, their scalar product will be denoted by $\langle v, w\rangle_{\Gamma}$, or just by $\langle v, w\rangle$ if there is no danger of confusion.

The *family-variable polytope* can be defined as the convex hull of the family-variable encodings of all possible $\mathsf{DAGS}\,(N)$:

$$\mathsf{F} := \mathrm{conv}\,(\{\, \eta_G \in \mathbb{R}^{\Upsilon} \,:\, G \in \mathsf{DAGS}\,(N)\,\})\,.$$

A general property of zero-one polytopes is that the only integer points in the family-variable polytope F are the codes (η_G) for directed acyclic graphs. It is not difficult to show that F contains precisely $n(2^{n-1}-1)$ affinely independent vectors, thus the dimension of F is $n(2^{n-1}-1)$. Moreover, none of the DAG-codes is a non-trivial convex combination of the others. In particular, the set of vertices (= extreme points) of F is just the set of DAG-codes. Lastly, since the vertices of F correspond to the graphs in $\mathsf{DAGS}\,(N)$, every face F of F can be identified with a subset of $\mathsf{DAGS}\,(N)$. That is, $S \subseteq \mathsf{DAGS}\,(N)$ is a face of F if $\mathrm{conv}\,\{\eta_G \in \mathbb{R}^{\Upsilon} \,|\, G \in S\}$ is a face of F.

3.2. Digraph and Cluster Polytopes. Consider the *digraph polytope* consisting the convex hull of all directed graphs, cyclic and acyclic:

$$\mathsf{P}_{digraph} := \mathrm{conv}\,\{\eta_G \in \mathbb{R}^{\Upsilon} : G \text{ a directed graph on nodes } N\}.$$

It is clear the only conditions on η_G for a digraph G is every component is non-negative and every node i has only one parent set. That is, from the definition of η_G the components are either 0 or 1, hence non-negative. Additionally, every node i of G has exactly one set of parents, which implies the components of η_G corresponding to the node i has one and only one entry which is 1. In fact, these two sets of inequalities fully describe the facets of $\mathsf{P}_{digraph}$, and hence yield a hyperplane representation.

THEOREM 3.1 (Proposition 1 [**18**]). *The facets of* $\mathsf{P}_{digraph}$ *are:*

(1) $\eta(a\,|\,B) \geq 0$, *for all* $(a\,|\,B) \in \Upsilon$.
(2) $\sum_{\emptyset \neq B \subseteq N \setminus a} \eta(a\,|\,B) \leq 1$ *for all* $a \in N$.

It follows that $\mathsf{F} \subsetneq \mathsf{P}_{digraph}$. Analogous to the family-variable polytope F above, one can define a face of $\mathsf{P}_{digraph}$ by a set S of digraphs on nodes N, provided the convex hull of the integer vectors of S define a face of $\mathsf{P}_{digraph}$.

The *cluster inequalities* in Υ were first introduced by [**28**], and are defined over clusters which are sets of nodes. Given $C \subseteq N$, the cluster inequality is

$$\sum_{a\in C} \sum_{\substack{B\cap C\neq\emptyset \\ a\notin B}} \eta(a\,|\,B) \leq |C| - 1. \tag{3.1}$$

The interpretation of the cluster inequalities is as follows: not all nodes in the cluster C can have a parent which is also in the cluster C. They will play a major role in family-variable polytope F, as described in §5.1.

REMARK 3.2. The cluster inequalities do not allow cycles: consider a digraph G which has a cycle $i_1 \to \cdots \to i_p \to i_1$ and note that the cluster inequality with $C := \{i_1, \ldots, i_p\}$ would be violated.

The *cluster polytope* [**18**] is defined in hyperplane representation form as:

$$\begin{aligned}\mathsf{P}_{cluster} := \Big\{\eta \in \mathbb{R}^{\Upsilon} \,|\, \forall\, (a\,|\,B) \in \Upsilon : & \qquad \eta(a\,|\,B) \geq 0, \\ \forall\, a \in N : & \qquad \sum_{B\subseteq N\setminus a} \eta(a\,|\,B) \leq 1, \\ \forall\, C \subseteq N,\ |C| > 1 : & \qquad \sum_{a\in C} \sum_{\substack{B\cap C \\ a\notin B}} \eta(a\,|\,B) \leq |C| - 1\Big\}.\end{aligned}$$

REMARK 3.3. Clearly one has: $\mathsf{F} \subsetneq \mathsf{P}_{cluster} \subsetneq \mathsf{P}_{digraph}$. It was shown [**57**, Lemma 4] [**18**, §3] [**28**] that the cluster polytope $\mathsf{P}_{cluster}$ indeed only contains the zero-one integer encodings of $\mathsf{DAGS}\,(N)$ as integer points. Thus both the family variable polytope F and cluster polytope $\mathsf{P}_{cluster}$ contain the integer vector encodings of directed acyclic graphs as their only integer points, that is: $\mathsf{F} \cap \mathbb{Z}^{\Upsilon} = \mathsf{P}_{cluster} \cap \mathbb{Z}^{\Upsilon}$.

The key difference here is that F currently only has a vertex representation, while $\mathsf{P}_{cluster}$ has a fully described hyperplane representation. Note the number of inequalities in $\mathsf{P}_{cluster}$ is exponential, as one needs a cluster inequality for each cluster $C \subseteq N$, $|C| > 1$.

3.3. Standard Imsets. The power set of N is $\mathcal{P}(N)$, which is the set of all subsets of N: $\mathcal{P}(N) := \{A : A \subseteq N\}$.

$$\Lambda := \{\, S \subseteq N \,:\, |S| \geq 2 \,\}.$$

Integer vectors in $\mathbb{R}^{\Lambda}$ are called **i**nteger valued **m**ulti**sets**, or *imsets* for short owing to [**52**]. Building up to the standard and characteristic imsets, the following notation is used: given $A \subseteq N$, the corresponding *basic vector* will be denoted by δ_A:

$$\delta_A(S) = \begin{cases} 1 & \text{if } S = A\,, \\ 0 & \text{if } S \subseteq N,\ S \neq A\,. \end{cases}$$

A special *semi-elementary imset* $\mathsf{u}_{\langle A,B|C\rangle}$ is associated with any (ordered) triplet of pairwise disjoint sets $A, B, C \subseteq N$:

$$\mathsf{u}_{\langle A,B|C\rangle} := \delta_C - \delta_{A\cup C} - \delta_{B\cup C} + \delta_{A\cup B\cup C}\,,$$

which, in the context of [**52**], encodes the corresponding conditional independence statement $A \perp\!\!\!\perp B \,|\, C$ (= A is independent of B given C) [**52**].

The *standard imset* of $G \in \mathsf{DAGS}\,(N)$ can then defined as an element of $\mathbb{R}^{\Lambda}$:

$$\mathsf{u}_G := \delta_N - \delta_\emptyset + \sum_{a \in N} \left\{ \delta_{pa_G(a)} - \delta_{\{a\} \cup pa_G(a)} \right\}.$$

The basic and fundamental important property of standard imsets is that they uniquely represent BN structure. That is, $\mathsf{u}_G = \mathsf{u}_H$ if and only if $G \sim H$ where $G, H \in \mathsf{DAGS}\,(N)$ [**52**, Corollary 7.1]. It is important to note that in [**62**] it was first proposed to study the *standard imset polytope*, defined as the convex hull of the set of vectors u_G, where G runs over all acyclic directed graphs with N as the set of nodes. Thus, sparking off one of the two main polyhedral approaches to learning BN.

3.4. Characteristic Imset. The *characteristic imset* (char imset) of $G \in \mathsf{DAGS}\,(N)$, introduced in [**26**] and denoted below by c_G, is an element of $\mathbb{R}^{\Lambda}$. The origins of the characteristic imset was first given by an affine transformation of the standards imsets, which was shown to be a bijection. The affine transformation is given by 1 minus the *portrait* p_G of the standard imset u_G:

$$\mathsf{p}(S) = \sum_{T,\, S \subseteq T \subseteq N,\, |S| \geq 1} \mathsf{u}(T) \qquad \text{for } S \subseteq N, \tag{3.2}$$

$$\mathsf{c}(S) = 1 - \mathsf{p}(S) \qquad \text{for } S \subseteq N,\ |S| \geq 1. \tag{3.3}$$

The mapping $\mathsf{u} \mapsto \mathsf{c}$ determined by (3.2)-(3.3) has a well-known inverse given by the Möbius inversion formula. The standard imset can be recovered via the Möbius inversion formula; see the inclusion-exclusion principle in § 2.1 of [**50**]:

$$\mathsf{u}(T) = \sum_{S,\, T \subseteq S \subseteq N} (-1)^{|S \setminus T|} \cdot \underbrace{[\,1 - \mathsf{c}(S)\,]}_{\mathsf{p}(S)} \qquad \text{for } T \subseteq N.$$

Since the transformation is one-to-one, for $G, H \in \mathsf{DAGS}\,(N)$ it follows that $G \sim H$ if and only if $\mathsf{c}_G = \mathsf{c}_H$. That is, since the standard imset uniquely represents Markov equivalent BN structure, so does the characteristic imset.

Characteristic imsets can also be defined directly on graphs [**26**, Theorem 1]:

THEOREM 3.4. *For $G \in \mathsf{DAGS}\,(N)$ where $S \subseteq N$, $|S| \geq 2$,*

$$\mathsf{c}_G(S) = 1 \iff \exists i \in S \text{ such that } S \setminus \{i\} \subseteq pa_G(i).$$

The *characteristic-imset polytope* (char-imset polytope) is then defined as follows:

$$\mathsf{C} := \operatorname{conv}\{\, \mathsf{c}_G \in \mathbb{R}^{\Lambda} \;:\; G \in \mathsf{DAGS}\,(N) \,\}.$$

It has been easily established that the char-imset polytope is full-dimensional in the space $\mathbb{R}^{\Lambda}$ [**57**], i.e. $\dim(\mathsf{C}) = |\Lambda| = 2^n - n - 1$.

Owing to the original standard imset to characteristic map, and a simple mapping of family variables to characteristics imsets, in [**57**, § 3.3.2] and [**4**, § 2] it was shown that c_G is a many-to-one linear function of η_G; the transformation is $\eta \mapsto \mathsf{c}_\eta$, where

$$\mathsf{c}_\eta(S) = \sum_{a \in S} \; \sum_{B : S \setminus \{a\} \subseteq B \subseteq N \setminus \{a\}} \eta(a \,|\, B) \qquad \text{for any } S \subseteq N,\ |S| \geq 2. \tag{3.4}$$

REMARK 3.5. C is the image of F by the linear map (3.4).

Again, a general property of zero-one polytopes is that the only lattice points in the characteristic imset polytope C are the codes for directed acyclic graphs c_G. Analogous to the family-variable polytope F above, one can define a face of C by a set S of digraphs on nodes N, provided the convex hull of the integer vectors of S define a face of C.

An interesting property of the characteristic imset is that the components corresponding to set $S \subseteq N$ where $|S| \leq 3$ completely characterize the higher order components where the relationship is non-linear [**59**]. Note that, while $\mathcal{Q}$ is an affine function of the characteristic imset, this is not true for the restriction of the characteristic imset to sets of cardinality two and three unfortunately.

4. Learning BN Structure

Learning BN structure is done on the basis of an observed (complete) *database* D, which is a sequence $x_1, \ldots, x_d$ of elements of the joint sample space $\mathsf{X}_N = \prod_{i \in N} \mathsf{X}_i$ ($d \geq 1$ is the length of the database). The *projection* of D to $A \subseteq N$ is the sequence of respective projections $x_A^1, \ldots, x_A^d$, denoted by the symbol D_A. Given a database D of length $d \geq 1$ and $y \in \mathsf{X}_A$, $A \subseteq N$ the special notation $d_{[y]} := |\{l : 1 \leq l \leq d,\ x_A^l = y\}|$ is used for the number of occurrences of y in the database projection D_A. In particular, for the empty list $y \in \mathsf{X}_\emptyset$, one always has $d_{[y]} = d$. The concatenation of two configurations $y \in \mathsf{X}_A$ and $z \in \mathsf{X}_B$ for disjoint A and B will be denoted by $[y, z]$; it belongs to $\mathsf{X}_{A \cup B}$.

Learning a BN structure in the context here is done by maximizing some *quality criterion*, also called a *score*, which is a bivariate real function $(G, D) \mapsto \mathcal{Q}(G, D)$, where $G \in \mathsf{DAGS}\,(N)$ and D a database. Informally, the value $\mathcal{Q}(G, D)$ expresses how well the statistical model defined by G explains the occurrence D. For a more formal definitions and coverage of related ideas see [**40**].

Given the observed database D, the goal is to maximize $G \mapsto \mathcal{Q}(G, D)$. A natural assumption is that the criterion $\mathcal{Q}$ being maximized should be *score equivalent*, meaning for every database D and DAGs G and H,

$$\mathcal{Q}(G, D) = \mathcal{Q}(H, D) \quad \text{whenever } G \text{ and } H \text{ are Markov equivalent.}$$

The crucial technical assumption, especially related to the polyhedral approach here, is that $\mathcal{Q}$ should be additively *decomposable*, which means, it has the form

$$(4.1) \qquad \mathcal{Q}(G, D) = \sum_{i \in N} q_D(i \,|\, \mathrm{pa}_G(i)),$$

where the summands $q_D(*|*)$ are called *local scores.*

Moreover, Chickering required [**9**] in his definition that each local score term $q_D(i|B)$, for $i \in N$ and $B \subseteq N \setminus \{i\}$, only depends on the database projection $D_{\{i\} \cup B}$. This is a natural requirement from the computational point of view, implying that the score for sparse graphs can be computed from low-dimensional projections of the database. One can re-write (4.1) in terms of η_G:

$$\mathcal{Q}(G, D) = \sum_{i \in N} \sum_{B \subseteq N \setminus \{i\}} q_D(i|B) \cdot \eta_G(i|B),$$

which allows one to interpret $\mathcal{Q}$ as (the restriction of) a linear function of η_G.

A well-known example of such a score is Schwarz's [**46**] *Bayesian information criterion* (BIC), whose local scores are given as follows:

$$(4.2)\quad \mathsf{bic}_D(i|B) = \underbrace{\sum_{y\in \mathsf{X}_B}\sum_{z\in \mathsf{X}_i} d_{[y,z]}\cdot \ln\frac{d_{[y,z]}}{d_{[y]}}}_{\mathsf{mll}_D(i|B)} - \frac{\ln d}{2}\cdot \underbrace{(r(i)-1)\cdot\prod_{j\in B} r(j)}_{\mathsf{dim}(i|B)}$$
$$\text{for } i\in N,\ B\subseteq N\setminus\{i\},$$

where $r(i) = |\mathsf{X}_i|$, $i \in N$ are the cardinalities of the individual sample spaces. In that formula, the conventions $0\cdot \ln\frac{0}{*} \equiv 0$ and $\prod_{j\in\emptyset} r(j) \equiv 1$ are applied. The first term is the local score of the *maximized log-likelihood* (MLL) criterion. The second term is a penalty determined by the length of the database d and the local contribution to the dimension. That is, the penalty term quantifies the complexity of the statistical model G.

An additional common score is the *Bayesian Dirichlet equivalence* (BDE) score [**19**], whose local scores are given by the following formula:

$$(4.3)\quad \mathsf{bde}_D(i|B) = \sum_{y\in \mathsf{X}_B}\left\{\ln\frac{\Gamma(\alpha_{[y]})}{\Gamma(\alpha_{[y]}+d_{[y]})} - \sum_{z\in \mathsf{X}_i}\ln\frac{\Gamma(\alpha_{[y,z]})}{\Gamma(\alpha_{[y,z]}+d_{[y,z]})}\right\}$$
$$\text{for } i\in N,\ B\subseteq N\setminus\{i\},$$

where Γ denotes the Gamma function and the α-terms are so-called *hyper-parameters*. The usual assumption is that the hyper-parameters are given by the formula $\alpha_{[y]} = \alpha^*\cdot(|\mathsf{X}_B|)^{-1} = \alpha^*\cdot(\prod_{j\in B} r(j))^{-1}$ for $y\in\mathsf{X}_B$, with some fixed $\alpha^* > 0$, called the *equivalent sample size*. In this special case, the criterion is *Bayesian Dirichlet equivalence uniform* score and abbreviated as BDEU or BDeu [**19**]. To avoid the troublesome problem of setting hyper-parameters, a modification has been proposed which simplifies parameter selection [**6**].

The fundamental point is that maximizing a score equivalent quality criteria $\mathcal{Q}(G, D)$ over the set of graphs is equivalent to maximizing a linear function over the family variable polytope F due to the decomposable property. Specifically $\mathcal{Q}$ has the form

$$\mathcal{Q}(G,D) = s_D^{\mathcal{Q}} - \langle v_D^{\mathcal{Q}}, \eta\rangle.$$

and both $s_D^{\mathcal{Q}}, v_D^{\mathcal{Q}} \in \mathbb{R}^{\Upsilon}$ depend only on the database D and the scoring criteria $\mathcal{Q}$. Thus, one can perform BN structure learning by linear or integer programming, depending on what description or relaxation of the family variable polytope F is available.

Following suit, it was shown in [**26**, Lemma 1] that a score equivalent and additively decomposable criteria $\mathcal{Q}$ has the form

$$\mathcal{Q}(G,D) = \mathcal{Q}(G^{\emptyset}, D) + \sum_{S\subseteq N,\,|S|\geq 2} r_D^{\mathcal{Q}}(S)\cdot \mathsf{c}_G(S),$$

where $G^{\emptyset}$ is the empty graph over N, and $r_D^{\mathcal{Q}}$ a uniquely determined vector only depending the database D, called the *revised data vector* (relative to $\mathcal{Q}$).

Formulas for the data vector corresponding to the BIC and BDE, with respect to the standard imset, were formulated in [**52**] and [**53**] respectively. Using the standard imset to characteristic imset one-to-one mapping given by (3.2) - (3.3) one can easily derive the revised data vector $r^{\mathcal{Q}}$ with respect to the characteristic imsets. Alternatively one is able to compute $r^{\mathcal{Q}}_D(S)$ for $S \subseteq N$, $|S| \geq 2$ directly from local scores.

LEMMA 4.1 (Lemma 3 [**58**]). *Let $\mathcal{Q}$ be a score equivalent and additively decomposable criterion with local scores $q_D(*|*)$. Then, for any $S \subseteq N$, $|S| \geq 2$, one has*

$$r^{\mathcal{Q}}_D(S) = \sum_{K \subseteq R} (-1)^{|R \setminus K|} \cdot q_D(j|K) \qquad \textit{where } j \in S \textit{ and } R = S \setminus \{j\}\,. \tag{4.4}$$

In particular, the right-hand side of (4.4) does not depend on the choice of $j \in S$.

Thus, just as in the family variable case, one can perform BN structure learning by linear or integer programming, depending on what description or relaxation of the characteristic imset polytope C is available.

5. Facets & Valid Inequalities of BN Polytopes

As of the writing of this chapter, the entire facet description is not fully known for the family variable polytope or the characteristic polytope. However, some facets are known explicitly, as well as properties on facets. Many technical proofs are omitted, although some are given if concise or to facilitate exposition.

Since F is a rational polytope, its facets are defined by inequalities with rational coefficients, that is, by $\langle o, \eta \rangle \leq u$ with $o \in \mathbb{Q}^{\Upsilon}$. Multiplying by a suitable positive factor one can get a unique integer vector objective $o \in \mathbb{Z}^{\Upsilon}$ whose components have no common prime divisor. Since the vertices of F are zero-one vectors, the tight upper bound in $\langle o, \eta \rangle \leq u$ must then be an integer as well: $u \in \mathbb{Z}$. Similar arguments hold for the characteristic imset polytope C.

5.1. Family Variable Polytope Faces & Facets. First, a polytope $P \subseteq \mathbb{R}^n_{\geq 0}$ is *monotone* if $\mathbf{x} \in P$ and $0 \leq \mathbf{y} \leq \mathbf{x}$ implies $\mathbf{y} \in P$. Here $\mathbf{y} \leq \mathbf{x}$ means component-wise, i.e. $y_i \leq x_i$ for all i. It is known that the family-variable polytope F is monotone (Proposition 2 [**18**]). One of the key properties, shown in [**25**], is that a polytope is monotone if and only if it can be represented in the form $\mathbf{x} \geq 0$, $A\mathbf{x} \leq \mathbf{b}$, where $A, \mathbf{b} \geq \mathbf{0}$.

THEOREM 5.1 (Theorem 2 [**18**]). *Every facet of F is either*

- *a lower bound on family variables:*

$$\eta(a \,|\, B) \geq 0 \tag{5.1}$$

- *an inequality of the form $\langle o, \eta \rangle \leq u$ where $o \geq 0$ and $u > 0$.*

Equation (5.1) is referred to as the *non-negativity constraints*. The *modified convexity* [**28**] constraints on the family-variable polytope F are given by: $\sum_{\emptyset \neq B \subseteq N \setminus i} \eta(i \,|\, B) \leq 1$ for all $i \in N$. It is easy to see that they enforce the notion that a node i in a graph G can have at most one parent set. One does not have equality because $\eta(a \,|\, \emptyset)$ is not allowed. Note, in the space $\mathbb{R}^{\Upsilon}$ the variables $(a \,|\, \emptyset)$ are not present, hence, the nomenclature 'modified convexity'. By the above reasoning it is clear the modified convexity constraints are valid inequalities, and in fact they are facets.

THEOREM 5.2 (Proposition 3 [**18**], Lemma 3 [**17**]). *If* $n := |N| > 2$ *then the modified convexity constraints*

$$\sum_{\emptyset \neq B \subseteq N \setminus i} \eta(i \,|\, B) \leq 1 \tag{5.2}$$

are facets of the family-variable polytope F.

The *cluster inequalities* (3.1) are valid for the family-variable polytope F since they do not allow cycles. These inequalities were crucial in certain IP approaches to learning BN structure, as will be discussed below (§8.2.2). Indeed in [**14**] a generalization of the cluster inequalities were proposed, which were used successfully in their IP approach to finding BN structure. For every cluster $C \subseteq N$, $|C| > 2$, and $k = 1, \ldots, |C| - 1$ the *generalized cluster inequality* is given by

$$\sum_{a \in C} \sum_{\substack{B \subseteq N \setminus a: \\ |B \cap C| \geq k}} \eta(a \,|\, B) \leq |C| - k. \tag{5.3}$$

The interpretation here is straightforward: for any $G \in \mathsf{DAGS}\,(N)$ the induced subgraph $G[C]$ is acyclic hence the first k nodes in C in a total order consistent with $G[C]$ have at most $k - 1$ parents in C. Note that for $k = |C|$ and $k = 0$ the inequalities are tight for all $G \in \mathsf{DAGS}\,(N)$ and are thus omitted from 5.3.

THEOREM 5.3 (Corollary 4 [**17**], Theorem 5 [**18**]). *The generalized cluster inequalities* (5.3) *are facets of the family-variable polytope* F.

The proof in [**17**, Corollary 4] differs greatly from that provided in [**18**, Theorem 5]. In the former, deeper connections to the characteristic imset polytope and extreme supermodular functions (discussed in §5.5.1) are used. In the latter, the authors first show that the generalized cluster inequality with $C = N$ and any $k \leq |C|$ is a facet [**18**, Theorem 3]. Next they show that a facet-defining inequality on the family-variable polytope F can be 'lifted' to a higher dimensional family-variable polytope with a new node inserted [**18**, Theorem 4, Corollary 1]. This lifting is then used to prove that all generalized cluster inequalities are facets.

In order to describe the next class of facets on the family-variable polytope F a few definitions are required. In fact, these definitions will be critical when discussing connections between the family-variable polytope F and the characteristic imset polytope C in §5.5.1.

DEFINITION 5.4 (standardized supermodular function). Any vector $m \in \mathbb{R}^{\mathcal{P}(N)}$ can be viewed as a real set function $m : \mathcal{P}(N) \to \mathbb{R}$. Such a set function is *standardized* if $m(S) = 0$ for $S \subseteq N$, $|S| \leq 1$, and *supermodular* if

$$\forall U, V \subseteq N \quad m(U) + m(V) \leq m(U \cup V) + m(U \cap V)\,.$$

The following (non-negative) characteristics are ascribed to any supermodular function m: for any $a, b \in N$, $a \neq b$ and $Z \subseteq N \setminus \{a, b\}$, we will denote

$$\Delta m(a, b \,|\, Z) := m(\{a, b\} \cup Z) + m(Z) - m(\{a\} \cup Z) - m(\{b\} \cup Z)\,.$$

It is easy to see that a set function m is supermodular if and only if $\Delta m(a, b \,|\, Z) \geq 0$ for any respective triplet $(a, b \,|\, Z)$; see, for example, [**36**, Theorem 24(iv)]. By its definition, the set of supermodular vectors in $\mathbb{R}^{\Lambda}$ is a polyhedral cone. Moreover, since it is pointed, it has finitely many extreme rays and motivates the following definition.

DEFINITION 5.5 (Extreme supermodular function). A standardized supermodular set function $m : \mathcal{P}(N) \to \mathbb{R}$ is *extreme* if it generates an extreme ray of the standardized supermodular cone.

The following mapping from $\mathbb{R}^{\Lambda}$ to $\mathbb{R}^{\Upsilon}$ will be crucial for describing the next set of facets of the family-variable polytope F. Given $m \in \mathbb{R}^{\Lambda}$, the map $\rho : \mathbb{R}^{\Lambda} \to \mathbb{R}^{\Upsilon}$ is given by

$$\rho(m)(a \,|\, B) := m(\{a\} \cup B) - m(B) \text{ for any } a \in N,\ B \subseteq N \setminus \{a\}. \tag{5.4}$$

This mapping is in fact one-to-one [**17**, Corollary 3]. With this definition in hand, it was shown that the extreme standardized supermodular functions correspond to facets of the family-variable polytope F. We say G is a *full graph* on nodes N if for every pair of nodes $i, j \in N$ either $i \to j$ or $j \to i$. That is, the induced undirected subgraph is the complete graph.

THEOREM 5.6 (Theorem 2 [**17**]). *An inequality $\langle o, \eta \rangle \leq u$ for $\eta \in \mathsf{F}$, where $o \in \mathbb{R}^{\Upsilon}$ and $u \in \mathbb{R}$, is facet-defining for F and tight at all full graphs over N if and only if there exists an extreme standardized supermodular set function m such that o is determined by* (5.4) *and u is the shared value of $\langle o, \eta_H \rangle$ for full graphs H over N.*

In fact, using Theorem 5.6, ideas of supermodular functions, and matroid theory, another class of facets can be described which specializes to the case of k cluster inequalities [**56**]. First recall some basic definitions on matroids [**43**].

5.2. Matroids. Matroids are an abstraction of independence and date far back to the 1930's. They have pervaded many aspects of computer science and mathematics and especially combinatorial optimization. A *matroid* is a pair $(C, \mathcal{I})$ where C is a finite set called the *ground set* and $\mathcal{I}$ is a non-empty set of subsets of C called the *independent sets* which is closed under subsets: $I \in \mathcal{I},\ J \subseteq I$ implies $J \in \mathcal{I}$ and satisfies the *independence augmentation axiom*:

$$\text{if } I, J \in \mathcal{I} \text{ and } |J| < |I| \text{ then } a \in I \setminus J \text{ exists with } J \cup \{a\} \in \mathcal{I}.$$

The *dependent* sets of the matroid $(C, \mathcal{I})$ are the subsets of C that are not independent. The *circuits* of the matroid are the inclusion-minimal dependent sets. The *rank function* r of the matroid $(C, \mathcal{I})$ is the integer valued function on $\mathcal{P}(C)$ defined as:

$$r(J) := \max \{|I| : I \subseteq J \text{ and } I \in \mathcal{I}\}.$$

The rank of the matroid $(C, \mathcal{I})$ is $r(C)$. A set $S \subseteq C$ is a *separator* of the matroid $(C, \mathcal{I})$ if $r(S) + r(C \setminus S) = r(C)$. A matroid is then called *connected* if it has no separators besides $S = \emptyset$ and $S = C$. Note, that one can equivalently define the matroid strictly in terms of circuits or the rank function. Finally, the related class of facets on the family-variable polytope F can be defined.

THEOREM 5.7 (Theorem 3 [**56**]). *Given a connected matroid $(C, \mathcal{I})$ on a cluster $C \subseteq N$, $|C| > 2$ of BN variables, the inequality*

$$\sum_{a \in C} \quad \sum_{\substack{B \subseteq N \setminus \{a\}: \\ \exists D \in \mathcal{E},\, a \in D \subseteq B \cup \{a\}}} \eta(a \,|\, B) \leq |C| - r(C)$$

where $\mathcal{E}$ the collection of its circuits, is a facet defining inequality for the family-variable polytope F.

The proof relies on the result that rank functions on connected matroids are extreme rays of non-decreasing submodular functions (mirrors to supermodular functions) [**41**]. Then in [**56**], it is shown with minor modification, the rank function is also a standardized extreme supermodular function.

The *uniform matroid* of rank k over ground set C is the independent sets all subsets of C of cardinality k. Then, one can see that in fact Theorem 5.7 implies Theorem 5.3. That is, the generalized cluster inequalities are the special case of Theorem 5.7 with uniform matroids and in fact the uniform matroids are in one-to-one correspondence with the k cluster inequalities. Two additional classes of faces of the family variable polytope F based on notions of total orders and sinks are provided in [**18**] and described briefly here.

5.3. Sink and Order Faces. An acyclic directed graph G is *consistent* with a total order $<$ on N if for all $i \leftarrow j \in G \implies j < i$. Then given $<$ it is not difficult to see that

$$\sum_{i,B \ s.t. \ \exists j \in B, \, i<j} \eta(i \,|\, B) \geq 0 \tag{5.5}$$

is a valid inequality [**18**]. In fact they define faces, which can be called *order faces.*

PROPOSITION 5.8 (§7.1 [**18**]).

(1) *For every total order* $<$, *the order face* (5.5) *is face* F *of the family variable polytope* F, *its dimension is* $2^n - n - 1$, *and* F *consists of all directed graphs* G *consistent with* $<$.
(2) *If all family variables not consistent with* $<$ *are removed, then the face* F *is full-dimensional in the space* $\mathbb{R}^{2^n-n-1}$.
(3) *Every extreme point in the family variable polytope* F *lies in some order face for some total order* $<$.
(4) *The full dimensional polytope defined by order face* F *with total order* $<$ *has as facets only lower bounds* $\eta(a \,|\, B) \geq 0$ *and the modified convexity constraints* $\sum_{a, \emptyset \neq B \subseteq N \setminus a} \eta(a \,|\, B) \leq 1$, *where the family variables not consistent with* $<$ *are removed.*

A node $j \in N$ is a *sink* of a directed graph G if it has no children, i.e. no nodes $i \in N$ such that $j \to i$. Every $G \in \mathsf{DAGS}\,(N)$ has at least one sink. For a node $j \in N$ the following

$$\sum_{i \neq j, \, j \in J} \eta(i \,|\, J) \geq 0 \tag{5.6}$$

is a valid inequality [**18**]. Again, these define faces, which can be called *sink faces.*

PROPOSITION 5.9 (§7.2 [**18**]).

(1) *For every* $j \in N$, *the sink face* (5.6) *is a face* F *of the family variable polytope* F, *its dimension is* $(n+1)2^{n-2} - n$, *and* F *consists of all graphs with sink* j.
(2) *If all family variables* $\eta(j \,|\, J)$ *where* $j \in J$ *are set to zero, then* F *defines a full dimensional polytope in* $R^{(n+1)2^{n-2}-n}$.
(3) *Every extreme point in the family variable polytope* F *lies in some sink face for some node* j.

(4) *The facets of the full dimensional polytope defined by F are by the facets of the family variable polytope where node j is removed and the modified convexity constraints for j:* $\sum_{j \in J, \emptyset \neq J \subseteq N \setminus j} \eta(j \mid J) \leq 1$.

One very useful implication of items (3) - (4) is that since every extreme point of the family variable polytope F appears in some sink face, and the sink faces themselves can be represented as smaller family variable polytopes (with the sink removed), then the family variable polytope F has an extended representation as the union of lower dimensional family variable polytopes with some additionally inequalities [**18**, §7.2.1]. By following a standard approach in [**12**, §2.11] this extended representation can be modeled with a few new variables and linking inequalities, leading to new interpretations of some previously unclassified facets.

5.4. Summary of Known Faces. The current (overlapping) list of known faces and facets of family-variable polytope F is thus summarized as:

- Order faces (5.5).
- Sink faces (5.6).
- Non-negative inequalities on family variables (5.1).
- Modified convexity constraints (5.2).
- Generalized cluster inequalities (5.3).
- Connected matroids on $C \subseteq N$, $|C| \geq 2$ (Theorem 5.7).
- Extreme supermodular set functions (Theorem 5.6).

In fact, the last item (extreme supermodular set functions) includes the case of connected matroid facets, which itself includes the case of generalized cluster inequalities. However, the class of facets corresponding to extreme supermodular set function is only implicitly defined in the general case.

Unfortunately it is not known if these are sufficient to fully characterize the family-variable polytope F. See §6 for examples of family variable polytopes. But, as stated before, these inequalities do form a linear relaxation of F, i.e. these inequalities contain only the integer points necessary for BN structure learning. In fact, all that are needed for a linear relaxation are the non-negative, modified convexity, and original cluster inequalities (§3.2). However, additional facets can be of great use in practice when learning BN structure.

Even though not all of the facets of the family-variable polytope F have been described, additional properties are known about general facets which may prove useful in discovering new facets, or for practical approaches to learning BN structure.

PROPOSITION 5.10 (Proposition 4 [**18**]). If $\langle o, \eta \rangle \leq u$ defines a facet of the family-variable polytope F and it is not a lower bound on a family variable, then for all $(a \mid B) \in \Upsilon$ there exists an extreme point $\eta \in \mathsf{F}$ such that $\eta(a \mid B) = 1$ and $\langle o, \eta \rangle = u$.

The proof follows by assuming the facet has $n(2^{n-1} - 1)$ affinely independent graphs which are zero on some $(a \mid B)$ and showing these graphs also lie in a lower bound facet, hence a contradiction.

PROPOSITION 5.11 (Proposition 5 [**18**]). Let $\langle o, \eta \rangle \leq u$ be a facet of the family-variable polytope F then $A \subseteq B \implies o(i \mid A) \leq o(i \mid B)$.

LEMMA 5.12 (Lemma 2 [**17**]). Let $\langle o, \eta \rangle \leq u$ be a valid inequality on F, where by Theorem 5.1 $u \geq 0$.

(1) $u = 0$ if and only if the corresponding face of F contains the empty graph.
(2) The facet-defining inequalities tight at the empty graph are just the non-negativity constraints $\eta(a \mid B) \geq 0$.
(3) If $\langle o, \eta \rangle \leq u$ is facet-defining on F with $u > 0$ then the objective coefficients are non-negative and increasing in the following sense:

$$o(a \mid B) \geq o(a \mid A) \geq 0 \quad \text{whenever } a \in N \text{ and } \emptyset \neq A \subseteq B \subseteq N \setminus \{a\}.$$

Note that item (3) is equivalent to Proposition 5.11. Then the following holds.

COROLLARY 5.13 (Corollary 2 [**17**]). Let S be a facet of F which does not contain the empty graph. Then S is closed under super-graphs in the sense: if $G \in \mathsf{S}$ is a subgraph of $H \in \mathsf{DAGS}(N)$ then $H \in \mathsf{S}$. Moreover, for every $(a \mid B) \in \Upsilon$, there exists $G \in \mathsf{S}$ with $pa_G(a) = B$.

5.5. Characteristic Imset Polytope Facets & Faces. Historically, the standard imsets [**52**] preceded the characteristic imsets [**26, 60**]. Due to the one-to-one correspondence the two representations are equivalent descriptors of conditional independence, as well as unique identifiers of Markov equivalence classes. However, characteristic imsets have the additional nice property that they are zero-one, and have a simple graphical interpretation (Theorem 3.4). Hence, this monograph focuses on characteristic imsets.

A key discovery was that in fact many of the faces of the family-variable polytope F correspond to faces of the characteristic imset polytope C via the concept of score equivalence.

5.5.1. *Score Equivalence & Connecting* F *and* C. In a score-based approach to learning BN structure one maximizes a function $G \in \mathsf{DAGS}(N) \mapsto \mathcal{Q}(G, D)$, where D is the database of observed values and $\mathcal{Q}$ a suitable scoring function which evaluates how the graph G fits the database, as described in §4. The criteria often used in practice turn out to be affine functions of the family-variable vector and thus BN structure learning turns into an IP problem, i.e. maximizing a linear function over the family-variable polytope F. Moreover, since the goal is typically to learn the BN structure described by Markov equivalence classes, scoring functions are selected such that they do not distinguish between Markov equivalence classes:

$$\mathcal{Q}(G, D) = \mathcal{Q}(H, D) \quad \text{whenever } G \text{ and } H \text{ are Markov equivalent.}$$

This motivates the following terminology on objectives in $\mathbb{R}^{\Upsilon}$.

DEFINITION 5.14 (score equivalent objective [**17**]). A vector $o \in \mathbb{R}^{\Upsilon}$ is a *score equivalent objective* (abbreviated as *an SE objective*) if it satisfies

$$\forall\, G, H \in \mathsf{DAGS}(N) \quad G \sim H \;\Rightarrow\; \langle o, \eta_G \rangle_{\Upsilon} = \langle o, \eta_H \rangle_{\Upsilon}\,.$$

Clearly, the set of SE objectives is a linear subspace of $\mathbb{R}^{\Upsilon}$.

Recall the faces and facets of any polytope can be defined in terms of normal vectors, which leads to the following concept.

DEFINITION 5.15 (SE face/facet, closed under Markov equivalence [**17**]). A face F of F is *score equivalent* (SE) if there exists an SE objective $o \in \mathbb{R}^{\Upsilon}$ and a constant $u \in \mathbb{R}$ such that it defines a face of F. An *SE facet* is a facet of F which is an SE face. In the terminology of graphs, one says a set $\mathsf{S} \subseteq \mathsf{DAGS}(N)$ of acyclic directed graphs is *closed under Markov equivalence* if

$$\forall\, G, H \in \mathsf{DAGS}(N) \;\text{ s.t. }\; G \sim H \qquad G \in \mathsf{S} \;\Rightarrow\; H \in \mathsf{S}.$$

REMARK 5.16. Note that an objective determining a face is not uniquely determined. Only in the case of a facet (of a full-dimensional polytope), is an objective unique up to a positive multiple. Therefore, one has to be careful when testing score equivalence of a face F which is not a facet, since one of the face-defining objectives for F could be SE and another objective for F need not be. Recall, the definition requires the *existence* of at least one SE objective defining the face.

From the above definition it follows directly that a set of graphs on an SE face is closed under Markov equivalence (Lemma 4 [**17**]). Interestingly the converse is still an open conjecture.

CONJECTURE 5.17 (Conjecture 1 [**17**]). *Every face* $\mathsf{S} \subseteq \mathsf{DAGS}\,(N)$ *of* F *closed under Markov equivalence is an SE face.*

In [**17**] the conjecture was shown to be true for facets (see Theorem 1 [**17**]), however, it has yet to be proven in the general cause. The conjecture was verified for $n = |N| = 3$ by exhaustive search, and additionally for $n = 4$ every inclusion-submaximal face among those closed under Markov equivalence is already an SE face [**17**].

The linear subspace of SE objectives have a few nice properties, and in fact connect nicely to the space $\mathbb{R}^{\Lambda}$ where the characteristic imset polytope C lives (see Lemma 5 [**17**]). In particular, the dimension of the linear subspace of SE objectives is the same as the one of C, that is $2^n - n - 1$.

What is an important theme throughout the comparison of the family variable polytope F and the characteristic imset polytope C are the faces of C which contain either the empty graph, or the Markov equivalence class of full graphs. For example, any face F of F containing the whole Markov equivalence class of full graphs is given by an SE objective. Hence, F is an SE face (Lemma 6 [**17**]).

It follows that no counter-example to Conjecture 5.17 is among the faces containing a full graph. Stronger relations can be established when considering facets instead of faces, and confirm Conjecture 5.17 for facets.

THEOREM 5.18 (Theorem 1 [**17**]). *The following conditions are equivalent for a facet* $\mathsf{S} \subseteq \mathsf{DAGS}\,(N)$*:*

(a) S *is closed under Markov equivalence,*
(b) S *contains the whole equivalence class of full graphs,*
(c) S *is SE.*

PROOF. To show (a)⇒(b) note, by Lemma 5.12(3), that S cannot contain the empty graph, since otherwise it is not closed under Markov equivalence. Clearly, S must be non-empty, because otherwise it is not a facet of F. Thus, $G \in \mathsf{S}$ exists and one can construct a full graph $H \in \mathsf{DAGS}\,(N)$ such that G is a subgraph of H. By Corollary 5.13, $H \in \mathsf{S}$. Since S is closed under Markov equivalence, all full graphs belong to S. The implication (b)⇒(c) is mentioned above (Lemma 6 [**17**]), and the implication (c)⇒(a) follows from Definition 5.15 (see also Lemma 4 [**17**]. □

In fact, such facets on the family variable polytope F tight on all full graphs correspond to standardized supermodular set functions (Definition 5.4).

LEMMA 5.19 (Lemma 7 [**17**]). An inequality $\langle o, \eta \rangle \leq u$, where $o \in \mathbb{R}^\Upsilon$ and $u \in \mathbb{R}$, is valid for all $\eta \in \mathsf{F}$ and tight at any full graph over N if and only if it corresponds to a standardized supermodular function m in the sense:

- o is given by $o(a \mid B) = m(\{a\} \cup B) - m(B)$ for $a \in N$, $B \subseteq N \setminus \{a\}$,
- u is the shared value $\langle o, \eta_H \rangle$ for full graphs H over N.

Moreover, the correspondence is one-to-one and preserves a conic combination.

As stated in §5.1 and following Definition 5.5, the set of standardized supermodular set functions is a pointed polyhedral cone leading to the following result.

LEMMA 5.20 (Lemma 8 [**17**]). Let $m_1, m_2 \in \mathbb{R}^{\mathcal{P}(N)}$ generate distinct extreme rays of the standardized supermodular cone. Then the faces of F determined by the corresponding inequalities, as described in Lemma 5.19, are inclusion-incomparable.

Piecing these SE results together, one is able to prove Theorem 5.6, which states that a facet of the family variable polytope F is tight at all full graphs if and only if it corresponds to an extreme supermodular set function. Note that a simple linear criterion for testing extremality of a standardized supermodular function has been recently proposed in [**36**]. Additionally, it was shown in §8 of [**17**] that the generalized k-cluster inequalities correspond to standardized extreme supermodular set functions, leading to Theorem 5.3. Finally, it was the same correspondence to extreme supermodular set functions which led to the generalization of the k-cluster inequalities to matroids and thus Theorem 5.7.

Returning now to faces and facets of the characteristic imset polytope C, a one-to-one correspondence exists between faces of the characteristic imset polytope C and SE faces of the family variable polytope F. Let $\langle z, \mathsf{c} \rangle_\Lambda \leq u$, where $z \in \mathbb{R}^\Lambda$ and $u \in \mathbb{R}$, be a valid inequality for c in the characteristic-imset polytope C. It defines a face of C: $\bar{F} = \{\mathsf{c} \in \mathsf{C} : \langle z, \mathsf{c} \rangle_\Lambda = u\}$. By substituting the η to c transformation given by (3.4) into the inequality $\langle z, \mathsf{c}_\eta \rangle_\Lambda \leq u$ and re-arranging terms after the components of η, one gets an inequality for $\eta \in \mathbb{R}^\Upsilon$ valid for any η_G, $G \in \mathsf{DAGS}(N)$. Indeed, this is because the image of η_G by (3.4) is just c_G. Moreover, the objective on the LHS of the obtained inequality is SE since whenever $G \sim H$, one has $\mathsf{c}_G = \mathsf{c}_H$ and, therefore, $\langle z, \mathsf{c}_G \rangle_\Lambda = \langle z, \mathsf{c}_H \rangle_\Lambda$. Thus, any face of C defines an SE face of F. Fortunately, the converse is also true. But, first an explicit mapping of SE objectives is required.

LEMMA 5.21 (Lemma 10 [**17**]). Given an SE objective $o \in \mathbb{R}^\Upsilon$, there exists unique $z_o \in \mathbb{R}^\Lambda$ such that the following holds:

$$\forall \eta \in \mathbb{R}^\Upsilon \qquad \langle o, \eta \rangle_\Upsilon = \langle z_o, \mathsf{c}_\eta \rangle_\Lambda \,.$$

Specifically, one has

$$z_o(T) \;:=\; \sum_{\emptyset \neq K \subseteq R} (-1)^{|R \setminus K|} \cdot o(b \mid K) \tag{5.7}$$

$$\text{for } T \in \Lambda, \text{ with any } b \in T \text{ and } R := T \setminus \{b\}.$$

In particular, the expression in (5.7) does not depend on the choice of $b \in T$.

Every face of the characteristic imset polytope C can be identified with a set of acyclic directed graphs closed under ME

$$\bar{F} \subseteq \mathsf{C} \text{ a face of } \mathsf{C} \longleftrightarrow \mathsf{S} = \{G \in \mathsf{DAGS}(N) : \mathsf{c}_G \in \bar{F}\}\,.$$

These are simply the graphs whose characteristic imsets belong to the face of C. It is easy to see that the correspondence preserves inclusion: $\bar{F}_1 \subseteq \bar{F}_2$ for faces of C if and only if $\mathsf{S}_1 \subseteq \mathsf{S}_2$ for the corresponding sets of graphs $\mathsf{S}_i \subseteq \mathsf{DAGS}\,(N)$.

COROLLARY 5.22 (Corollary 5 [**17**]). There is a one-to-one correspondence between SE faces of F and faces of C which preserves inclusion: given SE faces F_1, F_2 of F and the corresponding faces $\bar{F}_1, \bar{F}_2$ of C one has $F_1 \subseteq F_2$ if and only if $\bar{F}_1 \subseteq \bar{F}_2$. Specifically, the SE face of F given by an inequality $\langle o, \eta \rangle_\Upsilon \leq u$ corresponds to the face of C given the inequality $\langle z_o, \mathsf{c} \rangle_\Lambda \leq u$. This correspondence has the property that the sets of graphs identified with the faces coincide.

PROOF. It is easy to see that $\langle o, \eta \rangle_\Upsilon \leq u$ is valid for $\eta \in \mathsf{F}$ if and only if $\langle z_o, \mathsf{c} \rangle_\Lambda \leq u$ is valid for $\mathsf{c} \in \mathsf{C}$. Moreover, by Lemma 5.21, the set of $G \in \mathsf{DAGS}\,(N)$ such that $\langle o, \eta \rangle_\Upsilon \leq u$ is tight for η_G coincides with the set of $G \in \mathsf{DAGS}\,(N)$ such that $\langle z_o, \mathsf{c} \rangle_\Lambda \leq u$ is tight for c_G. Thus, an SE face of F and the corresponding face of C have the same sets of "belonging" graphs. This observation easily implies preserving the inclusion of faces. □

Returning to the important theme on the characteristic imset polytope C, there are two distinguished vertices: The vertex containing the 0-imset (zero vector in $\mathbb{R}^\Lambda$), which is the empty graph over N, and the vertex containing the 1-imset (vector of all ones in $\mathbb{R}^\Lambda$), which is the characteristic imset of the full graphs over N.

COROLLARY 5.23 (Corollary 6 [**17**]). SE facets of the family-variable polytope F correspond to those facets of the characteristic-imset polytope C that contain the 1-imset. None of those facets of C include the 0-imset. Moreover, these facets correspond to extreme supermodular functions.

The last part of Corollary 5.23 follows from Theorem 5.6 and 5.18 the fact that the facets in C contain the 1-imset. On the other hand, it follows from Corollaries 5.22 and 5.23 that the SE faces of F corresponding to facets of C not containing the 1-imset are *sub-maximal SE faces* with respect to inclusion, but *not SE facets*. That means, these are SE faces F of F such that there is no other SE face F' of F such that $F \subset F'$ except $F' = \mathsf{F}$ but F is not a facet of F since $\dim(F) < \dim(\mathsf{F}) - 1$.

Hence, extreme supermodular functions correspond to facets of the characteristic imset polytope C which contain the 1-imset. In the family variable polytope F form, some of the extreme supermodular functions were shown to correspond to the rank function of connected matroids, and gave facets of the F (Theorem 5.7). Fortunately, these also have an explicit form in the characteristic imset polytope C.

THEOREM 5.24 (Corollary 2 [**56**]). Given a connected matroid $(C, \mathcal{I})$ on a cluster $C \subseteq N$, $|C| > 2$ of BN variables, the inequality

$$\begin{aligned} &\sum_{T \in \Lambda,\, T \subseteq C} z(T) c(T) \leq |C| - k \quad \text{for } c \in \mathbb{R}^\Lambda \\ &\text{where} \quad z(t) = -\sum_{L \subseteq T} (-1)^{|T \setminus L|} r(L), \end{aligned} \tag{5.8}$$

and k is the rank of $(C, \mathcal{I})$, is a facet defining inequality for the characteristic imset polytope C.

5.5.2. *Sufficiency of Score Equivalent Faces.* As explained previously the goal of learning BN structure is equivalent to maximizing an SE objective over the family variable polytope F. The tractability of such a task is related to the number of inequalities required to describe the polytope. Given that the objective being maximized over F is SE, the natural question to ask is if all the facet describing inequalities of F are required when performing linear or integer programming with an SE objective. The correspondence between SE faces of F and faces of the characteristic imset polytope C described in Corollary 5.22 leads to the question of whether the non-SE faces of F are necessary when maximizing an SE objective. In fact, many of the SE faces of F are superfluous.

THEOREM 5.25. [**17**, Theorem 3] *Let o be an SE objective. Then the LP problem to*

$$\textit{maximize } \eta \mapsto \langle o, \eta \rangle_{\Upsilon} \textit{ over } \eta \in \mathsf{F}$$

has the same optimal value as the LP problem to maximize the same function over the polyhedron specified by

- *the inequalities defining SE faces that correspond to those facets of C that do not contain the 0-imset,*
- *the non-negativity ($\eta(a \mid B) \geq 0$) and modified convexity constraints* (5.2).

Again, one can ask if SE facets are enough in order to maximize an SE objective over the family variable polytope F. This is true for $|N| = 4$ (see [**17**, Corollary 7]). However, this does not hold in general as a facet for $|N| = 5$ (discovered in [**23**]) leads to a counter-example [**17**, § 12].

6. Example Family Variable Polytopes

The following section details example of the family variable polytope F for $|N| = 2, 3, 4$. These examples have been computed before, but see [**18**] for a most recent comprehensive list. To reduce clutter, the brackets denoting sets will be dropped, e.g. abc is used in place of $\{a, b, c\}$.

EXAMPLE 6.1 (F on 2 Nodes). $N = \{a, b\}$ and $\Upsilon = \{(a \mid b), (b \mid a)\}$. There are 3 acyclic directed graphs, and F has three facets: Two lower bounds and the score-equivalent 1-cluster constraint $\eta(a \mid B) + \eta(b \mid a) \leq 1$.

EXAMPLE 6.2 (F on 3 Nodes). $N = \{a, b, c\}$ and

$$\begin{aligned} \Upsilon = \{ &(a \mid b), (a \mid c), (a \mid bc), \\ &(b \mid a), (b \mid c), (b \mid ac), \\ &(c \mid a), (c \mid b), (c \mid ab) \}. \end{aligned}$$

There are 25 acyclic directed graphs and F consists of 17 facets:

- 9 lower bounds on the 9 family variables $(a \mid B)$.
- 3 modified convexity constraints, one for each variable $\{a, b, c\}$.
- 4 1-cluster (score-equivalent) constraints, one for each of the cluster $\{a, b\}$, $\{a, c\}$, $\{b, c\}$, $\{a, b, c\}$.
- 1 2-cluster (score-equivalent) constraint for the cluster $\{a, b, c\}$.

EXAMPLE 6.3 (F on 4 Nodes). [**17**, Example 2] [**18**, §6.3.3] $N = \{a, b, c, d\}$ and

$$\Upsilon = \{(a\,|\,b), (a\,|\,c), (a\,|\,d), (a\,|\,bc), (a\,|\,bd), (a\,|\,cd), (a\,|\,bcd),$$
$$(b\,|\,a), (b\,|\,c), (b\,|\,d), (b\,|\,ac), (b\,|\,ad), (b\,|\,cd), (a\,|\,acd),$$
$$(c\,|\,a), (c\,|\,b), (c\,|\,d), (c\,|\,ab), (c\,|\,ad), (c\,|\,bd), (a\,|\,abd),$$
$$(d\,|\,a), (d\,|\,b), (d\,|\,c), (d\,|\,ab), (d\,|\,ac), (d\,|\,bc), (a\,|\,abc)\}.$$

There are 543 acyclic directed graphs and F consists of 135 facets:

- 28 lower bounds on the 28 family variables $(a\,|\,B)$.
- 4 modified convexity constraints, one for each variable $\{a, b, c, d\}$.
- 6 1-cluster (score-equivalent) constraints for each of the $\binom{4}{2}$ size 2 clusters.
- 4 1-cluster (score-equivalent) constraints for each of the $\binom{4}{3}$ size 3 clusters.
- 1 1-cluster (score-equivalent) constraints for each of the $\binom{4}{4}$ size 4 clusters.
- 4 2-cluster (score-equivalent) constraints for each of the $\binom{4}{3}$ size 3 clusters.
- 1 2-cluster (score-equivalent) constraints for each of the $\binom{4}{4}$ size 4 clusters.
- 1 3-cluster (score-equivalent) constraints for each of the $\binom{4}{4}$ size 4 clusters.
- 86 other facets.

The other 86 facets, some score equivalent and some not, are of the following form. An alternate interpretation of the 86 inequalities in terms of sink-faces based on an extension representation is given in [**18**].

6 (score equivalent) inequalities of the form:

$$\eta(a\,|\,b) + \eta(a\,|\,bc) + \eta(a\,|\,bd) + \eta(a\,|\,cd) + \eta(a\,|\,bcd) +$$
$$\eta(b\,|\,a) + \eta(b\,|\,ac) + \eta(b\,|\,ad) + \eta(b\,|\,cd) + \eta(b\,|\,acd) +$$
$$\eta(c\,|\,ad) + \eta(c\,|\,bd) + \eta(c\,|\,abd) +$$
$$\eta(d\,|\,ac) + \eta(d\,|\,bc) + \eta(d\,|\,abc) \leq 2.$$

4 (score equivalent) inequalities of the form:

$$\eta(a\,|\,bc) + \eta(a\,|\,bd) + \eta(a\,|\,cd) + 2\eta(a\,|\,bcd) +$$
$$\eta(b\,|\,ac) + \eta(b\,|\,ad) + \eta(b\,|\,acd) +$$
$$\eta(c\,|\,ab) + \eta(c\,|\,ad) + \eta(c\,|\,abd) +$$
$$\eta(d\,|\,ab) + \eta(d\,|\,ac) + \eta(d\,|\,abc) \leq 2.$$

6 (score equivalent) inequalities of the form:

$$\eta(a\,|\,c) + \eta(a\,|\,d) + \eta(a\,|\,bc) + \eta(a\,|\,bd) + \eta(a\,|\,cd) + 2\eta(a\,|\,bcd) +$$
$$\eta(b\,|\,c) + \eta(b\,|\,d) + \eta(b\,|\,ac) + \eta(b\,|\,ad) + \eta(b\,|\,cd) + 2\eta(b\,|\,acd) +$$
$$\eta(c\,|\,a) + \eta(c\,|\,b) + \eta(c\,|\,d) + 2\eta(c\,|\,ab) + \eta(c\,|\,ad) + \eta(c\,|\,bd) + 2\eta(c\,|\,abd) +$$
$$\eta(d\,|\,a) + \eta(d\,|\,b) + \eta(d\,|\,c) + 2\eta(d\,|\,ab) + \eta(d\,|\,ac) + \eta(d\,|\,bc) + 2\eta(d\,|\,abc) \leq 4.$$

4 (score equivalent) inequalities of the form:

$$\eta(a\,|\,b) + \eta(a\,|\,c) + \eta(a\,|\,d) + 2\eta(a\,|\,bc) + 2\eta(a\,|\,bd) + 2\eta(a\,|\,cd) + 2\eta(a\,|\,bcd) +$$
$$\eta(b\,|\,a) + \eta(b\,|\,ac) + \eta(b\,|\,ad) + \eta(b\,|\,cd) + \eta(b\,|\,acd) +$$
$$\eta(c\,|\,a) + \eta(c\,|\,ab) + \eta(c\,|\,ad) + \eta(c\,|\,bd) + \eta(c\,|\,abd) +$$
$$\eta(d\,|\,a) + \eta(d\,|\,ab) + \eta(d\,|\,ac) + \eta(d\,|\,bc) + \eta(d\,|\,abc) \leq 3.$$

12 inequalities of the form:

$$\begin{aligned}&\eta(a\,|\,c)+\eta(a\,|\,b)+\eta(a\,|\,bc)+\eta(a\,|\,bd)+\eta(a\,|\,cd)+\eta(a\,|\,bcd)+\\&\eta(b\,|\,cd)+\eta(b\,|\,acd)+\\&\eta(c\,|\,ab)+\eta(c\,|\,bd)+\eta(c\,|\,abd)+\\&\eta(d\,|\,ab)+\eta(d\,|\,bc)+\eta(d\,|\,abc)\leq 2.\end{aligned}$$

12 inequalities of the form:

$$\begin{aligned}&\eta(a\,|\,b)+\eta(a\,|\,c)+\eta(a\,|\,d)+\eta(a\,|\,bc)+\eta(a\,|\,bd)+\eta(a\,|\,cd)+2\eta(a\,|\,bcd)+\\&\eta(b\,|\,a)+\eta(b\,|\,c)+\eta(b\,|\,d)+\eta(b\,|\,ac)+\eta(b\,|\,ad)+\eta(b\,|\,cd)+\eta(b\,|\,acd)+\\&\eta(c\,|\,a)+\eta(c\,|\,ab)+\eta(c\,|\,ad)+\eta(c\,|\,abd)+\\&\eta(d\,|\,a)+\eta(d\,|\,ab)+\eta(d\,|\,ac)+\eta(d\,|\,abc)\leq 3.\end{aligned}$$

24 inequalities of the form:

$$\begin{aligned}&\eta(a\,|\,cd)+\eta(a\,|\,bcd)+\\&\eta(b\,|\,c)+\eta(b\,|\,ac)+\eta(b\,|\,cd)+\eta(b\,|\,acd)+\\&\eta(c\,|\,b)+\eta(c\,|\,d)+\eta(c\,|\,ab)+\eta(c\,|\,ad)+\eta(c\,|\,bd)+2\eta(c\,|\,abd)+\\&\eta(d\,|\,a)+\eta(d\,|\,b)+\eta(d\,|\,c)+\eta(d\,|\,ab)+2\eta(d\,|\,ac)+\eta(d\,|\,bc)+2\eta(d\,|\,abc)\leq 3.\end{aligned}$$

12 inequalities of the form:

$$\begin{aligned}&\eta(a\,|\,cd)+\eta(a\,|\,bcd)+\\&\eta(b\,|\,c)+\eta(b\,|\,ac)+\eta(b\,|\,cd)+\eta(b\,|\,acd)+\\&\eta(c\,|\,b)+\eta(c\,|\,d)+\eta(c\,|\,ab)+\eta(c\,|\,ad)+\eta(c\,|\,bd)+2\eta(c\,|\,abd)+\\&\eta(d\,|\,a)+\eta(d\,|\,b)+\eta(d\,|\,c)+\eta(d\,|\,ab)+2\eta(d\,|\,ac)+\eta(d\,|\,bc)+2\eta(d\,|\,abc)\leq 2.\end{aligned}$$

6 inequalities of the form:

$$\begin{aligned}&\eta(a\,|\,c)+\eta(a\,|\,d)+\eta(a\,|\,bc)+\eta(a\,|\,bd)+\eta(a\,|\,cd)+2\eta(a\,|\,bcd)+\\&\eta(b\,|\,acd)+\\&\eta(c\,|\,ab)+\eta(c\,|\,abd)+\\&\eta(d\,|\,ab)+\eta(d\,|\,abc)\leq 3.\end{aligned}$$

7. Example Characteristic Imset Polytopes

EXAMPLE 7.1 (C on 3 nodes). $N=\{a,b,c\}$ and $|\Lambda|=4$. The 4-dimensional polytope C has 11 vertices and 13 facets, originally presented in [**57**, Examples 5,8]. There are five facet-defining inequalities tight for the 1-imset, and by Corollary 5.23 they correspond to SE facets of F mentioned in Example 6.2. The facets decompose into 3 permutation types, where their corresponding family variable facets is given if applicable.

- 3 inequalities of the form: $\mathsf{c}(ab)\leq 1$,
  in family variables $\eta(a\,|\,b)+\eta(a\,|\,bc)+\eta(b\,|\,a)+\eta(b\,|\,ac)\leq 1$,
- 1 inequality of the form: $\mathsf{c}(abc)\leq 1$,
  in family variables $\eta(a\,|\,bc)+\eta(b\,|\,ac)+\eta(c\,|\,ab)\leq 1$,

- 1 inequality of the form: $\mathsf{c}(ab) + \mathsf{c}(ac) + \mathsf{c}(bc) - \mathsf{c}(abc) \leq 2$
  in family variables
  $\eta(a\,|\,b) + \eta(a\,|\,c) + \eta(a\,|\,bc) + \eta(b\,|\,a) + \eta(b\,|\,c) + \eta(b\,|\,ac)$
  $+ \eta(c\,|\,a) + \eta(c\,|\,b) + \eta(c\,|\,ab) \leq 2.$

The remaining eight facet-defining inequalities of C are tight for the 0-imset and decompose into 4 permutation types:

- 3 inequalities of the form: $-\mathsf{c}(ab) \leq 0$,
  in family variables $-\eta(a\,|\,b) - \eta(a\,|\,bc) - \eta(b\,|\,a) - \eta(b\,|\,ac) \leq 0,$
- 1 inequality of the form: $-\mathsf{c}(abc) \leq 0$,
  in family variables $-\eta(a\,|\,bc) - \eta(b\,|\,ac) - \eta(c\,|\,ab) \leq 0,$
- 3 inequalities of the form: $-\mathsf{c}(ab) - \mathsf{c}(ac) + \mathsf{c}(abc) \leq 0$,
  in family variables $-\eta(a\,|\,b) - \eta(a\,|\,c) - \eta(a\,|\,bc) - \eta(b\,|\,a) - \eta(c\,|\,a) \leq 0,$
- 1 inequality of the form: $-\mathsf{c}(ab) - \mathsf{c}(ac) - \mathsf{c}(bc) + 2 \cdot \mathsf{c}(abc) \leq 0$,
  in family variables $-\eta(a\,|\,b) - \eta(a\,|\,c) - \eta(b\,|\,a) - \eta(b\,|\,c) - \eta(c\,|\,a) - \eta(c\,|\,b) \leq 0.$

Note that the previous eight inequalities define – in family variables – sub-maximal SE faces of F that are not facets: they are implied by the non-negativity constraints.

EXAMPLE 7.2. [**17**, Example 4] In case $N = \{a, b, c, d\}$ one has $|\Lambda| = 11$ and the 11-dimensional polytope C has 185 vertices and 154 facets. Unlike the case of three BN variables, there are facets of the characteristic-imset polytope C which neither contain the 0-imset nor the 1-imset. The characteristic polytope has 358 fewer vertices than the family-variable polytope F, but 19 more facets than F. Besides the 37 facets that correspond to SE facets of F (See Example 6.3) and contain the 1-imset (Corollary 5.23), there exist 117 facets of C that do not contain the 1-imset. With the exception of one permutation type all the 117 inequalities are tight for the 0-imset. The exception is

$$\begin{aligned}
& -\mathsf{c}(bc) - \mathsf{c}(bd) - \mathsf{c}(cd) \\
& +\mathsf{c}(abc) + \mathsf{c}(abd) + \mathsf{c}(acd) + 2 \cdot \mathsf{c}(bcd) - 2 \cdot \mathsf{c}(abcd) \;\leq\; 1, \qquad (7.1)
\end{aligned}$$

in family variables
$$\begin{aligned}
& +\eta(a\,|\,bc) + \eta(a\,|\,bd) + \eta(a\,|\,cd) + \eta(a\,|\,bcd) \\
& -\eta(b\,|\,c) - \eta(b\,|\,d) - \eta(c\,|\,b) - \eta(c\,|\,d) - \eta(d\,|\,b) - \eta(d\,|\,c) \;\leq\; 1,
\end{aligned}$$

consisting of 4 inequalities. The η-version of the inequality (7.1), therefore, defines an inclusion submaximal SE face of F which is not a facet. It is not difficult to see that (7.1) follows from the modified convexity constraint

$$\eta(a\,|\,b) + \eta(a\,|\,c) + \eta(a\,|\,d) + \eta(a\,|\,bc) + \eta(a\,|\,bd) + \eta(a\,|\,cd) + \eta(a\,|\,bcd) \leq 1$$

and the non-negativity constraints $-\eta(a\,|\,b) \leq 0, \ldots, -\eta(d\,|\,c) \leq 0$.

The remaining 117 are defined by means of the so-called *specific inequalities* discussed in [**57**, § 4.1.2]. Each of these inequalities corresponds to a clutter (= Sperner family) of non-empty subsets of N, that is, to a class of inclusion-incomparable subsets of N. The 117 specific facets decompose into 20 permutation types. Except for the 4 facets in (7.1) all the inequalities contain the 0-imset.

- $-\mathsf{c}(ab) \leq 0$ (6 inequalities of this type),
  Sperner family is $\mathcal{I} = \{ab\}$,
- $-\mathsf{c}(abc) \leq 0$ (4 inequalities of this type),
  Sperner family is $\mathcal{I} = \{abc\}$,
- $-\mathsf{c}(abcd) \leq 0$ (1 inequality of this type),
  Sperner family is $\mathcal{I} = \{abcd\}$,
- $-\mathsf{c}(ab) - \mathsf{c}(ac) - \mathsf{c}(bc) + 2 \cdot \mathsf{c}(abc) \leq 0$ (4 inequalities of this type),
  Sperner family is $\mathcal{I} = \{ab, ac, bc\}$,
- $-\mathsf{c}(ab) - \mathsf{c}(acd) - \mathsf{c}(bcd) + 2 \cdot \mathsf{c}(abcd) \leq 0$ (6 inequalities of this type),
  Sperner family is $\mathcal{I} = \{ab, acd, bcd\}$,
- $-\mathsf{c}(abc) - \mathsf{c}(abd) - \mathsf{c}(acd) + 2 \cdot \mathsf{c}(abcd) \leq 0$ (4 inequalities of this type),
  Sperner family is $\mathcal{I} = \{abc, abd, acd\}$,
- $-\mathsf{c}(abc) - \mathsf{c}(abd) - \mathsf{c}(acd) - \mathsf{c}(bcd) + 3 \cdot \mathsf{c}(abcd) \leq 0$ (1 inequality),
  Sperner family is $\mathcal{I} = \{abc, abd, acd, bcd\}$,
- $-\mathsf{c}(ab) - \mathsf{c}(ac) - \mathsf{c}(ad) - \mathsf{c}(bcd) + \mathsf{c}(abc) + \mathsf{c}(abd) + \mathsf{c}(acd) \leq 0$ (4 inequalities),
  Sperner family is $\mathcal{I} = \{ab, ac, ad, bcd\}$,
- $-\mathsf{c}(ab) - \mathsf{c}(ac) - \mathsf{c}(ad) - \mathsf{c}(bc) - \mathsf{c}(bd)$
  $+2 \cdot \mathsf{c}(abc) + 2 \cdot \mathsf{c}(abd) + \mathsf{c}(acd) + \mathsf{c}(bcd) - 2 \cdot \mathsf{c}(abcd) \leq 0$ (6 inequalities),
  Sperner family is $\mathcal{I} = \{ab, ac, ad, bc, bd\}$,
- $-\mathsf{c}(ab) - \mathsf{c}(ac) - \mathsf{c}(ad) - \mathsf{c}(bc) - \mathsf{c}(bd) - \mathsf{c}(cd)$
  $+2 \cdot \mathsf{c}(abc) + 2 \cdot \mathsf{c}(abd) + 2 \cdot \mathsf{c}(acd) + 2 \cdot \mathsf{c}(bcd) - 3 \cdot \mathsf{c}(abcd) \leq 0$ (1 inequality),
  Sperner family is $\mathcal{I} = \{ab, ac, ad, bc, bd, cd\}$,
- $-\mathsf{c}(ab) - \mathsf{c}(ac) + \mathsf{c}(abc) \leq 0$ (12 inequalities of this type),
  Sperner family is $\mathcal{I} = \{ab, ac\}$,
- $-\mathsf{c}(abc) - \mathsf{c}(abd) + \mathsf{c}(abcd) \leq 0$ (6 inequalities of this type),
  Sperner family is $\mathcal{I} = \{abc, abd\}$,
- $-\mathsf{c}(ab) - \mathsf{c}(acd) + \mathsf{c}(abcd) \leq 0$ (12 inequalities of this type),
  Sperner family is $\mathcal{I} = \{ab, acd\}$,
- $-\mathsf{c}(ab) - \mathsf{c}(cd) + \mathsf{c}(abcd) \leq 0$ (3 inequalities of this type),
  Sperner family is $\mathcal{I} = \{ab, cd\}$,
- $-\mathsf{c}(ab) - \mathsf{c}(ac) - \mathsf{c}(ad) + \mathsf{c}(abc) + \mathsf{c}(abd) + \mathsf{c}(acd) - \mathsf{c}(abcd) \leq 0$ (4 inequalities),
  Sperner family is $\mathcal{I} = \{ab, ac, ad\}$,
- $-\mathsf{c}(ab) - \mathsf{c}(ac) - \mathsf{c}(bd) + \mathsf{c}(abc) + \mathsf{c}(abd) \leq 0$ (12 inequalities of this type),
  Sperner family is $\mathcal{I} = \{ab, ac, bd\}$,
- $-\mathsf{c}(ab) - \mathsf{c}(ac) - \mathsf{c}(bcd) + \mathsf{c}(abc) + \mathsf{c}(abcd) \leq 0$ (12 inequalities of this type),
  Sperner family is $\mathcal{I} = \{ab, ac, bcd\}$,
- $-\mathsf{c}(ab) - \mathsf{c}(ac) - \mathsf{c}(bc) - \mathsf{c}(cd) + 2 \cdot \mathsf{c}(abc) + \mathsf{c}(acd) + \mathsf{c}(bcd) - \mathsf{c}(abcd) \leq 0$ (12 inequalities of this type),
  Sperner family is $\mathcal{I} = \{ab, ac, bc, cd\}$,

- $-\mathsf{c}(ab) - \mathsf{c}(ad) - \mathsf{c}(bc) - \mathsf{c}(cd)$
  $+\mathsf{c}(abc) + \mathsf{c}(abd) + \mathsf{c}(acd) + \mathsf{c}(bcd) - \mathsf{c}(abcd) \leq 0$ (3 inequalities),
  Sperner family is $\mathcal{I} = \{ab, ad, bc, cd\}$,
- $-\mathsf{c}(bc) - \mathsf{c}(bd) - \mathsf{c}(cd) + \mathsf{c}(abc) + \mathsf{c}(abd) + \mathsf{c}(acd) + 2\cdot\mathsf{c}(bcd) - 2\cdot\mathsf{c}(abcd) \leq 1$ (4 inequalities of this type),
  Sperner family is $\mathcal{I} = \{a, bc, bd, cd\}$. This is the exception which does not contain the 0-imset described by Equation (7.1).

8. Practical Approaches to Learning BN

A short history of practical polyhedral approaches to learning BN structure are briefly overviewed here. Since the full facet description of either the family variable polytope F or the characteristic imset polytope C are not known, practical approaches rely on linear relaxations and tools from integer and linear programming. As such, the approaches described below mostly rely on various linear relaxations of either F or C or extensions and hybridizations of those polytopes. A full description of each LP relaxation would be considerably lengthy, and are thus omitted. Instead the general approach, principles, and origins are described.

8.1. Search Space Reduction. One of the most efficient first steps for learning BN structure is a reduction of the search space initially discovered in [**63**, § 3.2]. Provided $i \in N$ and $C \subset B \subseteq N \setminus \{i\}$, the observed database D, and $q_D(i\,|\,C) > q_D(i\,|\,B)$ it follows from score-decomposability that no $G \in \mathsf{DAGS}\,(N)$ maximizing $\mathcal{Q}(G, D)$ has $pa_G(i) = B$. Thus, one does not need consider variable $\eta(i\,|\,B)$, that is, the variable $\eta(i\,|\,B)$ can be *pruned*. A similar statement can be made in the case of the characteristic imset polytope C [**58**, § 3.2]: If $\mathcal{Q}$ is a decomposable criterion and $S \subseteq N$, $|S| \geq 2$ such that

$$\forall i \in S \;\; \forall B,\, S \setminus \{i\} \subseteq B \subseteq N \setminus \{i\} \quad \exists C \subset B \;\, q_D(i|C) > q_D(i|B)\,,$$

then (3.4) implies that there is no optimal (acyclic directed) graph G with $\mathsf{c}_G(S) = 1$. In this case, the search for the optimal graph can be limited to graphs G with $\mathsf{c}_G(S) = 0$. Note that in general one needs to check exponentially many conditions.

A stronger form of pruning exists owing to the fact that typical scoring criteria prefer sparse solutions. For instance, the BIC criteria has a penalty for occurrences large parent sets. Two pruning conditions on BIC and one condition on the BDE criteria were given in [**20**, Theorem 2, 4, 9] [**21**, Theorem 1]. The important property of these pruning criteria is that once a $B \subseteq N$ is discovered which violates some inequality, not only can B be pruned, but any superset B' of B where $B \subseteq B' \subseteq N$ can be pruned. That is, one can *exponentially prune* many components of the η variables. A strengthening of Theorem 2 from [**20**] and Theorem 1 from [**21**] is given in [**55**, Corollary 7.1], allowing more components to be pruned. Moreover, Corollary 7.1 in [**55**] states that if a database is not long enough then one can exclude parent sets of large size. As stated above, pruning the family variables η can directly lead to pruning of the characteristic imset variables.

The pruning procedure can yield a dramatic reduction in the search space, typically offsetting the computational expense of pruning. In [**20**] a pruning procedure was applied to some databases from the UCI repository [**34**]. In most cases the parent sets were restricted to at most 5, and in only a few cases 6 and 7. Explicit algorithms for pruning in the family variable space were never published to the best knowledge of the authors. Some algorithms are proposed in [**55**] for the characteristic imset case.

In practice, often pruning is not performed at all and an upper bound is simply placed on the size of possible parent sets, leading to a dramatic reduction in the number of variables and hence speeding up the solver. However, it should be noted that in [**18**, § 5] it was shown that any general BN structure learning problem can be converted to a BN structure learning problem over more variables but with parent set size limited to 2 with the same set of optimal solutions. Hence, both the number of vertices and the parent set size must be considered when judging the difficulty of learning BN structure.

8.2. Family Variable Approaches.

8.2.1. *Jaakkola et. al.* Jaakkola et. al [**28**] first proposed to use the η encoding of directed graphs for the purpose of BN structure learning, independently and around the same time as Cussens et. al [**14, 16**]. In Jaakkola et. al's approach they used the family variable η encoding along with the non-negative inequalities (5.1), the convexity inequalities (5.2), as well as introducing the cluster inequalities (3.1). As stated in Remark 3.3 and originally shown in [**28**], these inequalities provide a linear relaxation of the family variable polytope F which allows one to use methods of integer programming (IP). Additionally, Jaakkola et. al employed pruning (§8.1) to reduce the solution space. The main computational challenge remaining was the fact that there are exponentially many cluster inequalities. To overcome this, Jaakkola et. al used iterative constraint adding: First only the non-negative and convexity constraints are added and an IP solver is applied. If the returned solution violates a cluster inequality, then this is not a valid BN structure solution, and the violated cluster inequality is added. Again, the IP solver is applied and the process of adding violated constraints is performed until the IP solver returns a solution which does not violate a cluster inequality. Hence, it is a solution to the BN structure learning problem. What was special in their approach is that they used the dual LP problem formulation as a tool for guiding which cluster constraint to add. Jaakkola et. al also combined iterative constraint adding with the branch-and-bound approach [**3, 31**]. In their original article, Jaakkola et. al outperformed both dynamic programming [**48**] and branch-and-bound approaches [**21**] on biological and UCI [**34**] data sets.

8.2.2. *Cussens et. al.* Originally Cussens et. al [**16, 47**] was interested in pedigree learning, in which case the parent set cardinality is bounded by 2. They too used the η encoding of directed acyclic graphs along with the non-negativity constraints (5.1) and convexity constraints (5.2). However, to ensure the acyclicity of the graph G they used another trick: adding additional components to the vector η_G which allows encoding of the total order of nodes consistent with the direction of arrows in the graph G. Additional constraints were added to this representation and Cussens et. al proved that it was indeed an LP relaxation. Moreover the number of total constraints added was polynomial in $|N|$. Overall this led to very fast solutions to the pedigree problem. For example, in [**16**] they were able to solve a pedigree problem with $|N| = 1614$ in about 3-42 minutes.

Cussens [**21**] later applied similar integer programming approaches to learning BN structure, independently using the family variable encoding used in Jaakkola et. al [**28**]. Thus, he used the non-negativity inequalities (5.1), the convexity inequalities (5.2), as well adopting the cluster inequalities (3.1) used in [**28**]. However, Cussens went even further and developed and used the generalized cluster inequalities [**28**]. SCIP [**2**] was employed as the IP solver (which performs branch-and-bound and addition of general cutting planes) and Cussens et. al developed the software GOBNILP [**15**] which incorporates all their BN learning methods. Their approach uses branch-and-bound, cutting planes (Gomory, etc.), and iterative constraint adding for the generalized cluster inequalities. In contrast to Jaakkola et. al's [**16**] search for violated cluster constraints in the dual space, Cussens performed the addition of generalized cluster inequalities in the primal space by the use of a special sub-IP problem which finds the most violated constraint. Even though the search for violated cluster inequalities appears quite fast in practice, Cussens et. al recently showed this separation problem is NP-hard [**18**, Theorem 6]. Many additional improvements have been added to GOBNILP [**4**, **18**], and GOBNILP remains under active development. For example GOBNILP incorporates results presented in [**4**], where a few key ideas are summarized as:

- The characteristic variables satisfy $\mathsf{c}(S) \leq 1$, and there is a corresponding (k-cluster) inequality on the family variables. All such non-redundant inequalities, called *set packing inequalities*, are added unconditionally. Adding the set packing inequalities has been shown to speed up solution times.
- GOBNILP creates additional binary variables representing directed edges and also undirected edges in the undirected skeleton. GOBNILP is then forced to branch on these additional edge variables and not on the family variables, leading to a more balanced search tree.
- GOBNILP has heuristics to generate a BN structure by "rounding" the current LP (possibly non-integer) solution. It has been shown that getting a good incumbent solution early can help speed up the branch-and-bound phase.
- Cussens et. al have also experimented with many hyper-parameters settings to improve solutions times. For example, GOBNILP gives up looking for cuts in the root node of branch-and-bound more easily. Also, a time limit is placed on the sub-IP which searches for violated inequalities. Both these measures prevent GOBNILP spending too long looking for (perhaps not very useful) cuts, and start branch-and-bound earlier.

GOBNILP currently has some of the best running times for exact BN structure learning methods, and also has the benefit that the search can be terminated at any time and a current best solution will be given. See [**15**] for the latest software and experimental results.

8.3. Characteristic Imset Polytope. Characteristic imsets have the useful benefit of encoding Markov equivalence classes of Bayesian networks as well as having a shorter encoding than the family variables. This is also true for their predecessor the standard imsets (which are in one-to-one correspondence with characteristic imsets [**26**]). However, the characteristic imsets have a nice graphical interpretation (Theorem 3.4) and are zero-one integer encodings. Thus, having these nice properties, efforts began towards integer programming approaches to learn BN structure at the inception of standard and characteristic imsets. The path towards practical applications of IP to learning BN using characteristic imsets is slightly longer than the family variable case.

Studený et. al [**62**] first abstractly defined the standard imset polytope and showed that BN structure learning using a decomposable and score-equivalent function is an affine function of this polytope. They also studied the edges of the standard-imset polytope and showed a failure of the GES algorithm [**9**, **39**] if the data are not generated from a perfectly Markovian distribution. This gave some evidence of the superiority of exact polyhedral approaches in such cases. Note that edges of the standard and characteristic imset are called *geometric neighbors* and some important sub-classes of the edges are known.

Shortly after, Studený et. al [**61**] defined three sets of valid inequalities over the standard imset polytope and conjectured that these defined the standard imset polytope, or more weakly, that they at least provided an LP relaxation. The three classes of inequalities given are (See [**61**, § 3]):

- Equality constraints, a linear number with respect to $|N|$,
- Non-specific constraints or constraints given by the supermodular functions (appearing in §5.1, and their extreme counterparts defining facets of the family variable polytope F and the characteristic imset polytope C). The number of these constraints unfortunately appear to be superexponential in $|N|$.
- Specific constraints, which correspond to clutters (= Sperner families), which are exponential in size in $|N|$.

The two main drawback to these sets of inequalities is that they are implicit, and also possibly super-exponential in $|N|$, although many of the inequalities do end up being superfluous.

Around the same time characteristic imsets were proposed in [**60**] and later the primary ground work for IP approaches was laid out by Hemmecke et. al in [**26**]. Therein, they gave a simple transition between the characteristic imset and the essential graph, extensions of the characteristic imset to chain graphs without flags, and an elegant geometric interpretation to a classic procedure for approximating probability distributions with trees [**11**], among other things. Two of the most important properties shown by Hemmecke et. al was that the characteristic imset is a zero-one vector (a very nice property for a polytope to have) and a close graphical description (See Theorem 3.4). Thus, efforts shifted away from the standard imset representation and focused on the characteristic imsets.

It wasn't until Studený et. al [**57**] that the weaker conjecture from [**61**] was proved, that is, the three classes of inequalities above (equality, non-specific, specific) indeed provide a linear relaxation of the standard imset polytope, and thus by the one-to-one mapping a linear relaxation of the characteristic imset polytope. They relied on Jaakkola et. al's [**28**] result that the family variable encoding therein is a linear relaxation of the family variable polytope F, and the mapping of family variables to characteristic imsets (Equation 3.4), to prove the weaker conjecture. However, an unpleasant finding was that, while there is a one-to-one correspondence for Jaakkola et. al's acyclicity-encoding inequalities, their basic non-negativity and equality constraints are transformed to much higher number of inequalities. Thus, direct transformation of their inequalities does not lead to a suitable ILP problem formulation in terms of characteristic imsets. Directly stemming from work in [**26**] and promising results in [**57**] suggesting linear relaxations of the characteristic imset were attainable, a few practical approaches to learning BN structure using characteristic imsets were developed.

Simultaneously, in her thesis, Lindner [**35**] provided a different linear relaxation, in this case using both characteristic imset variables as well as variables encoding the edges of directed graphs directly. She was able to show some promising results on some UCI [**34**] datasets (where she also employed pruning and parent size limits). Therein she also used branch-and-bound coupled with iterative constraint adding to deal with the overload of inequalities. Lindner also provided many useful related results on learning over chordal graphs, trees, the edges of the polytope, and complexity constraints. Some of these result were also published in [**59**] where theoretical and complexity results on learning undirected forests and chordal graphs were also provided.

Studený et. al [**58**] then came with a new provable linear relaxation where both characteristic imset and edge variables were used. In fact the relaxation contained encodings for chordal graphs, and involved two steps: computing an optimal IP solution, and then computing either the essential graph or a directed acyclic graph representative of the Markov equivalence class. The first set of inequalities provided basic constraints on the arrow variables and the characteristic imset variables $\mathsf{c}(S)$ where $|S| \leq 3$, and the total number of such inequalities was polynomial in $|N|$. Since the characteristic imset is uniquely determined by its variables corresponding to sets $S \subseteq N$ of cardinality three or less, the next a class of inequalities extended these to the variables of higher cardinality. These were exponential in size, but owing to pruning and parent set size limitations, were often polynomial in size in practice. The last set of inequalities were the now familiar cluster inequalities, which are exponential in size. Thus, a branch-and-bound approach combined with iterative constraint adding was used based on the ILOG CPLEX [**1**] solver. The experiments proved one can use IP and characteristic imsets to get the Markov equivalence class (essential graph). Though the goal was to verify the ability to learn BN structure using characteristic imsets, it was evident that the direct family variable polytope F encoding used by Cussens et. al [**21**], and contained in GOBNILP [**15**], provided faster solutions. For one, although the characteristic imsets provide a more compact encoding than the family variables, it often was the case that the family variable encoding was able to gain more benefit from the pruning procedure. Additionally, even though many of the inequalities used by Studený et. al [**58**] were polynomial in size, they still slowed down the computational time.

Characteristic imset approaches encode the Markov equivalence classes and theoretically have a shorter representation, but thus far have more complicated linear relaxations. Whereas the direct family variable encoding encodes all directed graphs but has relatively simple linear relaxations. However, as shown in §5.5.1, the faces of the family variable polytope F and the characteristic imset polytope C are in fact linked in many ways. Moreover, the objectives used for BN structure learning on the family variable polytope F are score equivalent, and thus not all the faces are required (Theorem 5.25). This leads to a suggestion to use a combined coding of BN structures. That is, one can encode a BN structure by a concatenation of the family variable and the characteristic imset and utilize the linear relation (3.4). Hence, linear constraints tight at the empty graph are better represented by simple non-negativity and modified convexity inequalities in the family variable part, while the other SE linear inequality constraints can be more naturally represented in the characteristic-imset part.

References

[1] *ILOG CPLEX*, http://www.ilog.com/products/cplex, 2013.

[2] Tobias Achterberg, *SCIP: solving constraint integer programs*, Math. Program. Comput. **1** (2009), no. 1, 1–41, DOI 10.1007/s12532-008-0001-1. MR2520442

[3] Tobias Achterberg, Thorsten Koch, and Alexander Martin, *Branching rules revisited*, Oper. Res. Lett. **33** (2005), no. 1, 42–54, DOI 10.1016/j.orl.2004.04.002. MR2091034

[4] M. Bartlett and J. Cussens, *Advances in Bayesian network learning using integer programming*, Uncertainty in Artificial Intelligence (Corvallis), vol. 29, AUAI Press, 2013, pp. 182–191.

[5] Alexander Barvinok, *A course in convexity*, Graduate Studies in Mathematics, vol. 54, American Mathematical Society, Providence, RI, 2002. MR1940576

[6] Andrés Cano, Manuel Gómez-Olmedo, Andrés R. Masegosa, and Serafín Moral, *Locally averaged Bayesian Dirichlet metrics for learning the structure and the parameters of Bayesian networks*, Internat. J. Approx. Reason. **54** (2013), no. 4, 526–540, DOI 10.1016/j.ijar.2012.09.003. MR3041116

[7] David Maxwell Chickering, *Learning Bayesian networks is NP-complete*, Learning from data (Fort Lauderdale, FL, 1995), Lecture Notes in Statist., vol. 112, Springer, New York, 1996, pp. 121–130, DOI 10.1007/978-1-4612-2404-4_12. MR1473013

[8] David Maxwell Chickering, *A transformational characterization of equivalent Bayesian network structures*, Uncertainty in artificial intelligence (Montreal, PQ, 1995), Morgan Kaufmann, San Francisco, CA, 1995, pp. 87–98. MR1615012

[9] David Maxwell Chickering, *Optimal structure identification with greedy search*, Journal of Machine Learning Research **3** (2002), 505–554.

[10] D.M. Chickering and C. Meek, *Finding optimal Bayesian networks*, Proceedings of the Eighteenth conference on Uncertainty in artificial intelligence, Morgan Kaufmann Publishers Inc., 2002, pp. 94–102.

[11] C.K. Chow and C.N. Liu, *Approximating discrete probability distributions with dependence trees*, Information Theory, IEEE Transactions on **14** (1968), no. 3, 462–467.

[12] Michele Conforti, Gérard Cornuéjols, and Giacomo Zambelli, *Integer programming*, Graduate Texts in Mathematics, vol. 271, Springer, Cham, 2014. MR3237726

[13] G.F. Cooper and E. Herskovits, *A Bayesian method for the induction of probabilistic networks from data*, Machine Learning **9** (1992), 309–347.

[14] J. Cussens, *Bayesian network learning with cutting planes*, Uncertainty in Artificial Intelligence (F. Cozman and Pfeffer A., eds.), vol. 27, AUAI Press, 2011, pp. 153–160.

[15] J. Cussens and M. Bartlett, *GOBNILP software*, www.cs.york.ac.uk/aig/sw/gobnilp/.

[16] J. Cussens, M. Bartlett, E.M. Jones, and N.A. Sheehan, *Maximum likelihood pedigree reconstruction using integer programming*, Genetic Epidemiology **37** (2012), no. 1, 69–83.

[17] J. Cussens, D. Haws, and M. Studený, *Polyhedral aspects of score equivalence in Bayesian network structure learning*, 2015.

[18] J. Cussens, M. Järvisalo, J.H. Korhonen, and M. Bartlett, *Polyhedral theory for Bayesian network structure learning*, 2015.

[19] D.M. Chickering D. Heckerman, D. Geiger, *Learning Bayesian networks: the combination of knowledge and statistical data*, Machine Learning **20** (1995), 194–243.

[20] Cassio P. de Campos and Qiang Ji, *Efficient structure learning of Bayesian networks using constraints*, J. Mach. Learn. Res. **12** (2011), 663–689. MR2786909

[21] C.P. de Campos, Z. Zeng, and Q. Ji, *Structure learning Bayesian networks using constraints*, Proceedings of the 26th International Conference on Machine Learning (ICML), 2009, pp. 113–120.

[22] Luis M. de Campos and Juan F. Huete, *A new approach for learning belief networks using independence criteria*, Internat. J. Approx. Reason. **24** (2000), no. 1, 11–37, DOI 10.1016/S0888-613X(99)00042-0. New perspectives on causal networks (Lisbon, 1998). MR1759736

[23] Orlinskaya G., *Linear constraints on standard and characteristic imsets for learning Bayesian network structures*, Ph.D. thesis, TU Munich, 2014.

[24] Paolo Giudici and Peter J. Green, *Decomposable graphical Gaussian model determination*, Biometrika **86** (1999), no. 4, 785–801, DOI 10.1093/biomet/86.4.785. MR1741977
[25] P. L. Hammer, E. L. Johnson, and U. N. Peled, *Facets of regular* $0-1$ *polytopes*, Math. Programming **8** (1975), 179–206. MR0384128
[26] Raymond Hemmecke, Silvia Lindner, and Milan Studený, *Characteristic imsets for learning Bayesian network structure*, Internat. J. Approx. Reason. **53** (2012), no. 9, 1336–1349, DOI 10.1016/j.ijar.2012.04.001. MR2994270
[27] E. Herskovits and G.F. Cooper, *An entropy-driven system for the construction of probabilistic expert systems from datasets*, Proceedings of the Sixth Conference on Uncertainty in Artificial Intelligence (Cambridge) (Bonissone P., ed.), 1990, pp. 54–62.
[28] T. Jaakkola, D. Sontag, A. Globerson, and M. Meila, *Learning Bayesian network structure using LP relaxations*, Journal of Machine Learning Research Workshop and Conference Proceedings (Teh Y.W. and Titterington M., eds.), vol. 9, AISTATS, 2010.
[29] Daphne Koller and Nir Friedman, *Probabilistic graphical models*, Adaptive Computation and Machine Learning, MIT Press, Cambridge, MA, 2009. Principles and techniques. MR2778120
[30] W. Lam and F. Bacchus, *Learning Bayesian belief networks: An approach based on the MDL principle*, Computational Intelligence **10** (1994), 269–293.
[31] A. Land and S. Powell, *Computer codes for problems of integer programming*, Ann. Discrete Math. **5** (1979), 221–269. Discrete optimization (Proc. Adv. Res. Inst. Discrete Optimization and Systems Appl., Banff, Alta., 1977), II. MR558573
[32] Steffen L. Lauritzen, *Graphical models*, Oxford Statistical Science Series, vol. 17, The Clarendon Press, Oxford University Press, New York, 1996. Oxford Science Publications. MR1419991
[33] Faming Liang and Jian Zhang, *Learning Bayesian networks for discrete data*, Comput. Statist. Data Anal. **53** (2009), no. 4, 865–876, DOI 10.1016/j.csda.2008.10.007. MR2657052
[34] M. Lichman, *UCI machine learning repository*, 2013.
[35] S. Lindner, *Discrete optimization in machine learning: learning Bayesian network structures and conditional independence implication*, Ph.D. thesis, TU Munich, 2012.
[36] Milan Studený and Tomáš Kroupa, *Core-based criterion for extreme supermodular functions*, Discrete Appl. Math. **206** (2016), 122–151, DOI 10.1016/j.dam.2016.01.019. MR3490436
[37] D. Madigan and E. Raftery, *Model selection and accounting for model uncertainty in graphical models using Occam's window*, Journal of American Statistical Association **89** (1994), 1535–1546.
[38] D. Madigan and J. York, *Bayesian graphical models for discrete data*, International Statistics Revie **63** (1995), 215–232.
[39] C. Meek, *Graphical models: Selecting causal and statistical models*, Ph.D. thesis, Carnegie Mellon University, 1997.
[40] R.E. Neapolitan, *Learning Bayesian networks*, Pearson Prentice Hall, Upper Saddle River, 2004.
[41] Hien Quang Nguyen, *Semimodular functions and combinatorial geometries*, Trans. Amer. Math. Soc. **238** (1978), 355–383. MR0491269
[42] M. Niepert, *Logical inference algorithms and matrix representations for probabilistic conditional independence*, Proceedings of the Twenty-Fifth Conference on Uncertainty in Artificial Intelligence, AUAI Press, 2009, pp. 428–435.
[43] James G. Oxley, *Matroid theory*, Oxford Science Publications, The Clarendon Press, Oxford University Press, New York, 1992. MR1207587
[44] Judea Pearl, *Probabilistic reasoning in intelligent systems: networks of plausible inference*, The Morgan Kaufmann Series in Representation and Reasoning, Morgan Kaufmann, San Mateo, CA, 1988. MR965765
[45] Alexander Schrijver, *Theory of linear and integer programming*, Wiley-Interscience Series in Discrete Mathematics, John Wiley & Sons, Ltd., Chichester, 1986. A Wiley-Interscience Publication. MR874114
[46] Gideon Schwarz, *Estimating the dimension of a model*, Ann. Statist. **6** (1978), no. 2, 461–464. MR0468014
[47] N. Sheehan, M. Bartlett, and J. Cussens, *Improved maximum likelihood reconstruction of complex multi-generational pedigrees*, Theoretical Population Biology **97** (2014), 11–19.

[48] T. Silander and P. Myllymäki, *A simple approach for finding the globally optimal Bayesian network structure*, Uncertainty in Artificial Intelligence (R. Dechter and T. Richardson, eds.), AUAI Press, 2006, pp. 445–452.

[49] Peter Spirtes, Clark Glymour, and Richard Scheines, *Causation, prediction, and search*, Lecture Notes in Statistics, vol. 81, Springer-Verlag, New York, 1993. MR1227558

[50] Richard P. Stanley, *Enumerative combinatorics. Vol. 1*, Cambridge Studies in Advanced Mathematics, vol. 49, Cambridge University Press, Cambridge, 1997. With a foreword by Gian-Carlo Rota; Corrected reprint of the 1986 original. MR1442260

[51] Milan Studený, *A recovery algorithm for chain graphs*, Internat. J. Approx. Reason. **17** (1997), no. 2-3, 265–293, DOI 10.1016/S0888-613X(97)00018-2. 1996 Uncertainty in AI (UAI '96) Conference. MR1462716

[52] Milan Studený, *Probabilistic conditional independence structures*, Information Science and Statistics, Springer, London, 2005. MR3183760

[53] Milan Studený, *Mathematical aspects of learning Bayesian networks: Bayesian quality criteria*, Research Report 2234, Institute of Information Theory and Automation, 2008.

[54] Milan Studený, *Integer linear programming approach to learning Bayesian network structure: towards the essential graph*, Proceedings of the 6th European Workshop on Probabilistic Graphical Models (PGM), 2012, pp. 307–314.

[55] Milan Studený, *LP relaxations and pruning for characteristic imsets*, (2012), no. research report n. 2323.

[56] Milan Studený, *How matroids occur in the context of learning Bayesian network structures*, 31st Conference on Uncertainty in Artificial Intelligence, 2015.

[57] Milan Studený and David C. Haws, *On polyhedral approximations of polytopes for learning Bayesian networks*, J. Algebr. Stat. **4** (2013), no. 1, 59–92, DOI 10.18409/jas.v4i1.19. MR3084211

[58] Milan Studený and David Haws, *Learning Bayesian network structure: towards the essential graph by integer linear programming tools*, Internat. J. Approx. Reason. **55** (2014), no. 4, 1043–1071, DOI 10.1016/j.ijar.2013.09.016. MR3178416

[59] Milan Studený, David Haws, Raymond Hemmecke, and Silvia Lindner, *Polyhedral approach to statistical learning graphical models*, Harmony of Gröbner bases and the modern industrial society, World Sci. Publ., Hackensack, NJ, 2012, pp. 346–372, DOI 10.1142/9789814383462_0020. MR2986888

[60] M. Studený, R. Hemmecke, and S. Lindner, *Characteristic imset: a simple algebraic representative of a Bayesian network structure*, Proceedings of the 5th European Workshop on Probabilistic Graphical Models (PGM) (P. Myllymäki, T. Roos, and T. Jaakkola, eds.), HIIT Publications, 2010, pp. 257–264.

[61] Milan Studený and Jiří Vomlel, *On open questions in the geometric approach to structural learning Bayesian nets*, Internat. J. Approx. Reason. **52** (2011), no. 5, 627–640, DOI 10.1016/j.ijar.2010.09.004. MR2787022

[62] Milan Studený, Jiří Vomlel, and Raymond Hemmecke, *A geometric view on learning Bayesian network structures*, Internat. J. Approx. Reason. **51** (2010), no. 5, 573–586, DOI 10.1016/j.ijar.2010.01.014. MR2644598

[63] M. Teyssier and D. Koller, *Ordering-based search: a simple and effective algorithm for learning Bayesian networks*, Proceedings of the 21st Conference on Uncertainty in Artificial Intelligence (UAI), 2005, pp. 584–590.

[64] T. Verma and J. Pearl, *Equivalence and synthesis of causal models*, Proceedings of the 6th Conference on Uncertainty in Artificial Intelligence (UAI), 1991, pp. 220–227.

[65] Nanny Wermuth and Steffen L. Lauritzen, *Graphical and recursive models for contingency tables*, Biometrika **70** (1983), no. 3, 537–552, DOI 10.2307/2336490. MR725368

[66] Laurence A. Wolsey, *Integer programming*, Wiley-Interscience Series in Discrete Mathematics and Optimization, John Wiley & Sons, Inc., New York, 1998. A Wiley-Interscience Publication. MR1641246

[67] Günter M. Ziegler, *Lectures on polytopes*, Graduate Texts in Mathematics, vol. 152, Springer-Verlag, New York, 1995. MR1311028

IBM Thomas J. Watson Research Center, Yorktown Heights, 1101 Kitchwan Road 134, New York
E-mail address: dchaws@gmail.com

University of York, Deramore Lane, York, YO10 5GE, United Kingdom
E-mail address: james.cussens@york.ac.uk

The Institute of Information Theory and Automation of the CAS, Pod Vodárenskou věží 4, Prague, 182 08, Czech Republic
E-mail address: studeny@utia.cas.cz

Contemporary Mathematics
Volume **685**, 2017
http://dx.doi.org/10.1090/conm/685/13814

Neural network coding of natural images with applications to pure mathematics

Christopher J. Hillar and Sarah E. Marzen

1. Introduction

The purpose of this chapter is to explain how a biologically-inspired decomposition of natural image patches [**39, 62**] can inspire interesting and even useful mathematics. In the process, we shall interact with several applied fields such as signal processing, neural memory modeling, statistical learning theory, and information theory. Our starting point is the problem of lossy digital image compression, a subject clearly having immediate practical applications to several industries but not usually associated with abstract mathematics or neural networks. Nonetheless, this engineering problem and a neural network approach to solving it can yield theorems (and open problems) at the interface of several mathematical and scientific disciplines.

To date, the field of lossy image compression has been mainly dominated by so-called *linear coding* methods. Typically, these schemes operate on 8×8 grayscale (8-bit) image patches $\mathbf{x} \in \{0, 1, \ldots, 255\}^{8\times 8} \subset \mathbb{R}^{64}$ and approximate them as:

$$\mathbf{x} \approx c_1\mathbf{b}_1 + \cdots + c_k\mathbf{b}_k, \tag{1}$$

using a particular basis $B = \{\mathbf{b}_1, \ldots, \mathbf{b}_{64}\} \subset \mathbb{R}^{64}$. The number $k = 1, \ldots, 64$ of real coefficients $\{c_1, \ldots, c_k\} \subset \mathbb{R}$ chosen to code $\mathbf{x}$ determines the compression level or quality, much as projecting data onto its top components in a principal components analysis (PCA) reduces its dimensionality. Commonly, the discrete cosine transform (DCT) basis is used for B (e.g., JPEG [**103**]), but others are also popular, such as wavelet bases (e.g., JPEG 2000 [**87**]) or those learned from natural data (e.g., PCA or ICA bases [**9**]). In the case of the DCT basis, the first term $c_1\mathbf{b}_1$ represents the mean ("DC component") of the patch, while the other terms correspond to weighted 2-D sinusoidal features $\mathbf{b}_k$ of increasing spatial frequency (see Fig. 5**a**).

2010 *Mathematics Subject Classification.* Primary 68T05, 82C32, 92B20, 94A08, 94A34.

The first author was funded by NSF Grant IIS-0917342 and an NSF All-Institutes Postdoctoral Fellowship administered by the Mathematical Sciences Research Institute through core grant DMS-0441170.

The second author was supported by an NSF Graduate Research Fellowship.

Here, we explore a recent alternative[1] to linear coding that uses a *discrete recurrent neural network* (DRNN) trained on natural data to compress grayscale images. The main idea is deceptively simple: decompose an image patch $\mathbf{x}$ into a mean μ, a standard deviation σ, and a discrete structure $\mathbf{D}$ computed from the attractor of a recurrent network dynamics; that is,

$$\mathbf{x} \approx \mu + \sigma \mathbf{D}. \tag{2}$$

In Fig. 1, we display two example codings using the scheme, while Fig. **4b** summarizes in more detail all of its main components (elaborated upon in Section 3 below). Whereas many lossy coders utilize linear algebra to represent the vast number 2^{512} of possible 8×8 image patches, the DRNN model gets its expressive power from an exponential repertoire of structures $\mathbf{D}$.

Although able to compress images at high perceptual quality with competitive rates, this new DRNN codingstrategy faces a number of conceptual challenges that arise from conventional intuition for discrete networks and image processing. Most of these concerns are summarized in Table 1 and resolving them required either new results or better understanding of old ones. In fact, this chapter represents an attempt to rigorously address each of the puzzling issues that are raised in this table.

Our first hurdle for approximating patches using a formula such as (2) is that the classical auto-associative neural network used to determine the discrete structure $\mathbf{D}$ has long been believed to have a capacity limitation of at most 2 memories per neuron [**16**]. Such a constraint represents a major obstacle for the method since it limits the number of possible expressions (2) for reconstructing image patches. Contrary to popular belief, however, robust storage of an exponential number of attractors is not only possible in such architectures but also can be learned directly from exceptionally small amounts of data [**41**]. This surprising mathematical finding is described in Section 2.d (specifically, Theorem 1) and provides intuition for how special connectivity structures and pattern collections allow networks to violate known capacity bounds.

Second, we confront several concerns involving the modeling of structure in natural image data. One issue is that the network (seemingly) should only be able to capture "second-order" structure since the underlying maximum entropy model is determined by its mean and covariance, whereas "higher-order" statistics appear to be fundamental for human perception. Another is that while the DRNN scheme performs well when a standard perceptual similarity measure [**105**] is used to quantify reconstruction quality (Fig. **6bd**), it is usually worse than JPEG when using a mean-squared pixel error (Fig. **6a**).

To better understand how the DRNN utilizes its statistical model of natural data to compress at such high perceptual fidelity, we performed a rate-distortion (RD) analysis in the more manageable case of coding 2×2 image patches. Using this tool from information theory, we find that the deterministic poset structure of the DRNN dynamics, rather than precise statistical model fit, is key to understanding its high-quality lossy coding of images. In particular, we demonstrate that despite all of the issues formulated in Table 1, a DRNN coding of 2×2 discretized natural image patches is near-optimal (Fig. 9).

---

[1] The literature also contains other diverse approaches to lossy image compression; e.g., fractal [**48**], neural networks [**19**], compressed sensing [**32**].

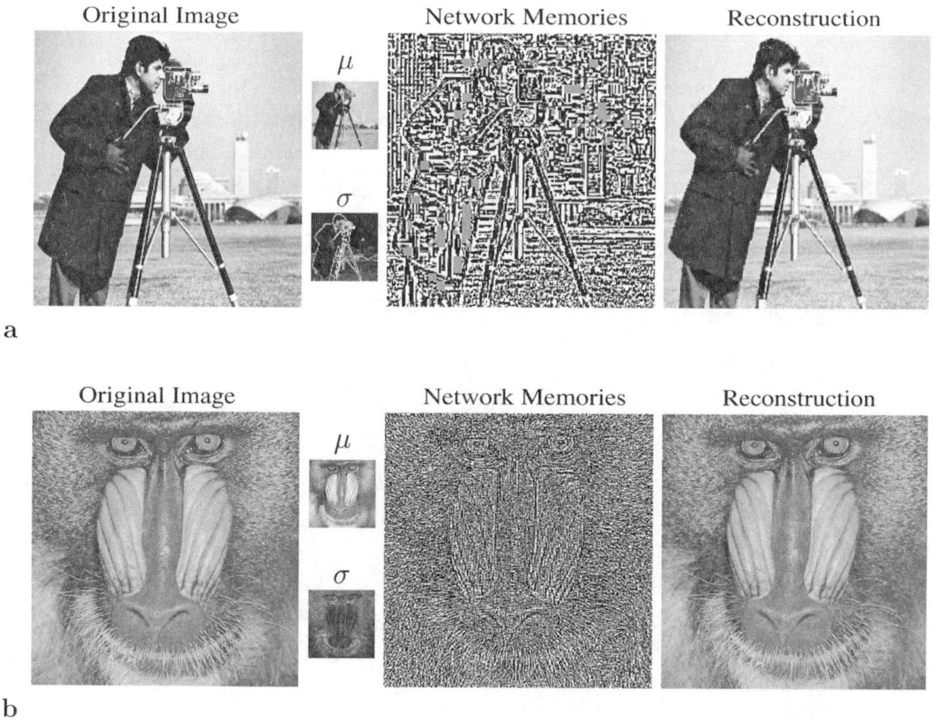

FIGURE 1. **Compressing grayscale images with a discrete recurrent neural network (DRNN) trained on natural data [39, 62].** A digital image to compress is first tiled into non-overlapping 4×4 patches and then means μ and standard deviations σ are extracted per patch. The remainder of each patch is represented by an ON/OFF ternary (white, gray, black) attractor of the neural network, which is entropy coded. To reconstruct the image, each memory is replaced with a continuous representative (e.g., Fig. 5**b**), and means and standard deviations are restored. **a)** 256×256 "cameraman"; 64×64 patch means μ and standard deviations σ. **b)** 512×512 "baboon" with additive Gaussian white noise (s.d. 5); 128×128 μ, σ. Coding quality of this image matches JPEG as shown in Fig. 6**b** ("baboon noisy").

Finally, most compression methodologies view natural images as continuous objects, which are to be approximated by sparse linear combinations of continuous "basis functions" with real-valued coefficients that require quantization. On the other hand, the DRNN approach reconstructs an image patch using a single binary structure and two patch statistics, thereby taking a fundamentally discrete view of natural images. Although rather uncommon in image processing applications, this perspective is partially justified by Theorem 3 in Section 7, where we prove quite generally that optimal codings (in a rate-distortion sense) of continuous input signals usually have discrete reproduction distributions. Discovered decades ago in various contexts [**26, 76, 90**] but largely ignored, this striking observation opens up several opportunities for new mathematics and algorithms. Here, we correct small gaps in previous proofs of the theorem and also generalize it somewhat. We remark that it is already a very difficult math problem to determine the finite number of points that optimally code these distributions given a desired distortion. For an

Conventional Wisdom	Conceptual Issue	§
The collection of possible natural images is vast	Neural networks typically store a small number of memories	2
DRNN coding is worse than JPEG on mean-squared error (MSE)	DRNN fit to distribution of input rather than minimizing MSE	3,4
Natural images have "higher-order" statistical structure	Lenz-Ising models are determined by mean and covariance	4
Good lossy compression should require good models of data	Lenz-Ising imperfectly models discretized natural images	5,6
Natural images are continuous objects	Discrete neural networks code for discrete alphabets	5,7

TABLE 1. Issues facing DRNN codings of natural images. (see Table 2 for resolutions).

example of how surprisingly complex such codes can be, see Fig. 10 where we display optimal output alphabets of size less than ten for coding the uniform distribution with a mean-squared error distortion.

The organization of this chapter is as follows. In Sections 2 and 3, we provide the necessary background for understanding the DRNN coding scheme of [**39**]. The next sections then go on to present in-depth investigations of the concerns outlined in Table 1. A final summary of our resolutions to Table 1 can be found at the end of this chapter in Table 2.

2. Discrete memory models in neuroscience

The key object in this chapter will be a symmetrically-connected McCulloch-Pitts [**59**] linear threshold binary recurrent neural network, otherwise known in the literature as a Hopfield network [**43**] or a single-layer auto-associative perceptron [**77**] with symmetric weight matrix. This classical discrete model of neural computation had a significant impact on several domains other than neuroscience. For instance, it was the single work cited (twice) in computing pioneer John von Neumann's design of the first electronic stored-program computer architecture (EDVAC) [**102**].

The original intent of the authors of [**59**], though, was primarily to model the computation of synapses and spiking neurons in a brain (e.g., Fig. 4**a**). Thus, their ability to simulate a calculus for propositional logic using a "theory of nets" or to emulate Turing machines [**98**], a mathematical model of computation, was mainly a proof-of-concept for their approach to neuroscience theory. See [**72**] for a close analysis of this work and its implications for natural philosophy and the history of science.

We first describe the underlying model of data in these neural networks and then explain some of their properties.

2.a. Lenz-Ising model. Our starting point is the non-ferromagnetic[2] *Lenz-Ising model* [**47**] from statistical physics, more generally called a *Markov random field* in the machine learning literature, and the underlying probability distribution of a fully observable *Boltzmann machine* [**1**] in artificial intelligence. This discrete distribution has as states all length n binary column vectors, with the probability $p(\mathbf{x})$ of a particular state $\mathbf{x} = (x_1, \ldots, x_n) \in \{0,1\}^n$ determined by the parameterization:

$$p(\mathbf{x}) := \frac{1}{Z} \exp\left(\sum_{i<j} W_{ij} x_i x_j - \sum_{i=1}^{n} \theta_i x_i \right) = \frac{1}{Z} \exp\left(-E_{\mathbf{x}}\right), \tag{3}$$

in which $W = W^\top \in \mathbb{R}^{n \times n}$ is a symmetric matrix with zero diagonal (the *weight matrix*), the column vector $\theta \in \mathbb{R}^n$ is a bias or *threshold* term, and $Z = \sum_{\mathbf{x}} \exp(-E_{\mathbf{x}})$ is the *partition function* (which normalizes p to sum to 1).

It is often useful to view the expression inside the exponential of (3) as the negative of an energy function:

$$E_{\mathbf{x}} := -\frac{1}{2}\mathbf{x}^\top W \mathbf{x} + \theta^\top \mathbf{x}. \tag{4}$$

States $\mathbf{x}$ with low energy (4) appear often when sampling from (3). We also remark (see [**17**, Theorem 12.1.1] and Section 4) that (3) is the maximum entropy[3] distribution on binary vectors given its first- and second-order statistics (mean and covariance).

2.b. Hopfield networks. Discovered first by Pastur and Figotin [**70**] as a simplified spin glass [**21**] in statistical physics, the Hopfield model [**43**] is a recurrent network of n McCulloch-Pitts [**59**] binary neurons that can learn to store at least $n/(4 \ln n)$ distributed memories [**61**]. This model and its earlier incarnations (e.g., [**2**]) have been studied intensely in neuroscience theory and physics [**93**].

Formally, a *Hopfield network* on n nodes $\{1, \ldots, n\}$ is a Lenz-Ising model equipped with recurrent dynamics on states defined as follows. A *dynamics update* of $\mathbf{x}$ consists of replacing (in some fixed order through all nodes) each x_i in $\mathbf{x}$ with:

$$x_i = \begin{cases} 1 & \text{if } \sum_{j \neq i} W_{ij} x_j > \theta_i, \\ 0 & \text{otherwise.} \end{cases} \tag{5}$$

We call $F_i := \sum_{j \neq i} W_{ij} x_j$ the *feedforward input* to node i. Update (5) was inspired by consideration of communication possible between elements of nervous tissue, and such a computational unit is usually called a linear threshold *McCulloch-Pitts neuron.* We write the relation $\mathbf{x} \preceq \mathbf{y}$ for network states $\mathbf{x}, \mathbf{y}$ if m dynamics updates take $\mathbf{y}$ to $\mathbf{x}$ for some $m \geq 0$. It is important to note that as defined above, updates are asynchronous, and thus depend on an ordering of the nodes.

A fundamental property of Hopfield networks is that dynamics does not increase energy $E_{\mathbf{x}}$, evident from

$$\Delta E_i = -\Delta x_i (F_i - \theta_i),$$

[2] In the literature, "non-ferromagnetic" refers to the fact that all-to-all and positive or negative connectivity is allowed in the network, unlike the classical model which has "nearest-neighbor" connectivity for nodes on a square lattice.

[3] Maximum entropy models [**49**] such as the Lenz-Ising model appear often in science; e.g., [**64, 80, 83**]. Importantly, few samples are needed to determine them with high fidelity [**13, 29, 42**].

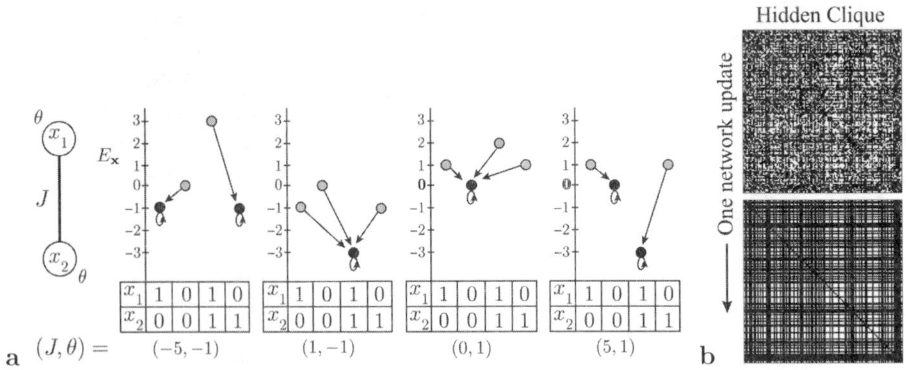

FIGURE 2. **Hopfield networks. a) Energy landscape and poset structure**. Consider a symmetric McCulloch-Pitts network of two neurons with weight J and threshold θ. Each binary state $\mathbf{x} = (x_1, x_2)$ has an energy $E_{\mathbf{x}}$. Arrows represent iteration of network dynamics (5) through node x_1 and then x_2; resulting memories (attractors) $\mathbf{x}^*$ are indicated by black circles. **b) Exponential storage and hidden cliques** [**41**]. (Bottom) One 64-clique on 128 vertices (of $\approx 10^{37}$) is described by its symmetric binary adjacency matrix. (Top) The same graph with 20% of its edges/non-edges randomly flipped (corrupted). One iteration of dynamics uncovers the original clique.

where ΔE_i (resp., $\Delta x_i = 0, \pm 1$) is the energy (resp., bit) change at node i after one step (5) in a dynamics update. Hence, after a finite number of updates, each initial state $\mathbf{x}$ converges to its *attractor* $\mathbf{x}^*$ (or *memory*), which is a fixed-point of the dynamics. Sometimes this property is expressed by saying that the energy $E_{\mathbf{x}}$ is a "Lyapunov function" for the network dynamics. Another common and useful intuition is that the dynamics is an inference technique, "freezing" a noisy version of a memory into a probable nearby state. See Fig. **2a** for energy landscapes of some small networks.

A natural mathematical structure emerges when studying the deterministic dynamics in these DRNNs.

LEMMA 1 (DRNN posets). *The dynamics relation* $\mathbf{x} \preceq \mathbf{y}$ *makes* $\{0,1\}^n$ *into a partially ordered set.*

PROOF. The relation $\preceq$ is reflexive since $m = 0$ takes $\mathbf{y}$ to itself. It is also antisymmetric because if m updates take $\mathbf{y}$ to $\mathbf{x}$ and m' updates $\mathbf{x}$ to $\mathbf{y}$, then $m+m'$ of them take $\mathbf{x}$ to itself; hence, $\mathbf{x}$ is a fixed-point, implying that $\mathbf{x} = \mathbf{y}$. And finally, if $\mathbf{x} \preceq \mathbf{y}$ and $\mathbf{y} \preceq \mathbf{z}$, then it is easy to see that $\mathbf{x} \preceq \mathbf{z}$, so that the relation is transitive. □

Even though Hopfield networks are well-studied, many basic questions about them remain open, such as the average time for dynamics to converge. We state one folklore experimental finding as a conjecture.

CONJECTURE 1. *Dynamics in Hopfield networks converge to attractors in time that is usually logarithmic in network size.*

2.c. Learning in DRNNs. A basic problem is to construct Hopfield networks with a given set X (or multiset – so that X forms a probability distribution q) of

Local learning rule	**Inception**	**Principle**
Outer-Product (OPR)	1933 [**55**], 1949 [**36**]	Spike correlations as weights
Perceptron (PER)	1958 [**77**]	Supervised pattern storage
Least Mean Square (LMS)	1960 [**106**]	Descend error function objective
Min Probability Flow (MPF)	2011 [**40, 91**]	Descend density estimation objective

FIGURE 3. Deterministic learning rules to train auto-associative networks of binary linear threshold neurons.

binary patterns as memories (i.e., local minima of $E_{\mathbf{x}}$). Such networks are useful for memory denoising since corrupted versions of patterns in X converge through the dynamics to the originals. Note that there are already necessary restrictions for attractors X; for instance, as the reader can verify, no two attractors can be a single bit apart.

Traditional approaches to this problem consist of iterating over X a *learning rule* [**37**] that updates the network's weights and thresholds given a training pattern $\mathbf{x} \in X$. In [**43**], the *outer-product learning rule* (OPR) for finding such networks was used to store at most $n/(4 \ln n)$ patterns without errors in an n-node network [**61**]. This "one-shot" method for storing collections of patterns essentially assigns weights between neurons to be their correlation, an early idea in neuroscience [**36, 55**] that is usually called "Hebbian learning". The *perceptron learning rule* (PER) [**77**] provides an alternative method to store patterns in a Hopfield network[4]. Unlike OPR, it achieves *optimal* storage capacity, in that if it is possible for a collection of patterns X to be memories of a Hopfield network, then PER will find parameters for which all of X are memories. Another common (optimal) method is the "Least Mean Square" (LMS) or "Delta" rule, attributed to an early paper by Woodrow-Hoff [**106**] but applied in many other contexts (e.g., psychology [**75**]). A list of these methods can be found in Fig. 3.

To estimate Hopfield networks from data, instead we use an objective [**40**] inspired by the recently developed *minimum probability flow* (MPF) technique [**91**] that avoids computation of the partition function Z. The *MPF objective* we use here is:

$$K_X(W, \theta) := \sum_{\mathbf{x} \in X} \sum_{\mathbf{x}' \in \mathcal{N}(\mathbf{x})} \exp\left(\frac{E_{\mathbf{x}} - E_{\mathbf{x}'}}{2}\right), \tag{6}$$

in which the *neighborhood* $\mathcal{N}(\mathbf{x})$ consists of those vectors which are Hamming distance 1 away from $\mathbf{x}$ (i.e., having exactly one bit different from $\mathbf{x}$). The function (6) is smooth, consists of only $n|X|$ terms, is jointly convex in the parameters, and can be minimized using standard gradient methods (e.g., the limited-memory Broyden-Fletcher-Goldfarb-Shanno algorithm [**66**]).

[4] In PER, the change in weights is proportional to the desired node value minus the updated node value.

We call a descent down the gradient of (6) given a single pattern $X = \{\mathbf{x}\}$ the *MPF learning rule*. Parameter changes are:

$$\Delta W_{ij} \propto -x_j \Delta x_i e^{\frac{1}{2}\Delta x_i (F_i - \theta_i)} \quad \text{and} \quad \Delta \theta_i \propto \Delta x_i e^{\frac{1}{2}\Delta x_i (F_i - \theta_i)}. \tag{7}$$

After learning, the weights are symmetrized to $(W + W^\top)/2$, which preserves the energy E and endows the DRNN with only fixed-point attractors. As these directions descend the gradient of a smooth convex function, traversing them can be very fast [**35**]. Note that setting $e^x \approx 1$ gives an outer-product type rule; thus, (7) can be seen as a generalization of OPR learning.

The following proposition is elementary, but important for linking this MPF objective to optimal memory storage. We say that $\mathbf{x}$ is a *strict local minimum* if every $\mathbf{x}'$ in $\mathcal{N}(\mathbf{x})$ has larger energy:

$$E_{\mathbf{x}} - E_{\mathbf{x}'} < 0. \tag{8}$$

From (5) and (8), it follows that strict local minima are fixed-points of the dynamics.

PROPOSITION 1. *If a set of binary vectors X can be stored as strict local minima (8) of a Hopfield network, then minimizing the convex MPF objective (6) will find such a network.*

PROOF. We first claim that X can be stored as strict local minima of a Hopfield network if and only if the MPF objective (6) satisfies $K_X < 1$ for some W, θ. If parameters W and θ have (8) hold for each $\mathbf{x} \in X$ and $\mathbf{x}' \in \mathcal{N}(\mathbf{x})$, then a uniform scaling in the parameters will make the energy differences in (6) arbitrarily large and negative; thus, K_X can be made less than 1. Conversely, if $K_X < 1$ for some choice of parameters, then each term in the sum of positive numbers (6) is less than 1. This implies that each $\mathbf{x} \in X$ and $\mathbf{x}' \in \mathcal{N}(\mathbf{x})$ satisfies (8) so that X are all strict local minima. Suppose now that X can be stored as strict local minima of a Hopfield network. By the claim, any method producing parameters with objective (6) arbitrarily close to the infimum of K_X produces a network with MPF objective less than 1 and therefore memorizes X. □

Typically, networks trained on input patterns X will have many more memories than elements of X. Nonetheless, in practice these additional *spurious minima* usually have shallow basins of attraction and are not easily reached (but see Theorem 1).

2.d. Exponential memory and error-correcting codes. Eq. (2) suggests that the number of different structures $\mathbf{D}$ coding for image patches should be large for sufficient expressive power, but [**43**] showed experimentally that $\approx .15n$ dense binary patterns (generated uniformly at random) can be stored with OPR in an n-node network if some fixed percentage of errors in a recovered pattern were tolerated. Since then, improved approaches for fitting Hopfield networks have been developed; however, independent of method, arguments of Cover [**16**] can be used to show that the number of such generic patterns storable in a Hopfield network (or more generally, a perceptron) with n nodes is at most $2n$. Interestingly, if the generic patterns to memorize are sparse (i.e., have few nonzero entries), then nearly a quadratic number of them can be stored [**3**].

It is still possible for certain Hopfield networks to have exponentially large collections of highly structured ("non-random") memories. For instance, choosing weights for the model from a standard normal distribution makes n-node networks

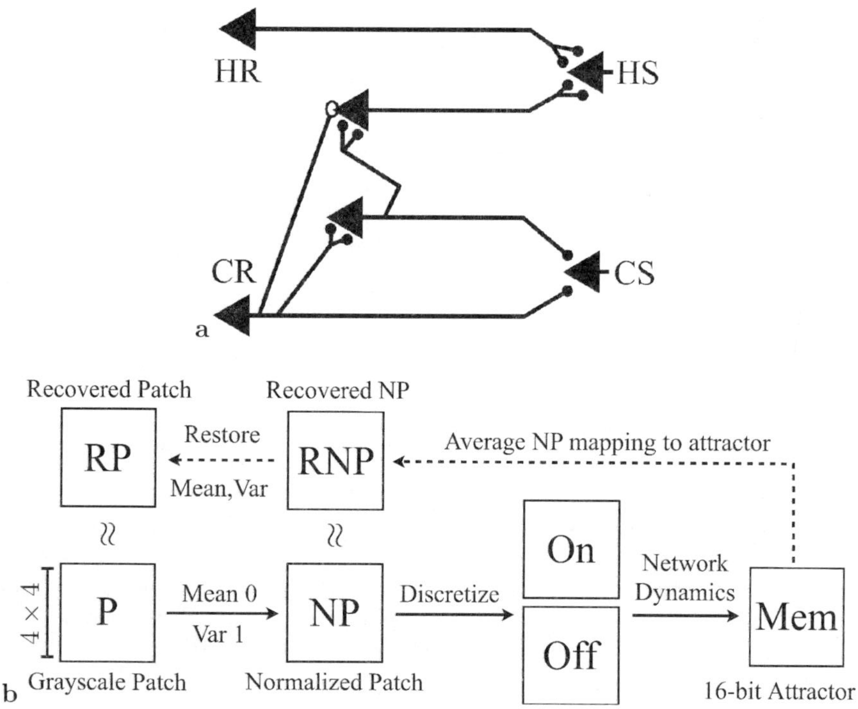

FIGURE 4. **DRNNs coding for natural signals. a) McCulloch-Pitts network** modeling somatosensory "hot and cold illusion" [**59**]: A heat sensation (HS) neuron fires if and only if the heat reception (HR) neuron fires or cold reception (CR) fires once and then immediately ceases firing. On the other hand, when CR fires twice in succession, the cold sensation (CS) neuron activates. **b) ON/OFF DRNN** trained on ternarized natural image patches performs image compression [**39**]; see Section 3 for a detailed summary of the scheme.

with $\approx 1.22^n$ memories asymptotically [**34, 60, 94**]. Although a generic network has exponential capacity, its basins of attraction are shallow and difficult to predetermine from network parameters, leading researchers to speculate that such spurious minima are to be avoided. The surprising finding below that special connectivity structures allow for DRNNs with robust memories in an exponential number of combinatorial configurations (e.g., cliques in graphs) opens up new possibilities.

THEOREM 1. *([**41**]) McCulloch-Pitts DRNNs on n nodes can learn to robustly*[5] *store* $\approx \frac{2^{\sqrt{2n}+\frac{1}{4}}}{n^{1/4}\sqrt{\pi}}$ *memories.*

The robustness aspect of Theorem 1 says that these networks have large attraction basins around an exponential number of memories. Fig. 2**b** shows the

[5]More precisely, we say that a sequence B_n of binary pattern collections is *robustly stored* by an n-node DRNN if a pattern in B_n having αn of its bits altered at random can be recovered (with probability limiting to 1 as $n \to \infty$) by converging the network dynamics, where $0 < \alpha < 1$ is a constant independent of n. In this sense, randomly generated networks do not have robust storage because the number of bits of corruption tolerated in memory recovery does not even increase with the number of nodes.

denoising of an example corrupted memory using only one iteration of the network dynamics.

Another lens with which to view this result is the field of coding theory. Namely, Theorem 1 provides an efficient set of Shannon optimal [**82**] low-density error-correcting codes that can be obtained from training on very small amounts of data (i.e., randomly generated cliques). See [**25**] for a related approach to error-correcting codes using DRNNs, [**57**] for a reference on interactions between learning and coding theory, [**14**] for a review of neural memory theory more generally, and [**101**] for a classical yet still poignant look at the problem of the mathematics underlying brain computation. We also remark that estimating DRNNs can aid in the unsupervised discovery of reoccurring spatiotemporal patterns in noisy neural data [**38**], [**22**].

3. Natural image coding with a discrete neural network

Inspired by the application of maximum entropy modes in computational neuroscience (e.g., in retina [**80**], [**83**]), a DRNN consisting of "ON" and "OFF" neurons was developed in [**39**] for compressing natural images. The main idea there was that image patches appear to be well-described perceptually by a decomposition (2) into three components: a patch mean, a patch variance, and a pattern computed using a Hopfield network of ON/OFF neurons trained on natural data. Although neural networks have been used for various image processing tasks by many previous researchers [**23**], we note that there are only a few papers using Hopfield DRNNs for lossy compression of digital images [**54, 65**].

The first step of the process is to ternarize millions of grayscale $L \times L$ patches from an image database (later, we will approximate the resulting data distribution with a Lenz-Ising model). To do this, we first remove the mean of each patch, and then we discretize each pixel. Next, we discretize each pixel x in such a patch to a pair of abstract neurons $(ON, OFF) \in \{0,1\}^2$ according to a parameter $\alpha \geq 0$. When $x > \alpha$, the discretized pixel is assigned $(ON, OFF) = (1,0)$; similarly, when $x < -\alpha$, we have $(ON, OFF) = (0,1)$; and finally, when $x \in [-\alpha, \alpha]$, we set $(ON, OFF) = (0,0)$. (The parameter α is chosen such that $2\alpha = 1$ is the smallest possible absolute difference in intensity values.) We can thus convert any grayscale $L \times L$ image patch into a binary vector of length $2L^2$. When no pixel has $(ON, OFF) = (0,0)$ or when all of them do, we call the patch *binary*. Also, we call a patch with zero mean pixel intensity and variance 1 *normalized* (NP).

EXERCISE 1. *By minimizing probability flow (6) (e.g., with constant thresholds $\theta_i = 1$), determine network parameters storing all binary patterns in an $L \times L$ network of ON/OFF neurons. (Compare with Fig. 7***a***, "Constructed (MPF)".)* □

The $2L^2$-bit DRNN is then trained using MPF estimation on millions of discretized patches. After learning, the network acts deterministically on ternary patches by converging its recurrent dynamics (each ON/OFF pair is viewed as a single unit for the asynchronous dynamics). In Fig. 7**b**, we display examples of these dynamics landscapes in the $L = 2$ case. It was found that such networks have all binary patches as memories, and that these primitives are good feature labels for image compression. Specifically, a converged, ternarized NP (i.e., binary patch) is remapped to a continuous one by averaging over all natural patches those NPs with that binary patch as DRNN output. Examples are shown in Fig. 5**b** for $L = 4$.

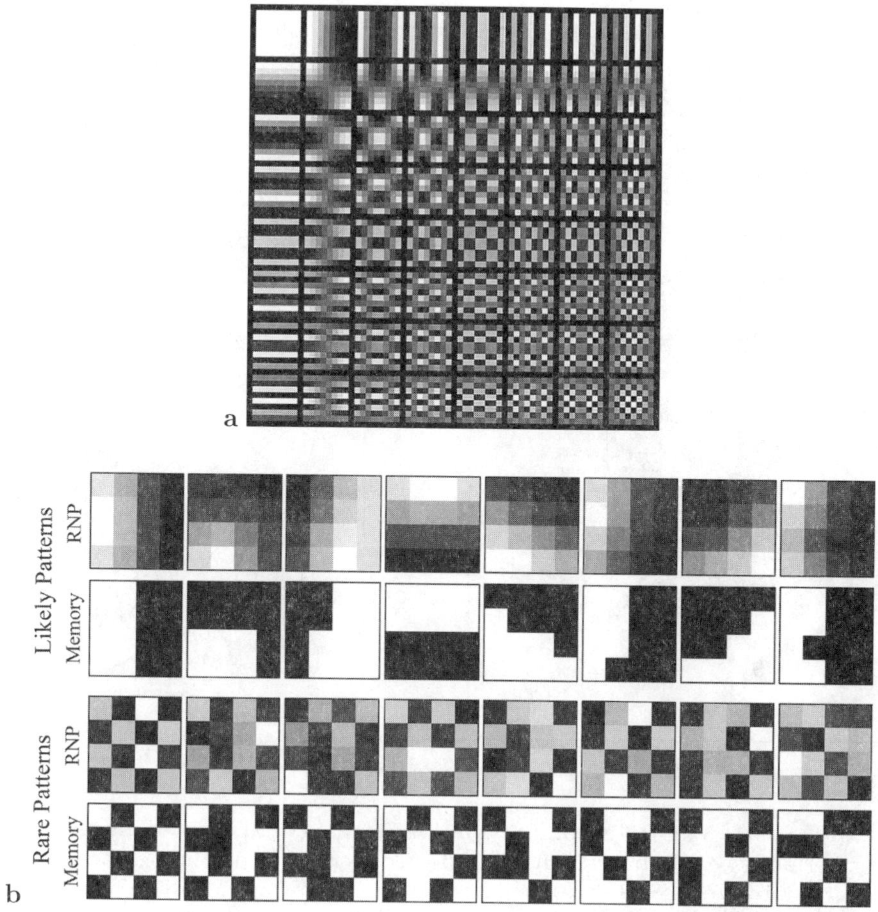

FIGURE 5. **Example "bases" for image compression. a) Linear:** The DCT basis $\{\mathbf{b}_1, \ldots, \mathbf{b}_{64}\}$ commonly used in linear coding, **b) Discrete:** Likely (Top) / Rare (Bottom) memories (Mems) and recovered normalized patches (RNPs) in a 4×4 ON/OFF net; see Fig. 4**b**.

To perform compression, a grayscale digital image is broken up into non-overlapping $L \times L$ patches, and then each patch is independently decomposed into a mean, a standard deviation, and L^2 (ON, OFF) neurons. Means and variances are saved losslessly as "Portable Network Graphics" (PNG) files, while converged discretized patches are encoded with an entropy coder (e.g., a Huffman code [**17**]) to get bytes on disk. For reconstruction, each binary patch is replaced by its continuous normalized representative RNP and then means and variances are restored. The mechanics of the scheme are illustrated in Fig. 4**b**. In addition, several codings and their performance can be found in Figs. 1 and 6, respectively.

Focusing on small natural image patches necessarily ignores larger-scale structure in the full image [**11, 51**], but nonetheless makes for a compression algorithm that matches JPEG perceptually (Fig 6**bd**). The theory of rate-distortion (RD) can validate (at least in the $L = 2$ case) this scheme [**62**], as we shall see in Section 6. We note that many other workers have also developed models of the structure in small natural image patches (e.g., [**12, 33, 52, 71, 79**]).

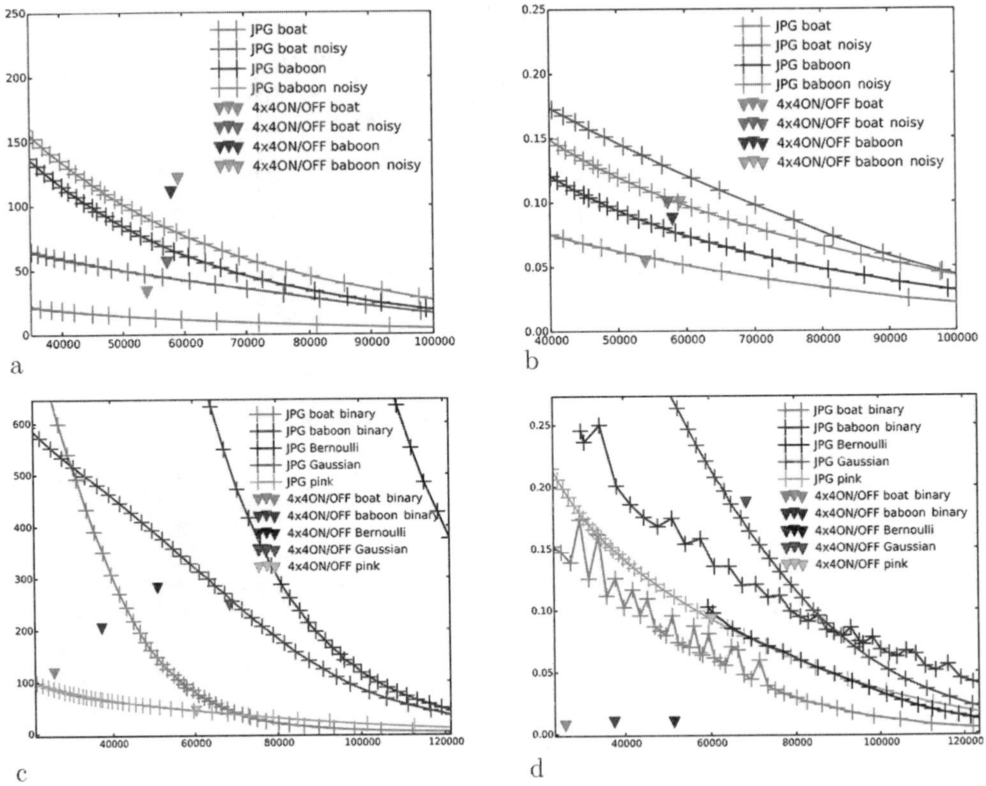

FIGURE 6. **DRNN compression performance. Standard images "boat" / "baboon" and low noised versions: a)** Mean-squared error vs bytes on disk, **b)** Mean structural similarity SSIM [**105**]. **Non-standard images: c)** MSE, **d)** Mean SSIM. "Noisy" versions of images refer to additive Gaussian noise (boat, baboon: $\sigma = 7.5, 5$), "Binary" versions refer to binarization about the mean intensity of the entire image, the "Bernoulli" image was generated using fair coin flips at each pixel, the "Gaussian" image was generated by draws of a standard normal at each pixel, and the "pink" noise image was generated to have the same correlation structure as continuous natural images.

4. Natural image patch modeling

A cat raised in a deprived visual environment will have differently tuned neurons than a cat raised in the wild [**45**]. One common hypothesis for such findings in neuroscience and psychology research is that brains adapt to efficiently process input from stimuli. Formal definitions of such "efficiency" have led to a variety of normative principles of brain function and behavior (see, for instance, [**5**, **6**, **15**, **28**, **53**, **68**, **85**, **86**, **95**, **99**]). In particular, there is evidence that the visual pathway in humans is adapted to efficiently process natural images [**84**]. Here, we explain how RD theory can be applied to validate such ideas, and then show its practical application to evaluating DRNN codings of 2×2 ON/OFF patches.

Natural images are a good dataset for experimenting with this hypothesis since they are highly structured in ways that are surprisingly consistent across contexts.

The best known of these regularities is the behavior of the orientation-averaged power spectrum $S(f)$ at different spatial frequencies f. For any given natural image, this statistic closely follows a power law: $S(f) \sim 1/f^2$ [**24, 78**]. Equivalently, amplitudes of Fourier coefficients fall off as $\sim 1/f$, and so truncated Fourier expansions (1) of natural images have diminishing error. This empirical fact provides a loose explanation for the success of linear coding. We remark that more generally characterizing the higher-order statistics – i.e., higher than pairwise correlations – of natural images is an active research area (e.g., [**7, 44, 46, 67, 96, 100**]).

It is clear that recognizing objects requires more than a power spectrum; thus, in some sense, it is obvious that the higher-order statistics of natural images should have affected our visual pathways. In fact, there is evidence that this is even true locally: humans are sensitive to higher-order statistics in small patches [**30**]. Accordingly, the commonly used perceptual distortion measure SSIM [**105**] also depends on the higher-order statistics of patches through its structural similarity comparator. One would therefore expect that a perceptually useful model of natural images should be effective at explaining higher-order statistics in natural images. In particular, one would expect the model underlying the coding in Section 3 should capture these statistics.

Recall that a discretized $L \times L$ image patch is identified by a binary vector $\mathbf{x} = (x_1, \ldots, x_n)$ of length $n = 2L^2$. A model $p(\mathbf{x})$ of these discretized patches will summarize $q(\mathbf{x})$, the frequency at which $\mathbf{x}$ is observed in a large ensemble of natural images, and better statistical models will have smaller distances between these two probability distributions. Of course, zero distance between model and data is trivially possible by setting the model to be a histogram of the data. However, this is not very useful, as a model should summarize the data in some way. Thus, researchers commonly set $p(\mathbf{x}) = f_\Theta(\mathbf{x})$ to come from a parametrized family of distributions, and then search for a Θ^* that minimizes a distance between q and f_Θ.

One common choice for this distortion measure is the (asymmetric) *Kullback-Leibler divergence*,

$$D_{KL}[q \| p] := \sum_{\mathbf{x}} q(\mathbf{x}) \log \frac{q(\mathbf{x})}{p(\mathbf{x})}, \tag{9}$$

so that Θ^* would satisfy:

$$\Theta^* = \arg \inf_{\Theta} D_{KL}[q \| f_\Theta]. \tag{10}$$

We note that D_{KL} has an interpretation in coding theory as the number of extra bits required to code the data when the code has been optimized for the model [**31**]. Finding such Θ^* also maximizes $\mathcal{L} := \prod_{\mathbf{x}} p(\mathbf{x})^{q(\mathbf{x})}$, the *likelihood* of the data.

Recall in Section 3 that we set p to be a Lenz-Ising distribution with weights W and thresholds θ; see (3). This distribution captures first- and second-order statistics of the data, but no higher-order dependencies. For completeness, we next sketch this well-known connection relating maximum likelihood estimation, minimization of D_{KL}, and the matching of statistics.

The Kullback-Leibler distance between data and model can be written (Shannon entropy H is defined in Section 5):

$$D_{KL}[q \| p] = -H[q] + \sum_{\mathbf{x}} q(\mathbf{x}) E_{\mathbf{x}} + \log Z.$$

Since $\frac{\partial E_{\mathbf{x}}}{\partial W_{ij}} = -x_i x_j$ and $\frac{\partial E_{\mathbf{x}}}{\partial \theta_i} = x_i$, maximum (log) likelihood estimates of weights W_{ij} come from minimizing D_{KL}:

$$0 = \frac{\partial D_{KL}[q\|p]}{\partial W_{ij}} = -\langle x_i x_j \rangle_q + \frac{\partial \log Z}{\partial W_{ij}} \implies \frac{\partial \log Z}{\partial W_{ij}} = \langle x_i x_j \rangle_q,$$

where $\langle \cdot \rangle_r$ denotes an average with respect to a distribution $r(\mathbf{x})$. Similarly, maximum likelihood estimates of θ_i come from:

$$\frac{\partial \log Z}{\partial \theta_i} = \langle x_i \rangle_q.$$

Finally, notice that

$$\frac{\partial \log Z}{\partial W_{ij}} = \frac{1}{Z}\frac{\partial Z}{\partial W_{ij}} = \frac{1}{Z}\sum_{\mathbf{x}} \frac{\partial (e^{-E_{\mathbf{x}}})}{\partial W_{ij}} = \sum_{\mathbf{x}} p(\mathbf{x})\, x_i x_j = \langle x_i x_j \rangle_p,$$

and similar manipulations show that $\frac{\partial \log Z}{\partial \theta_i} = \langle x_i \rangle_p$.

Together with previous expressions, it follows that optimal D_{KL} parameters are precisely those for which the model has the same first- and second-order statistics as the data:

$$\langle x_i x_j \rangle_p = \langle x_i x_j \rangle_q \quad \text{and} \quad \langle x_i \rangle_p = \langle x_i \rangle_q.$$

Similar results hold for other maximum entropy models (see also [**20**, Proposition 2.1.5]).

The minimum probability flow distance from Section 2.c is related to Kullback-Leibler divergence, though parameters from maximum likelihood estimation may not be exactly equal to those obtained from MPF (in practice, they are quite close).

The above analysis illuminates one of the concerns from Table 1. The model underlying the DRNN compression scheme in Section 3 should capture perceptually relevant higher-order dependencies in natural image patches, given its performance. However, as we have just seen, the model only captures second-order structure. It follows that the correlations in ternarized patches (and of the estimated Lenz-Ising model) should contain information about higher-order statistics in natural images. To test this hypothesis, one can whiten[6] natural images and then retrain the DRNN on this uncorrelated input, as was done in [**62**]. The finding is that the DRNN is similar and the scheme's performance is unaffected, largely confirming the hypothesis.

Another issue, as can be seen from Fig. **7a**, is that the resulting Lenz-Ising model when fit to natural data is not an exceptional match. If the two distributions were very similar, then the matrices in the top row should match the models displayed in the bottom row. Instead, we see that the model distributions correctly estimate the locations of the local maxima of the data, but fail to precisely match the frequencies (i.e., underestimating high empirical probabilities and overestimating lower ones). We shall explain a partial resolution to this second issue next in Section 6, after a short digression in rate-distortion theory.

[6]By Fourier transforming the original image and flattening the amplitude spectrum; this sets correlations between disparate pixels to be zero.

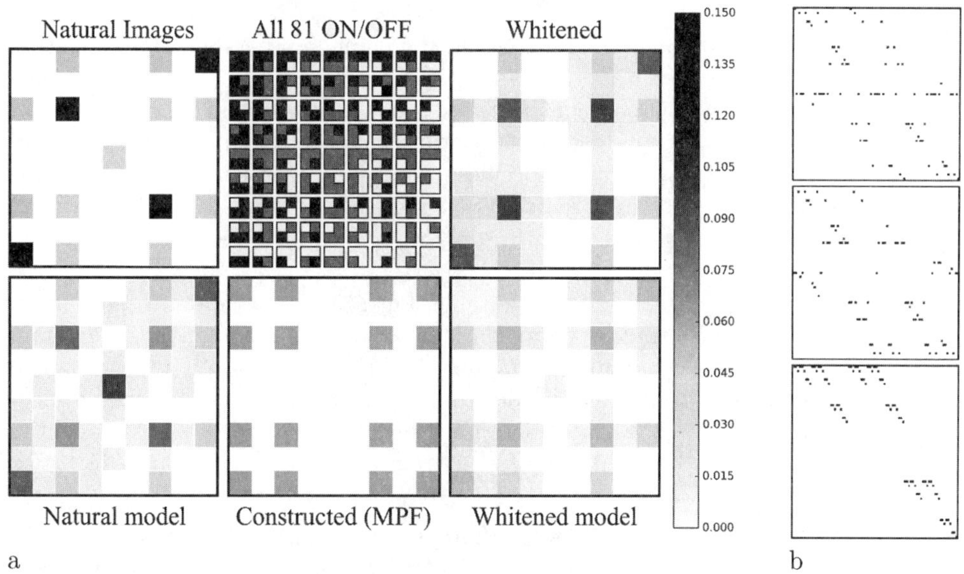

FIGURE 7. **DRNN coding of ternarized patches. a) Lenz-Ising model vs. data.** The top middle template image shows how 2×2 ON/OFF ternary patches $\mathbf{x}$ are ordered in the other images in **a**; e.g., the all-OFF (all-black) patch is shown at top left template corner. All other 5 images show either $p(\mathbf{x})$ (bottom row) or $q(\mathbf{x})$ (top row) via the shading on a given patch's location. For instance, the frequency of the all-black patch for the "constructed" model is ≈ 0.075 according to the colorbar at top right. The top left and bottom right ON/OFF patches are zero except for the constructed network because they do not occur in the data (consider how the scheme in Fig. 4 would produce such a pattern). **b) DRNN deterministic codebooks**. In each matrix, rows index $\tilde{\mathbf{x}}$ and columns index $\mathbf{x}$; ordering of the 81 states is left-to-right down rows of 2×2 ON/OFF key in **a**. Black (resp. white) pixels represent a value of 1 (resp. 0) in the matrix. **Natural image:** (Top) $p(\tilde{\mathbf{x}}|\mathbf{x}) = \delta_{\tilde{\mathbf{x}},\mathbf{x}^*}$ from Hopfield networks fit to natural image patches have coding rate $R = 3.63$ bits and expected distortion $D = 0.0549$. **Whitened natural image:** (Middle) have coding rate $R = 3.63$ and distortion $D = .0549$ (relative to natural data). **Constructed:** (Bottom) fit to all non-gray binary patterns have rate $R = 3.61$ and $D = 0.172$ (relative to natural data).

5. Rate-distortion theory

Information theory is often thought of as a framework for transmitting and decoding signals without losing any information. Nevertheless, one can also speak about the regime in which information is lost, and the subfield of rate-distortion theory quantifies the tradeoff between the size of a representation and its fidelity. We shall use RD theory to test the quality of the proposed deterministic codebook from Section 3, and we hope that its brief explanation below is adequate for the reader.

More precisely, consider the following classical setup: an *information source* sends N symbols $x_1, \ldots, x_N$ to an *encoder*, where x_i belong to an *alphabet* $\mathcal{X}$ and

N can be arbitrarily large. The encoder, upon receiving N symbols, sends one of M codewords to a *decoder*. As N grows larger, the number of possible codewords that could be sent grows larger as well. The *absolute rate* r of the encoding is defined as $(\log_2 M)/N$ bits per input symbol.

Given a codeword, the decoder then turns it into a reproduction sequence $\tilde{x}_1, \ldots, \tilde{x}_N$, with $\tilde{x}_i$ in $\tilde{\mathcal{X}}$. A user specifies a *distortion* function $d : \mathcal{X} \times \tilde{\mathcal{X}} \to \mathbb{R}_{\geq 0}$ that quantifies the error in a reproduction sequence. With such a function, we can compute an average distortion $\bar{d} := \frac{1}{N}\sum_{i=1}^{N} d(x_i, \tilde{x}_i)$ between the input and decoder output.

Larger average distortion is undesirable, but so is larger absolute rate. Intuitively, there must be a tradeoff between the two. Achieving zero distortion requires copying the input, with a minimal rate of $H[X]$ (defined below) bits per symbol. If there are strong constraints on encoder complexity, then we will necessarily incur large distortion. The problem of determining optimal encoders and decoders at rates and distortions between these two extremes gives rise to the rate-distortion function.

DEFINITION 1. *A* ***rate-distortion pair*** (R, D) *is one such that for any* $\varepsilon > 0$, *there is an encoding with* $r \leq R + \varepsilon$ *and* $\mathbb{P}(\bar{d} > D + \varepsilon) \leq \varepsilon$. *The* ***rate-distortion region*** *in* $\mathbb{R}^2$ *is the subset of all such pairs.*

The subset of $\mathbb{R}^2$ defined above is closed and convex (e.g., [**107**, Theorem 8.12], [**8**, Theorem 2.4.1]). Thus, the rate-distortion region is typically described by its boundary, which is called the *rate-distortion function* $R(D)$.

At first glance, determining $R(D)$ seems hopeless. The rate-distortion theorem (Theorem 2 below) equates this optimization problem with another seemingly disparate one. Instead of viewing $x_1, x_2, \ldots$ as a string of symbols, we view them as a realization of a stochastic process generated by an information source $\{X_k\}_{k=1}^{\infty}$. The output of the decoder is also viewed as an information source $\{\tilde{X}_k\}_{k=1}^{\infty}$ for which $\tilde{X}_i$ has realizations $\tilde{x}_k$ in the reproduction alphabet $\tilde{\mathcal{X}}$. For now, we assume that X_k and $\tilde{X}_k$ are discrete random variables with marginal probability distributions $p_k(x) = \mathbb{P}(X_k = x)$ and $p_k(\tilde{x}) = \mathbb{P}(\tilde{X}_k = \tilde{x})$, and joint probability distribution $p_k(x, \tilde{x}) = \mathbb{P}(X_k = x, \tilde{X}_k = \tilde{x})$. We also make two strong simplifying assumptions: first, that X_i are i.i.d. copies of random variable X; and second, that the decoder is memoryless, so that $\tilde{X}_i$ are i.i.d. copies of a random variable $\tilde{X}$. In this case, $p_k(x, \tilde{x})$ is independent of k, and so we speak of $p(x, \tilde{x})$ without fear of confusion.

To understand the rate-distortion theorem, we must briefly define certain information measures. The *Shannon entropy* [**82**] of a random variable X, denoted

$$H[X] := -\sum_{x \in \mathcal{X}} p(x) \log_2 p(x),$$

in units of bits, is colloquially known as a measure of uncertainty, and is operationally the expected number of "yes/no" questions needed to identify a realization of X. Accordingly, entropy $H[X]$ is maximal ($\log_2 |\mathcal{X}|$) when $p(x)$ is uniform on $\mathcal{X}$ and 0 when $p(x)$ has support on only one element in $\mathcal{X}$. A practical interpretation of entropy is given by the *source coding theorem*. Transmitting N symbols from an information source using a string of 0s and 1s requires at least $NH[X]$ bits, and this lower bound is asymptotically ($N \to \infty$) achievable.

The *conditional entropy* of $\tilde{X}$ given X quantifies our uncertainty about $\tilde{X}$ given knowledge of X and is defined as $H[\tilde{X}|X] := -\sum_{x \in \mathcal{X}, \tilde{x} \in \tilde{\mathcal{X}}} p(x, \tilde{x}) \log_2 p(\tilde{x}|x)$, where

$p(\tilde{x}|x)$ is the conditional probability of observing $\tilde{x}$ given x. The shared or *mutual information* between X and $\tilde{X}$ quantifies our information gain about X knowing $\tilde{X}$, or our reduction in X's uncertainty:

$$(11) \qquad I[X;\tilde{X}] := H[\tilde{X}] - H[\tilde{X}|X] = \sum_{x\in\mathcal{X}} \sum_{\tilde{x}\in\tilde{\mathcal{X}}} p(x)p(\tilde{x}|x)\log_2 \frac{p(\tilde{x}|x)}{p(\tilde{x})}.$$

We choose the convention that $0\log_2 0 = 0$ and, when both $p(\tilde{x}|x)$ and $p(\tilde{x})$ are 0, we set the term $p(\tilde{x}|x)\log_2 \frac{p(\tilde{x}|x)}{p(\tilde{x})}$ to also be 0. Later, we shall relax the assumption that X and $\tilde{X}$ are discrete random variables in defining these quantities.

THEOREM 2. *(The rate-distortion theorem) The rate-distortion function R(D) can be computed as:*

$$R(D) = \min_{p(\tilde{x}|x):\,\mathbb{E}[d]=D} I[X;\tilde{X}].$$

For a proof of Theorem 2, see e.g. Chapter 8 in [**107**]. Notice here that the distortion inequality constraint has become the equality $D=\mathbb{E}[d]:=\sum_{x,\tilde{x}} p(x,\tilde{x})d(x,\tilde{x})$. The optimal stochastic codebooks $p(\tilde{x}|x)$ which achieve this minimum are often more interesting than the rate-distortion function itself. For instance, they are thought of as "soft clusters" in the machine learning literature (e.g., [**88**, **97**]). See Figs. 8, 9, and 10 for some examples of optimal codebooks.

This still leaves us with the problem of effectively computing the rate-distortion function and optimal stochastic codebooks given an input distribution $p(x)$ and distortion $d(x,\tilde{x})$. The mutual information $I[X;\tilde{X}]$ is a convex function of $p(\tilde{x}|x)$, the equality constraint $\mathbb{E}[d] = D$ is linear in $p(\tilde{x}|x)$, and the implied normalization constraint $\sum_{\tilde{x}} p(\tilde{x}|x) = 1$ is also linear. Therefore, calculating the rate-distortion function reduces to the convex program of minimizing

$$(12) \qquad \mathcal{L}_\beta := I[X;\tilde{X}] + \beta\sum_{x,\tilde{x}} p(x,\tilde{x})d(x,\tilde{x}),$$

with respect to $p(\tilde{x}|x)$ for each β, and appropriate constraints.

The first term in $\mathcal{L}_\beta$ is the coding rate that we wish to minimize, and the second term constrains the expected distortion $\mathbb{E}[d]$. (There are additional inequality constraints, $p(\tilde{x}|x) \geq 0$, restricting solutions to a product Δ of simplices.) Once we obtain

$$(13) \qquad p_\beta(\tilde{x}|x) := \text{argmin}_{p(\tilde{x}|x)\in\Delta}\mathcal{L}_\beta,$$

we can determine our desired quantities:

$$(14) \qquad R_\beta = \sum_{x,\tilde{x}} p(x)p_\beta(\tilde{x}|x)\log_2\frac{p_\beta(\tilde{x}|x)}{p_\beta(\tilde{x})} \quad\text{and}\quad D_\beta = \sum_{x,\tilde{x}} p(x)p_\beta(\tilde{x}|x)d(x,\tilde{x}),$$

which parametrically trace out the rate-distortion function $R(D)$ (whose slope is $-\beta$ at differentiable $D = D_\beta$ [**8**, Theorem 2.5.1]).

There are several different numerical approaches to calculating $p_\beta(\tilde{x}|x)$, but most are based on gradient descent of the Lagrangian in (12). Specifically, we seek a matrix $p_\beta(\tilde{x}|x)$ for which

$$(15) \qquad \frac{\partial\mathcal{L}_\beta}{\partial p(\tilde{x}|x)}|_{p_\beta(\tilde{x}|x)} = 0.$$

After some straightforward algebra [**8**, Ch. 2] manipulating (15), one finds that $p_\beta(\tilde{x}|x)$ satisfies the following equation:

$$p_\beta(\tilde{x}|x) = p_\beta(\tilde{x})e^{-\beta d(x,\tilde{x})}/Z(x,\beta), \tag{16}$$

where the normalization constraint $\sum_{\tilde{x}} p_\beta(\tilde{x}|x) = 1$ implies that $Z(x,\beta) = \sum_{\tilde{x}} p_\beta(\tilde{x})e^{-\beta d(x,\tilde{x})}$. Expression (16) is then used to construct a dynamical system which converges to the desired $p_\beta(\tilde{x}|x)$.

For instance, the *Blahut-Arimoto (BA) algorithm* [**4, 10**] corresponds to iterating (16) until convergence ($t = 0, 1, \ldots, T$):

$$\begin{aligned} p^{(t+1)}(\tilde{x}|x) &= p^{(t)}(\tilde{x})e^{-\beta d(x,\tilde{x})}/Z^{(t+1)}(x,\beta), \\ p^{(t+1)}(\tilde{x}) &= \sum_x p(x)p^{(t+1)}(\tilde{x}|x), \\ Z^{(t+1)}(x,\beta) &= \sum_{\tilde{x}} p^{(t)}(\tilde{x})e^{-\beta d(x,\tilde{x})}. \end{aligned} \tag{17}$$

If $p^{(0)}(\tilde{x}|x) \in \Delta$ (i.e., $p^{(0)}(\tilde{x}|x) \geq 0$ and $\sum_{\tilde{x}} p^{(0)}(\tilde{x}|x) = 1$), then it is easy to see that $p^{(t)}(\tilde{x}|x)$ will remain in Δ.[7] Moreover, one can prove that as $t \to \infty$, the codebook $p^{(t)}(\tilde{x}|x)$ converges to the optimal one $p_\beta(\tilde{x}|x)$ [**18**].

The discussion in this section can be extended to general random variables. In particular, when X and $\tilde{X}$ are both continuous, and a joint probability density function $p(x, \tilde{x})$ exists, the extended definitions are essentially equivalent to those here except that sums are replaced with integrals. Furthermore, the derivation above can be replaced with a calculus of variations argument [**8**, Ch. 4.2], and the resulting Euler-Lagrange equations also take the form (16).

Still, rate-distortion theorists generally face the difficult problem of how to choose a distortion function, as optimal stochastic codebooks can change drastically with d. This occurs because choice of distortion implicitly corresponds to a decision of what is being computed, in some limited sense of the word "computed". The issue is particularly important for image compression, as it is well-known that commonly-used mean-squared error correlates poorly with perceptual distortion [**104**].

Example: Lossy addition by probabilistic hash table. We illustrate RD theory by constructing a code that approximately adds two numbers. Though elementary, this example (distributed functional source coding [**63**]) shows how to use RD theory for constructing imperfect computational devices.

Consider the following distortion measure on binary vectors $\mathbf{x} = (x_1, x_2)$ and $\mathbf{y} = (y_1, y_2)$:

$$d(\mathbf{x}, \mathbf{y}) = |(x_1 + x_2) - (y_1 + y_2)|.$$

Notice that $(0, 1)$ and $(1, 0)$ are equivalent as far as d is concerned, whereas both $(0, 0)$ and $(1, 1)$ are distinguishable from each other and from $(0, 1)$ and $(1, 0)$.

The structure of d has strong implications for the $\beta \to \infty$ limit of optimal codebooks p_β. To understand why, we temporarily consider a more general setting. Let $p(x) = \frac{1}{|\mathcal{X}|}$ be uniform, and assume that d is symmetric: $d(x, \tilde{x}) = d(\tilde{x}, x)$. We

[7]Note that in other applications, such as the information bottleneck method [**97**], the distortion $d(x, \tilde{x})$ might depend on $p(\tilde{x}|x)$ and will therefore need to be recomputed at each timestep t.

claim:

$$\lim_{\beta\to\infty} p_\beta(\tilde{x}|x) = \begin{cases} 0 & d(x,\tilde{x}) > 0, \\ \frac{1}{|\{y:d(x,y)=0\}|} & d(x,\tilde{x}) = 0. \end{cases} \tag{18}$$

To verify this informally, one can check that the solution satisfies equation (16). In other words, the $\beta \to \infty$ limit implicitly clusters inputs x according to the equivalence relation $x \sim \tilde{x}$ if and only if $d(x,\tilde{x}) = 0$.[8]

Applied to our example, we have $\lim_{\beta\to\infty} p_\beta(\tilde{\mathbf{x}}|\mathbf{x}) = \delta_{\mathbf{x},\tilde{\mathbf{x}}}$ when $\mathbf{x} = (0,0)$ or $(1,1)$ (here, δ is the Kronecker delta), and

$$\lim_{\beta\to\infty} p_\beta(\tilde{\mathbf{x}}|\mathbf{x}) = \begin{cases} 0 & \tilde{\mathbf{x}} = (0,0),\ (1,1) \\ \frac{1}{2} & \tilde{\mathbf{x}} = (0,1),\ (1,0) \end{cases}, \quad \text{when } \mathbf{x} = (0,1) \text{ or } (1,0).$$

Thus, binary vectors with identical sum are clustered together. These codebooks, however, provide no direction for how to assign function output to the soft clusters given by $p_\beta(\tilde{\mathbf{x}}|\mathbf{x})$. One straightforward procedure is to output $\tilde{x}_1 + \tilde{x}_2$ with probability $p_\beta(\tilde{\mathbf{x}}|\mathbf{x})$ when presented with $\mathbf{x}$, similar in spirit to the idea of a "soft heap".

Fig. 8 shows estimated $p_\beta(\tilde{\mathbf{x}}|\mathbf{x})$ and $R(D)$ for the lossy adder defined above. The different curves are estimations of the rate-distortion function at different numbers of iterating (17). Comparison of the curves resulting from 50 and 5000 BA iterations reveals that the estimated $R(D)$ has essentially converged after 50 steps. Inspection of $p_\beta(\tilde{\mathbf{x}}|\mathbf{x})$ also suggests that no solution, for any β, differentiates between inputs $(0,1)$ and $(1,0)$. For small enough β, the codebook $p_\beta(\tilde{\mathbf{x}}|\mathbf{x})$ is represented by a single "cluster", a solution with zero rate. If a lossy adder were constructed based on this solution, one would simply output 1 regardless of input. This zero-rate solution can be seen as the flat plateau in Fig. 8**b** (black line). For $\beta > 1$, the codebooks $p_\beta(\tilde{\mathbf{x}}|\mathbf{x})$ begin to differentiate between $(0,0)$, $(1,1)$, and the input pair $(0,1)$, $(1,0)$. When $\beta = 10$, we recover a solution $p_\beta(\tilde{\mathbf{x}}|\mathbf{x})$ that is indistinguishable from the $\beta \to \infty$ limiting codebook in the paragraph above. □

In computing applications, deterministic codebooks are often preferable to the usually stochastic ones $p_\beta(\tilde{\mathbf{x}}|\mathbf{x})$. Even then, rate-distortion theory provides a guide for choosing between algorithms: $R(D)$ functions can validate clustering performance. An example of this for 2×2 DRNN image coding will be presented in the next section.

6. Near-optimal DRNN coding of ternarized natural image patches

We next quantify the quality of the ON/OFF DRNN patch coding described in Section 2, but emphasize that modeling and compression are not equivalent goals [**96**]. Our estimated Lenz-Ising model is not perfect at matching the probabilities of discretized natural image patches. However, matching the local maxima of the data distribution nonetheless allows one to construct a near-optimal deterministic codebook for natural images in a rate-distortion sense, as we shall now explain.

Consider the Hamming distance $d(\mathbf{x}, \tilde{\mathbf{x}})$ between two ON/OFF patches $\mathbf{x}$ and $\tilde{\mathbf{x}}$, and an input source $p(\mathbf{x})$ given by the empirical distribution $q(\mathbf{x})$ of ternarized 2×2 patches as displayed in Fig. 7**a**. The deterministic codebook derived from

[8] This can be extended to informational distortions (i.e., the information bottleneck method [**97**]), in which case it can be shown that the $\beta \to \infty$ limit of rate-distortion objectives recover minimal sufficient statistics [**81**].

the corresponding Hopfield network is given by $p(\tilde{\mathbf{x}}|\mathbf{x}) = \delta_{\tilde{\mathbf{x}},\mathbf{x}^*}$, where $\mathbf{x}^*$ is the attractor of the dynamics initialized at $\mathbf{x}$. The resulting output distribution places a probability $p(\tilde{\mathbf{x}}) = \sum_{\mathbf{x}:\mathbf{x}^*=\tilde{\mathbf{x}}} q(\mathbf{x})$ at every binary vector $\tilde{\mathbf{x}}$.

Examples of these codebooks are shown in Fig. **7b** for networks fit to natural image patches, whitened patches, and artificially constructed patches. These diagrams are partial representations of the poset structure of ON/OFF patches subject to the DRNN dynamics. A full representation of this poset structure might look like Fig. 2, showing exactly which patterns follow single updates. The deterministic codebooks displayed in Fig. **7b** can be compared to the optimal stochastic ones depicted in Fig. 9**c**. Note that the optimal codebook $p_\beta(\tilde{\mathbf{x}}|\mathbf{x})$ (III; $\beta \approx 3$) resembles the natural deterministic one shown in Fig. **7a**. Also, as shown in Fig. 9**b**, this optimal p_β is close to a transition between codebooks that copy the input and low-rate, high-distortion solutions.

We suspect that the similarity between the optimal stochastic codebook and the DRNN one allows the network to nearly achieve performance bounds imposed by rate-distortion theory. In Fig. 9**a**, we compare the rate and expected distortion of the deterministic (natural image) coder to the asymptotically achievable rates and distortions. The coding rate $I[X;\tilde{X}]$ and expected distortion $\mathbb{E}[d]$ of each of these codebooks is calculated using formulae given in Section 5.[9] Interestingly, networks trained on natural image patches and whitened natural patches have nearly identical rates and expected distortions (Fig. **7b**).

7. Optimal codings of continuous variables are usually discrete

Almost all compression in signal processing for continuous-valued signals involves "quantization". For example, many methods for compressing an image patch first Fourier transform it and then discretize the resulting coefficients. From this traditional mindset, quantization is usually seen as a necessary but cumbersome step to actually code images on a computer (with more lossy settings corresponding to ignoring higher frequency components). However, a theorem discovered in [**26**, **76**] for mean-squared error distortions states that optimal reproduction variables $\tilde{X}$ are often discrete even when the original information source emits continuous-valued signals. Viewed from the perspective of this mathematical fact, discretization of the original signal (or optimization of an agent's actions [**58**]) is key to compression rather than a post-transform nuisance.

Previous proofs of this theorem, however, were slightly incomplete. The equations used to show that the reproduction alphabet is discrete were derived by assuming the existence of a well-defined probability density function $p_\beta(\tilde{x}|x)$, which, the proofs then go on to argue, usually does not exist. This section is devoted to outlining a measure-theoretic formulation of the problem that circumvents this technical difficulty. We refer to [**27**] for requisite background in analysis and probability.

Let μ be a random variable (i.e., a standard Lebesgue-Rokhlin probability measure on $\mathbb{R}$) such that the support of μ lies in $R := [-A, A]$ for some positive A. We define a *coding* of μ to be a family of Borel probability measures $\{\nu_x\}_{x\in R}$ on R, each of which has the property that $x \mapsto \nu_x(E)$ is μ-measurable for each

[9] Since this transducer is deterministic, so that $H[\tilde{X}|X] = 0$, the coding rate is simply the entropy of the probability distribution over fixed-points in the network: $I[X;\tilde{X}] = H[\tilde{X}] - H[\tilde{X}|X] = H[\tilde{X}]$. Note that it is not always the case that $H[\tilde{X}] = I[X;\tilde{X}]$ for stochastic codebooks.

Borel set E. Given this data, there is a construction [**50**] (usually overlooked as being necessary in the literature) of a measure λ on the product space $R^2 = R \times R$ satisfying:

$$\int \left(\int f d\nu_x(\tilde{x}) \right) d\mu(x) = \int f d\lambda, \quad \text{for } f \in L^+(R \times R).$$

Given such a measure λ, we have a marginal distribution:

$$\nu(E) = \int_R \nu_x(E) d\mu(x) = \lambda(R \times E), \quad \text{for each Borel set } E.$$

Also, set $\pi = \mu \times \nu$ to be the standard (independent) product distribution on R^2.

With these basic notions in place, we can define the distortion and rate of a coding $\{\nu_x\}_{x \in R}$. The *expected distortion* for a nonnegative d is

$$\mathbb{E}[d] := \int d(x, \tilde{x}) \, d\lambda,$$

and the *coding rate* is

$$I[\lambda] := \sup_{\{G_i\}} \sum_{i=1}^{N} \lambda(G_i) \log \frac{\lambda(G_i)}{\pi(G_i)},$$

where $\{G_i\}_{i=1}^N$ is a finite[10] partition of $R \times R$ into rectangles. When $I[\lambda] < \infty$, the Radon-Nikodym derivative $\frac{d\lambda}{d\pi}$ exists and we have [**73**, Theorem 2.4.2] (compare with definition (11)):

$$I[\lambda] = \int \log \frac{d\lambda}{d\pi} d\lambda.$$

Let Λ_D denote the set of measures on R^2 with expected distortion $\mathbb{E}[d] = D$, and with the proper marginal distribution $\mu(E) = \lambda(E \times R)$, for all Borel sets E. The original optimization problem from Section 5 now reads:

$$R(D) = \inf_{\lambda \in \Lambda_D} I[\lambda].$$

We note that [**89, 90**] studied a related optimization problem asking for the input probability measure μ that maximizes $I[\lambda]$ given a fixed coding $\{\nu_x\}_{x \in R}$.

Before stating the precise form of the continuous-to-discrete coding theorem that we shall prove here, we need to extend optimal coding to that of a continuous variable. First, we have the following measure-theoretic analogue to the limit of (17).

LEMMA 2. *(Optimization lemma [**74, 89, 90**]) For all $D > 0$, there is a unique coding and corresponding joint probability measure λ attaining the infimum of $I[\lambda]$ on Λ_D.*

PROOF. As Λ_D is a compact, convex topological space under the Lévy-Prokhorov metric [**56**] and $I[\lambda]$ is a strictly convex function of λ, the conclusion follows. □

From Lemma 2, there is a unique $\lambda^* \in \Lambda_D$ which minimizes $I[\lambda]$:

$$\lambda^* := \arg \min_{\lambda \in \Lambda_D} I[\lambda].$$

The uniqueness and existence of this optimal coding λ^* are crucial for proving the main result of this section.

---

[10]In [**73**], there are countably many rectangles, but since R is bounded we can restrict to a finite set.

THEOREM 3. *Let μ be any random variable with bounded support R, and let ρ be any nonnegative analytic function. If d is any distortion on $R \times R$ given by $d(x, \tilde{x}) = \rho(\tilde{x} - x)$, then the optimal coding of μ with distortion D has finite support.*

Much of the legwork for this result takes place in verifying the following lemma and its main implication, Eq. (20).

LEMMA 3. *(First order conditions) Assume that μ has bounded support R and that d is normal (i.e., $d(x, x) = 0$ and $d(x, \tilde{x}) < \infty$ for all $x, \tilde{x}$). Define λ^* as above, let ν^* be the corresponding marginal distribution, and let $\pi^* = \mu \times \nu^*$. The Radon-Nikodym derivative $\frac{d\lambda^*}{d\pi^*}$ exists and satisfies for some $\beta \geq 0$:*

$$\frac{d\lambda^*}{d\pi^*} = \frac{e^{-\beta d(x,\tilde{x})}}{Z(x, \beta)}, \tag{19}$$

for all $(x, \tilde{x})$ in the support of λ^*, where $Z(x, \beta) = \int e^{-\beta d(x,\tilde{x})} d\nu^*$. Additionally, for all $\tilde{x}$ in the support of ν^*, we have:

$$1 = \int \frac{1}{Z(x, \beta)} e^{-\beta d(x,\tilde{x})} d\mu(x). \tag{20}$$

PROOF. Given any $\lambda \in \Lambda_D$, define

$$\lambda_\varepsilon := (1 - \varepsilon)\lambda^* + \varepsilon\lambda.$$

Also, denote the weak derivative of I by I':

$$I'[\lambda^*] := \lim_{\varepsilon \to 0^+} \frac{I[\lambda_\varepsilon] - I[\lambda^*]}{\varepsilon}.$$

Since λ^* is the global minimum of $I[\lambda]$ in Λ_D, we have $I'[\lambda^*] \geq 0$. But from [**90**], a necessary and sufficient condition for $I[\lambda^*] = \inf_{\lambda \in \Lambda_D} I[\lambda]$ is that $I'[\lambda^*] \leq 0$; so together, we search for $I'[\lambda^*] = 0$.

Consider the coding $\nu_x = \frac{1}{m(R)}$, where m is standard Lebesque measure. By construction this coding has finite expected distortion and rate; hence, for all nonzero expected distortions D, we have $I[\lambda^*] = R(D) < \infty$. As stated earlier, this implies that $\frac{d\lambda^*}{d\pi^*}$ exists and that $I[\lambda^*] = \int \log \frac{d\lambda^*}{d\pi^*} d\lambda^*$. Continuity of $I[\lambda]$ in λ [**90**] implies that $I[\lambda_\varepsilon]$ is also finite, so that $\frac{d\lambda_\varepsilon}{d\pi_\varepsilon}$ exists where $\pi_\varepsilon = (1 - \varepsilon)\pi^* + \varepsilon\pi$ and $I[\lambda_\varepsilon] = \int \log \frac{d\lambda_\varepsilon}{d\pi_\varepsilon} d\lambda_\varepsilon$.

We restrict ourselves to perturbations λ such that $\lambda_\varepsilon \ll \lambda^* \ll \pi^* \ll \pi_\varepsilon$ so that the chain rule can be applied to Radon-Nikodym derivatives. This still provides us with enough necessary conditions on $\frac{d\lambda^*}{d\pi^*}$. The chain rule gives:

$$\frac{d\lambda_\varepsilon}{d\pi_\varepsilon} = \frac{d\lambda_\varepsilon}{d\lambda^*} \frac{d\lambda^*}{d\pi^*} \frac{d\pi^*}{d\pi_\varepsilon},$$

so that

$$\log \frac{d\lambda_\varepsilon}{d\pi_\varepsilon} = \log \frac{d\lambda_\varepsilon}{d\lambda^*} + \log \frac{d\lambda^*}{d\pi^*} + \log \frac{d\pi^*}{d\pi_\varepsilon}.$$

It is straightforward to show that $\lambda \ll \lambda^*$ under the assumption that $\lambda_\varepsilon \ll \lambda^*$, since if $\lambda^*(E) = 0$ then $\lambda_\varepsilon(E) = 0$, giving $\lambda(E) = \frac{\lambda(E)-(1-\varepsilon)\lambda^*(E)}{\varepsilon} = 0$. Then, $\frac{d\lambda}{d\lambda^*}$ exists and $\frac{d\lambda_\varepsilon}{d\lambda^*} = (1-\varepsilon) + \varepsilon\frac{d\lambda}{d\lambda^*}$. Similar logic implies $\frac{d\pi^*}{d\pi_\varepsilon} = \frac{1}{1-\varepsilon}\left(1 - \varepsilon\frac{d\pi}{d\pi_\varepsilon}\right)$.

Together, this allows us to express the difference $I[\lambda_\varepsilon] - I[\lambda^*] = \int \log\frac{d\lambda_\varepsilon}{d\pi_\varepsilon}d\lambda_\varepsilon - \int \log\frac{d\lambda^*}{d\pi^*}d\lambda^*$ as follows:

$$\begin{aligned} I[\lambda_\varepsilon] - I[\lambda^*] &= \int \log\frac{d\lambda_\varepsilon}{d\lambda^*}d\lambda_\varepsilon + \int \log\frac{d\lambda^*}{d\pi^*}d\lambda_\varepsilon + \int \log\frac{d\pi^*}{d\pi_\varepsilon}d\lambda_\varepsilon - \int \log\frac{d\lambda^*}{d\pi^*}d\lambda^* \\ &= (1-\varepsilon)\int \log\left((1-\varepsilon)+\varepsilon\frac{d\lambda}{d\lambda^*}\right)d\lambda^* + \varepsilon\int \log\left((1-\varepsilon)+\varepsilon\frac{d\lambda}{d\lambda^*}\right)d\lambda^* + (1-\varepsilon)\int \log\frac{d\lambda^*}{d\pi^*}d\lambda^* \\ &\quad + \varepsilon\int \log\frac{d\lambda^*}{d\pi^*}d\lambda + \int \log\left(\frac{1}{1-\varepsilon}\left(1-\varepsilon\frac{d\pi}{d\pi_\varepsilon}\right)\right)d\lambda_\varepsilon - \int \log\frac{d\lambda^*}{d\pi^*}d\lambda^* \\ &= (1-\varepsilon)\int \log\left(1+\frac{\varepsilon}{1-\varepsilon}\frac{d\lambda}{d\lambda^*}\right)d\lambda^* + \varepsilon\int \log\left(1+\frac{\varepsilon}{1-\varepsilon}\frac{d\lambda}{d\lambda^*}\right)d\lambda^* - \varepsilon\int \log\frac{d\lambda^*}{d\pi^*}d\lambda^* \\ &\quad + \varepsilon\int \log\frac{d\lambda^*}{d\pi^*}d\lambda + \int \log\left(1-\varepsilon\frac{d\pi}{d\pi_\varepsilon}\right)d\lambda_\varepsilon. \end{aligned}$$

Since R is bounded, each of the Radon-Nikodym derivatives $\frac{d\lambda}{d\lambda^*}$ and $\frac{d\pi}{d\pi_\varepsilon}$ is too; hence, the integrands $\frac{(1-\varepsilon)\log\left(1+\frac{\varepsilon}{1-\varepsilon}\frac{d\lambda}{d\lambda^*}\right)}{\varepsilon}$, $\log\left(1+\frac{\varepsilon}{1-\varepsilon}\frac{d\lambda}{d\lambda^*}\right)$, and $\frac{\log\left(1-\varepsilon\frac{d\pi}{d\pi_\varepsilon}\right)}{\varepsilon}$ above are bounded for small ε. From an application of L'Hopitals Rule, we find that $\lim_{\varepsilon\to 0^+}\frac{(1-\varepsilon)\log\left(1+\frac{\varepsilon}{1-\varepsilon}\frac{d\lambda}{d\lambda^*}\right)}{\varepsilon} = \frac{d\lambda}{d\lambda^*}$ and $\lim_{\varepsilon\to 0^+}\frac{\log\left(1-\varepsilon\frac{d\pi}{d\pi_\varepsilon}\right)}{\varepsilon} = -\frac{d\pi}{d\pi_\varepsilon}$. Thus,

$$I'[\lambda^*] \quad = \quad \int \frac{d\lambda}{d\lambda^*}d\lambda^* + \int \log\frac{d\lambda^*}{d\pi^*}d(\lambda-\lambda^*) - \int \frac{d\pi}{d\pi^*}d\lambda^*.$$

It follows from these calculations that $\int \frac{d\lambda}{d\lambda^*}d\lambda^* = 1$. Note also that $\frac{d\pi}{d\pi^*} = \frac{d\nu}{d\nu^*}$, and so $\int \frac{d\pi}{d\pi^*}d\lambda^* = 1$. Hence,

$$I'[\lambda^*] = \int \log\frac{d\lambda^*}{d\pi^*}d(\lambda-\lambda^*).$$

Since $\lambda \in \Lambda_D$, we have $\int d(x,\tilde{x})d(\lambda-\lambda^*) = 0$ and $\int \gamma(x)d(\lambda-\lambda^*) = 0$ for any $\gamma(x) \in L^+$. Thus, a necessary condition for λ^* to have $I[\lambda^*] = R(D)$ is

$$0 = \int \left(\log\frac{d\lambda^*}{d\pi^*} + \beta d(x,\tilde{x}) + \gamma(x)\right)d(\lambda-\lambda^*),$$

for some scalar β. Since this holds for any $\lambda \in \Lambda_D$ with $\lambda \ll \lambda^*$, we must have $\log\frac{d\lambda^*}{d\pi^*} + \beta d(x,\tilde{x}) + \gamma(x) = 0$ or Eq. 19 holds for all $(x,\tilde{x})$ in the support of λ^*.

Finally, note that for all Borel sets E:

$$\int_E \left(\int_R \frac{d\lambda^*}{d\pi^*} d\nu^*\right) d\mu = \int_{E\times R} \frac{d\lambda^*}{d\pi^*} d\pi^* = \lambda^*(E \times R) = \mu(E),$$

which implies that $\int_R \frac{d\lambda^*}{d\pi^*} d\nu^* = 1$ for all $x \in R$, giving $Z(x,\beta) = \int e^{-\beta d(x,\tilde{x})} d\nu^*$.

Similarly, for all Borel sets E, we have

$$\int_E \left(\int_R \frac{d\lambda^*}{d\pi^*} d\mu\right) d\nu^* = \int_{R\times E} \frac{d\lambda^*}{d\pi^*} d\pi^* = \lambda^*(R \times E) = \nu^*(E),$$

which implies that $\int_R \frac{d\lambda^*}{d\pi^*} d\mu = 1$ or, as is easily seen, identity (20) for all $\tilde{x}$ in the support of ν^*. □

We next show that optimal reproduction alphabets of continuous-valued input sources are usually discrete. We require only one more well-known basic fact from analytic function theory.

FACT 1. *If F is analytic on a bounded interval R and $F(x) = 0$ for an infinite number of $x \in R$, then $F = 0$ on R.*

PROOF OF THEOREM 3. Let S be the support of ν^*. Recall from Lemma 2 that ν^* exists, is unique, and that (20) holds for $\tilde{x} \in S$. For notational ease, set $F(\tilde{x}) := \int_{-\infty}^{\infty} \frac{p(x)}{Z(x,\beta)} e^{-\beta d(x,\tilde{x})} dx - 1$ and the related function $G(\tilde{x}) := \int_{-\infty}^{\infty} \frac{p(x)}{Z(x,\beta)} e^{-\beta\rho(\tilde{x}-x)} dx - 1$, in which $p(x) = \frac{d\mu}{dx}$. By construction, $F(\tilde{x}) = G(\tilde{x})$ for $\tilde{x} \in R$, so $G(\tilde{x}) = 0$ for $\tilde{x} \in S$. We work with G rather than F in what follows, as the analyticity of ρ and [**58**, Lemma III.2] imply that G is analytic. By Fact 1, if S has an accumulation point, then $G(\tilde{x}) = 0$ for almost all $\tilde{x} \in R$, and not only $\tilde{x} \in S$. For ease below, we also set $d = \infty$ outside of R^2.

From above, we have that either S has no accumulation point or $\int_{-\infty}^{\infty} \frac{p(x)}{Z(x,\beta)} e^{-\beta\rho(\tilde{x}-x)} dx = 1$ for all $\tilde{x} \in R$. Suppose the latter holds. We claim that the solution $Z(x,\beta)$ to this integral equation is unique up to a set of μ-measure 0. If there were some other solution $p(x)/Z'(x,\beta)$, then subtraction would yield $\int_{-\infty}^{\infty} p(x)(\frac{1}{Z(x,\beta)} - \frac{1}{Z'(x,\beta)}) e^{-\beta\rho(\tilde{x}-x)} dx = 0$. We recognize this as a convolution of $p(x)(\frac{1}{Z(x,\beta)} - \frac{1}{Z'(x,\beta)})$ and $e^{-\beta\rho(x)}$, and as the Fourier transform of $e^{-\beta\rho(x)}$ is nonzero almost everywhere, we have that the Fourier transform of $p(x)(\frac{1}{Z(x,\beta)} - \frac{1}{Z'(x,\beta)})$ is 0 almost everywhere, implying that $Z(x,\beta) = Z'(x,\beta)$ up to a set of μ-measure 0. It is straightforward to show that this equation is satisfied by $p(x)/Z(x,\beta) = 1/\int_{-\infty}^{\infty} e^{-\beta\rho(z)} dz$. However, for $x \notin R$, the left-hand side is 0 while the right-hand side is some positive number. Hence, we have $\int_{-\infty}^{\infty} \frac{p(x)}{Z(x,\beta)} e^{-\beta\rho(\tilde{x}-x)} dx = 1$ for all $\tilde{x} \in R$ cannot hold. In particular, the set S has no accumulation point. As R is bounded, we must have that S is a finite set. The statement of the theorem now follows. □

For other distortion measures, discrete support is not optimal when the so-called Shannon lower bound can be attained. Out of all codings with expected distortion D, consider the coding with maximal conditional entropy $H[\tilde{X}|X]$. This coding then places a lower bound on $I[X;\tilde{X}]$ for codings with expected distortion D, a lower bound called the Shannon lower bound. This lower bound is, for example,

TABLE 2. Issues and partial resolutions for DRNN coding of natural images.

Issue	Partial resolution
Neural networks typically store a linear number of memories	Special connectivity and patterns give exponential storage (Theorem 1)
DRNN coding matches model density to input rather than minimizing MSE	Competitive with JPEG using perceptual distortion (Figs. 6**bd**)
2nd-order Lenz-Ising models are determined by their 2nd-order statistics	ON/OFF correlations capture higher-order continuous structure (Section 4)
2nd-order Lenz-Ising model is not an excellent fit to data	Similar local extrema; near-optimal RD for 2×2 patches (Fig. 9)
Discrete networks code for discrete alphabets	Optimal codings of continuous variables are discrete (Theorem 3)

achieved by a Gaussian coding when input is normally distributed and the distortion measure is mean-squared error. In Theorem 3, we avoided discussion of the Shannon lower bound by way of a special distortion measure. However, more generally, we should expect the Shannon lower bound to be an important exception to the optimality of discretely supported codings.

QUESTION 1. *What is the most general formulation of Theorem 3?*

Even given this theory, rate-distortion functions are hard to compute when the input is a continuous variable. Naive implementation of the iteration (17) would be to mesh R and approximate $p(\tilde{x}|x)$ by a very large matrix. Alternatively, we can use that $\tilde{X}$ is a discrete variable to simplify computation of the rate-distortion function and optimal codebooks [**76**].

Example: Uniform continuous source and mean-squared error distortion. To illustrate some of the complexity of continuous RD, consider an input source that is uniform in $[-1, 1]$ with squared-error distortion: $d(x, \tilde{x}) = (x - \tilde{x})^2$. To find the optimal coding in this case, we numerically minimize the functional

$$\mathcal{L}_\beta = -\frac{1}{2}\int_{-1}^{1} \log\left(\sum_{\tilde{x}\in\mathcal{X}} p(\tilde{x})e^{-\beta(x-\tilde{x})^2}\right) dx,$$

with respect to $p(\tilde{x})$ and the elements of a finite set $\mathcal{X}$ (subject to the constraint that $p(\tilde{x}) \geq 0$ and $\sum_{\tilde{x}} p(\tilde{x}) = 1$) using the programming language Python's integrate and optimization packages (Scipy).

From Fig. 10, we see that there are certain critical β at which the reproduction alphabet $\tilde{\mathcal{X}}$ increases in size. These are markers of bifurcations in the solution space $p(\tilde{x}|x)$ with respect to β. At some critical β, mass is "created" at $\tilde{x} = 0$; at others, mass "splits" so that a single point mass at $\tilde{x} = 0$ becomes two points near 0. Examination of these optimal solutions hint at deep connections between rate-distortion objectives and bifurcation theory, some of which are explored in [**69**]. □

We close this section with a very general open problem whose study should bear (mathematical) fruit.

PROBLEM 1. *Construct optimal (discrete) codings of continuous input distributions subject to analytic distortion measures.*

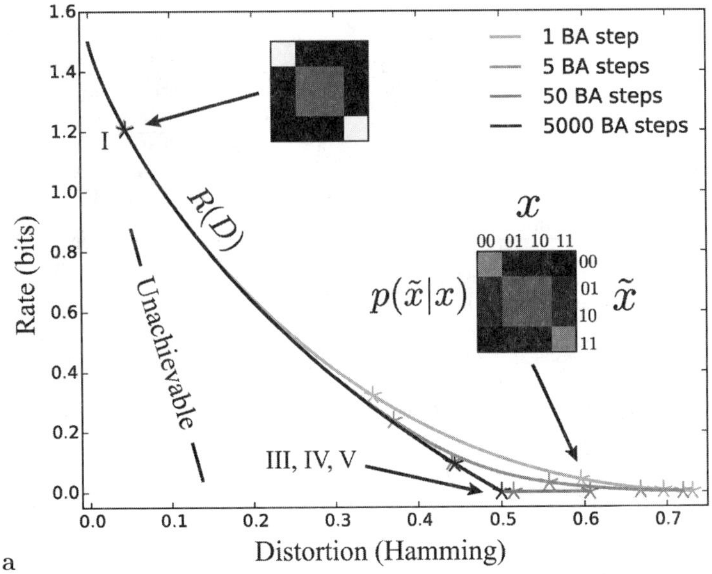

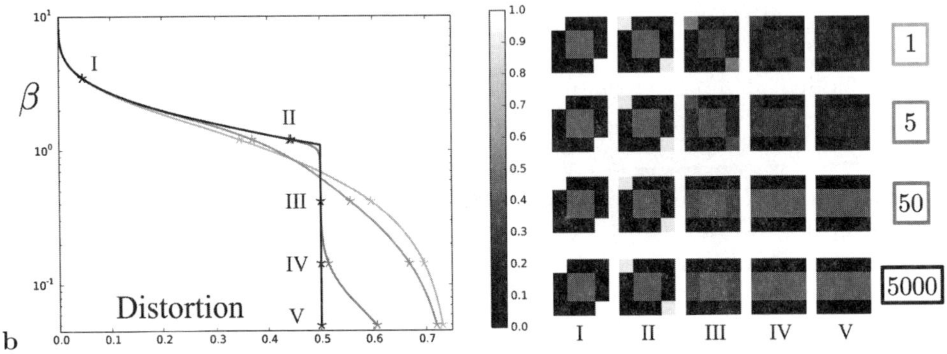

FIGURE 8. **RD for lossy addition. a)** Rate-distortion function $R(D)$. **b)** (Top) Distortion D_β defined in (14). To find optimal stochastic codebooks and the rate-distortion function, we iterate Eq. 17 for some number of iterations T at various β between 0 and 10. The plot shows corresponding estimates of the $R(D)$ function for several steps T, as in the legend, with rate in bits on the x-axis and expected distortion on the y-axis. At $T = 50$ iterations, the estimate of the rate-distortion function is nearly the same as its estimate at $T = 5000$ iterations, illustrating convergence. The stars $(*)$ on the curves indicate β's at which the estimated solutions $p^{(T)}(\tilde{x}|x)$ were visually examined. Arrows point to particular curve (and so T), rate, and distortion. Each box is the image of the matrix representing $p^{(T)}(\tilde{x}|x)$ and shading translates probability according to the colorbar shown in **b**. Rows and columns are both ordered by $(0,0), (0,1), (1,0), (1,1)$. Left-to-right, marked points have $\beta = 3.8, 1.3, .43, .15, .05$.

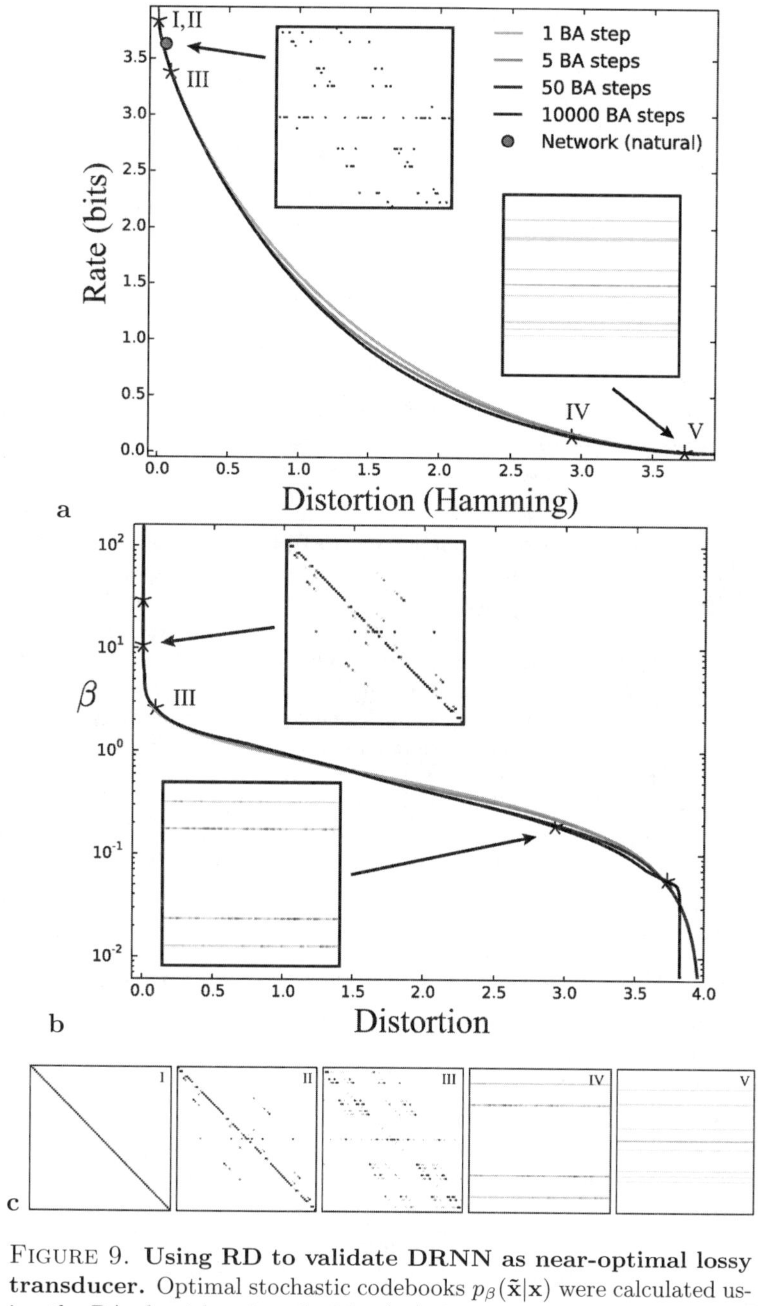

FIGURE 9. **Using RD to validate DRNN as near-optimal lossy transducer.** Optimal stochastic codebooks $p_\beta(\tilde{\mathbf{x}}|\mathbf{x})$ were calculated using the BA algorithm described by (17), Section 5, with the number of iterations indicated by the shade of the curve. **a) Rate-distortion function**, with rate in bits on the y-axis and the expected distortion on the x-axis. Note that DRNN coding (circle) of natural patches is near-optimal (compare with III in **c**). **b)** D_β on the x-axis vs. "inverse temperature" β on the y-axis; see Eq. (14). **c) Optimal stochastic codebooks** at corresponding star $(*)$ markers (left-to-right) in **ab**, evaluated using the BA algorithm with 10,000 iterations; ordering of 81 states is left-to-right down the rows in ON/OFF key in Fig. 7**a**; entries are shaded by reverse of colorbar in Fig. 7**a**.

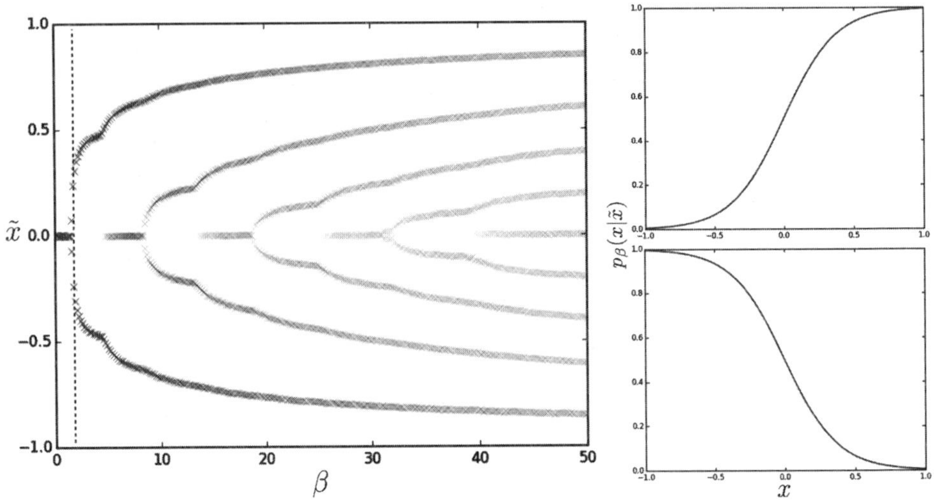

FIGURE 10. **Bifurcation structure in optimal stochastic codebooks for compressing uniform source with mean-squared error distortion.** The plot at left shows the values in the reproduction alphabet $\tilde{\mathcal{X}}$ as a function of β. The transparency of the point corresponds to $p_\beta(\tilde{x})$; lighter points have smaller probability. The size of the reproduction alphabet increases by 1 at critical βs. At some of these critical β, the values in the reproduction alphabet appear to change discontinuously; at others, the change is just nondifferentiable. For the former, see the transition from $|\tilde{\mathcal{X}}| = 2$ to $|\tilde{\mathcal{X}}| = 3$; for the latter, see the transition from $|\tilde{\mathcal{X}}| = 1$ to $|\tilde{\mathcal{X}}| = 2$. The plot at right shows $p_\beta(x|\tilde{x})$ at $\beta = 3$, illustrating the stochastic nature of the codebook.

8. Conclusion

The new image compression scheme described in Section 3 seems, at first glance, to violate many principles of statistical modeling of natural image patches (Table 1). As such, its near-optimal compression performance seems surprising. However, modeling is not compression: an imperfect statistical model can give rise to a high-quality lossy compression scheme. And as suggested by Sections 4 and 7, statistical models of discretized natural image patches may be more natural for compression efforts, as they do capture information about higher-order local structure while also being true to mathematical results such as Theorem 3 on quantization. A summary of these resolutions can be found in Table 2. As a closing remark, we hope that this works answers the call [**92**] for new problems and theorems arising from theoretical investigations into biology.

Acknowledgment

The authors thank Bill Bialek, Jim Crutchfield, Charles Garfinkle, Kilian Koepsell, Urs Köster, Jesse Livezey, Ram Mehta, Bruno Olshausen, Susanna Still, and Lav Varshney for useful comments and references.

References

[1] D. Ackley, G. Hinton, and T. Sejnowski. A learning algorithm for Boltzmann machines. *Cognitive science*, 9(1):147–169, 1985.

[2] S.-I. Amari, *Learning patterns and pattern sequences by self-organizing nets of threshold elements*, IEEE Trans. Computers **C-21** (1972), 1197–1206. MR0366123

[3] S.-I. Amari. Characteristics of sparsely encoded associative memory. *Neural Networks*, 2(6):451–457, 1989.

[4] S. Arimoto, *An algorithm for computing the capacity of arbitrary discrete memoryless channels*, IEEE Trans. Information Theory **IT-18** (1972), 14–20. MR0403796

[5] F. Attneave. Some informational aspects of visual perception. *Psychological review*, 61(3):183–193, 1954.

[6] H. Barlow. Sensory mechanisms, the reduction of redundancy, and intelligence. In *NPL Symposium on the Mechanization of Thought Process*, number 10, pages 535–539. HM Stationery Office, London, 1959.

[7] A. Bell and T. Sejnowski. The "independent components" of natural scenes are edge filters. *Vision research*, 37(23):3327–3338, 1997.

[8] T. Berger, *Rate distortion theory: Mathematical basis for data compression*, Prentice-Hall, Inc., Englewood Cliffs, N. J., 1971. MR0408988

[9] M. Bethge, *Factorial coding of natural images: how effective are linear models in removing higher-order dependencies*, J. Opt. Soc. Amer. A **23** (2006), no. 6, 1253–1268, DOI 10.1364/JOSAA.23.001253. MR2307154

[10] R. E. Blahut, *Computation of channel capacity and rate-distortion functions*, IEEE Trans. Information Theory **IT-18** (1972), 460–473. MR0476161

[11] P. Burt and E. Adelson. The Laplacian pyramid as a compact image code. *Communications, IEEE Transactions on*, 31(4):532–540, 1983.

[12] G. Carlsson, T. Ishkhanov, V. de Silva, and A. Zomorodian. On the local behavior of spaces of natural images. *International journal of computer vision*, 76(1):1–12, 2008.

[13] S. Chatterjee, P. Diaconis, and A. Sly, *Random graphs with a given degree sequence*, Ann. Appl. Probab. **21** (2011), no. 4, 1400–1435, DOI 10.1214/10-AAP728. MR2857452

[14] R. Chaudhuri and I. Fiete. Computational principles of memory. *Nature neuroscience*, 19(3):394–403, 2016.

[15] D. Chklovskii and A. Koulakov. Maps in the brain: what can we learn from them? *Annu. Rev. Neurosci.*, 27:369–392, 2004.

[16] T. Cover. Geometrical and statistical properties of systems of linear inequalities with application in pattern recognition. *IEEE Trans. Electronic Computers*, 14(3):326–334, 1965.

[17] T. M. Cover and J. A. Thomas, *Elements of information theory*, 2nd ed., Wiley-Interscience [John Wiley & Sons], Hoboken, NJ, 2006. MR2239987

[18] I. Csiszár, *On the computation of rate-distortion functions*, IEEE Trans. Information Theory **IT-20** (1974), 122–124. MR0368936

[19] R. Dony and S. Haykin. Neural network approaches to image compression. *Proceedings of the IEEE*, 83(2):288–303, 1995.

[20] M. Drton, B. Sturmfels, and S. Sullivant, *Lectures on algebraic statistics*, Oberwolfach Seminars, vol. 39, Birkhäuser Verlag, Basel, 2009. MR2723140

[21] S. Edwards and P. Anderson. Theory of spin glasses. *J. Phys. F: Met. Phys.*, 5(5):965, 1975.

[22] F. Effenberger and C. Hillar. Discovery of salient low-dimensional dynamical structure in neuronal population activity using Hopfield networks. In *Similarity-Based Pattern Recognition*, pages 199–208. Springer, 2015.

[23] M. Egmont-Petersen, D. de Ridder, and H. Handels. Image processing with neural networks – a review. *Pattern recognition*, 35(10):2279–2301, 2002.

[24] D. Field. Relations between the statistics of natural images and the response properties of cortical cells. *JOSA A*, 4(12):2379–2394, 1987.

[25] I. Fiete, D. Schwab, and N. Tran. A binary Hopfield network with $1/\log(n)$ information rate and applications to grid cell decoding. *arXiv preprint arXiv:1407.6029*, 2014.

[26] S. L. Fix, *Rate distortion functions for squared error distortion measures*, Sixteenth Annual Allerton Conference on Communication, Control, and Computing, Univ. Illinois, Urbana, IL, 1978, pp. 704–711. MR691536

[27] G. B. Folland, *Real analysis*, Pure and Applied Mathematics (New York), John Wiley & Sons, Inc., New York, 1984. Modern techniques and their applications; A Wiley-Interscience Publication. MR767633

[28] K. Friston. The free-energy principle: a unified brain theory? *Nature Reviews Neuroscience*, 11(2):127–138, 2010.

[29] S. Geman and D. Geman. Stochastic relaxation, Gibbs distributions, and the bayesian restoration of images. *IEEE Trans. Pattern Analysis and Machine Intelligence*, 6(6):721–741, 1984.

[30] H. Gerhard, F. Wichmann, and M. Bethge. How sensitive is the human visual system to the local statistics of natural images? *PLoS computational biology*, 9(1):e1002873, 2013.

[31] E. N. Gilbert, *Codes based on inaccurate source probabilities*, IEEE Trans. Information Theory **IT-17** (1971), 304–314. MR0312983

[32] V. Goyal, A. Fletcher, and S. Rangan. Compressive sampling and lossy compression. *Signal Processing Magazine, IEEE*, 25(2):48–56, 2008.

[33] L. Griffin. The second order local-image-structure solid. *Pattern Analysis and Machine Intelligence, IEEE Transactions on*, 29(8):1355–1366, 2007.

[34] D. J. Gross and M. Mézard, *The simplest spin glass*, Nuclear Phys. B **240** (1984), no. 4, 431–452, DOI 10.1016/0550-3213(84)90237-2. MR766359

[35] E. Hazan, A. Agarwal, and S. Kale. Logarithmic regret algorithms for online convex optimization. *Machine Learning*, 69(2-3):169–192, 2007.

[36] D. Hebb. The organization of behavior. *New York, Wiley*, 1949.

[37] J. Hertz, A. Krogh, and R. G. Palmer, *Introduction to the theory of neural computation*, Santa Fe Institute Studies in the Sciences of Complexity. Lecture Notes, I, Addison-Wesley Publishing Company, Advanced Book Program, Redwood City, CA, 1991. With forewords by Jack Cowan and Christof Koch. MR1096298

[38] C. Hillar and F. Effenberger. Robust discovery of temporal structure in multi-neuron recordings using Hopfield networks. *Procedia Computer Science*, 53:365–374, 2015.

[39] C. Hillar, R. Mehta, and K. Koepsell. A Hopfield recurrent neural network trained on natural images performs state-of-the-art image compression. In *Image Processing (ICIP), 2014 IEEE International Conference on*, pages 4092–4096. IEEE, 2014.

[40] C. Hillar, J. Sohl-Dickstein, and K. Koepsell. Efficient and optimal binary Hopfield associative memory storage using minimum probability flow. In *4th Neural Information Processing Systems (NIPS) workshop on Discrete Optimization in Machine Learning (DISCML)*, 2012.

[41] C. Hillar and N. Tran. Robust exponential memory in Hopfield networks. *ArXiv e-prints: nlin.AO 1411.4625*, 2014.

[42] C. Hillar and A. Wibisono. Maximum entropy distributions on graphs. *arXiv preprint arXiv:1301.3321*, 2013.

[43] J. J. Hopfield, *Neural networks and physical systems with emergent collective computational abilities*, Proc. Nat. Acad. Sci. U.S.A. **79** (1982), no. 8, 2554–2558. MR652033

[44] J. Huang, *Statistics of natural images and models*, ProQuest LLC, Ann Arbor, MI, 2000. Thesis (Ph.D.)–Brown University. MR2701355

[45] D. Hubel and T. Wiesel. Effects of monocular deprivation in kittens. *Naunyn-Schmiedeberg's Archives of Pharmacology*, 248(6):492–497, 1964.

[46] J. Hurri, A. Hyvärinen, J. Karhunen, and E. Oja. Image feature extraction using independent component analysis. In *Proc. NORSIG*, 1996.

[47] E. Ising. Beitrag zur Theorie des Ferromagnetismus. *Zeitschrift fur Physik*, 31:253–258, 1925.

[48] A. Jacquin. Image coding based on a fractal theory of iterated contractive image transformations. *Image Processing, IEEE Transactions on*, 1(1):18–30, 1992.

[49] E. T. Jaynes, *Information theory and statistical mechanics*, Phys. Rev. (2) **106** (1957), 620–630. MR0087305

[50] G. W. Johnson, *An unsymmetric Fubini theorem*, Amer. Math. Monthly **91** (1984), no. 2, 131–133, DOI 10.2307/2322111. MR729555

[51] J. J. Koenderink, *The structure of images*, Biol. Cybernet. **50** (1984), no. 5, 363–370, DOI 10.1007/BF00336961. MR758126

[52] A. Lee, K. Pedersen, and D. Mumford. The nonlinear statistics of high-contrast patches in natural images. *International Journal of Computer Vision*, 54(1-3):83–103, 2003.

[53] T. Lee and D. Mumford. Hierarchical bayesian inference in the visual cortex. *JOSA A*, 20(7):1434–1448, 2003.

[54] J.-S. Lin and S.-H. Liu. A competitive continuous Hopfield neural network for vector quantization in image compression. *Engineering Applications of Artificial Intelligence*, 12(2):111–118, 1999.

[55] R. Lorente de Nó. Vestibulo-ocular reflex arc. *Archives of Neurology & Psychiatry*, 30(2):245–291, 1933.

[56] D. G. Luenberger, *Optimization by vector space methods*, John Wiley & Sons, Inc., New York-London-Sydney, 1969. MR0238472

[57] D. J. C. MacKay, *Information theory, inference and learning algorithms*, Cambridge University Press, New York, 2003. MR2012999

[58] F. Matějka and C. Sims. Discrete actions in information-constrained tracking problems. *CERGE-EI Working Paper Series*, (441), 2011.

[59] W. S. McCulloch and W. Pitts, *A logical calculus of the ideas immanent in nervous activity*, Bull. Math. Biophys. **5** (1943), 115–133. MR0010388

[60] R. McEliece and E. Posner. The number of stable points of an infinite-range spin glass. *JPL Telecomm. and Data Acquisition Progress Report*, 42-83:209–215, 1985.

[61] R. J. McEliece, E. C. Posner, E. R. Rodemich, and S. S. Venkatesh, *The capacity of the Hopfield associative memory*, IEEE Trans. Inform. Theory **33** (1987), no. 4, 461–482, DOI 10.1109/TIT.1987.1057328. MR901677

[62] R. Mehta, S. Marzen, and C. Hillar. Exploring discrete approaches to lossy compression schemes for natural image patches. In *European Signal Processing Conference (EUSIPCO 2015)*, 2236–2240, 2015.

[63] V. Misra, V. K. Goyal, and L. R. Varshney, *Distributed scalar quantization for computing: high-resolution analysis and extensions*, IEEE Trans. Inform. Theory **57** (2011), no. 8, 5298–5325, DOI 10.1109/TIT.2011.2158882. MR2849118

[64] T. Mora, A. Walczak, W. Bialek, and C. Callan. Maximum entropy models for antibody diversity. *Proceedings of the National Academy of Sciences*, 107:5405–5410, 2010.

[65] M. Naillon and J.-B. Theeten. Neural approach for TV image compression using a Hopfield type network. In *Advances in neural information processing systems 1*, pages 264–271. Morgan Kaufmann Publishers Inc., 1989.

[66] J. Nocedal, *Updating quasi-Newton matrices with limited storage*, Math. Comp. **35** (1980), no. 151, 773–782, DOI 10.2307/2006193. MR572855

[67] B. Olshausen and D. Field. Emergence of simple-cell receptive field properties by learning a sparse code for natural images. *Nature*, 381(6583):607–609, 1996.

[68] S. Palmer, O. Marre, M. Berry II, and W. Bialek. Predictive information in a sensory population. *Proceedings of the National Academy of Sciences*, page 201506855, 2015.

[69] A. E. Parker, A. G. Dimitrov, and T. Gedeon, *Symmetry breaking in soft clustering decoding of neural codes*, IEEE Trans. Inform. Theory **56** (2010), no. 2, 901–927, DOI 10.1109/TIT.2009.2037045. MR2724414

[70] L. Pastur and A. Figotin. Exactly soluble model of a spin glass. *Sov. J. Low Temp. Phys*, 3:378–383, 1977.

[71] K. Pedersen and A. Lee. Toward a full probability model of edges in natural images. In *Computer Vision?ECCV 2002*, pages 328–342. Springer, 2002.

[72] G. Piccinini, *The first computational theory of mind and brain: a close look at W. S. McCulloch and W. Pitts's "A logical calculus of ideas immanent in nervous activity" [Bull. Math. Biophys.* **5** *(1943), 115–133; MR 0010388]*, Synthese **141** (2004), no. 2, 175–215, DOI 10.1023/B:SYNT.0000043018.52445.3e. MR2101909

[73] M. S. Pinsker, *Information and information stability of random variables and processes*, Translated and edited by Amiel Feinstein, Holden-Day, Inc., San Francisco, Calif.-London-Amsterdam, 1964. MR0213190

[74] E. C. Posner, *Random coding strategies for minimum entropy*, IEEE Trans. Information Theory **IT-21** (1975), 388–391. MR0689214

[75] R. Rescorla and A. Wagner. *A theory of Pavlovian conditioning: Variations in the effectiveness of reinforcement and nonreinforcement*, chapter 3, pages 64–99. Appleton-Century-Crofts, New York, 1972.

[76] K. Rose. A mapping approach to rate-distortion computation and analysis. *IEEE Transactions on Information Theory*, 40(6):1939–1952, 1994.

[77] F. Rosenblatt, *Principles of neurodynamics. Perceptrons and the theory of brain mechanisms*, Spartan Books, Washington, D.C., 1962. MR0135635

[78] D. Ruderman and W. Bialek. Statistics of natural images: Scaling in the woods. *Physical review letters*, 73(6):814, 1994.

[79] S. Saremi and T. J. Sejnowski, *Hierarchical model of natural images and the origin of scale invariance*, Proc. Natl. Acad. Sci. USA **110** (2013), no. 8, 3071–3076, DOI 10.1073/pnas.1222618110. MR3045555

[80] E. Schneidman, M. Berry II, R. Segev, and W. Bialek. Weak pairwise correlations imply strongly correlated network states in a neural population. *Nature*, 440(7087):1007–1012, 2006.

[81] O. Shamir, S. Sabato, and N. Tishby, *Learning and generalization with the information bottleneck*, Theoret. Comput. Sci. **411** (2010), no. 29-30, 2696–2711, DOI 10.1016/j.tcs.2010.04.006. MR2666286

[82] C. Shannon. A mathematical theory of communication. *ACM SIGMOBILE Mobile Computing and Communications Review*, 5(1):3–55, 2001.

[83] J. Shlens, G. Field, J. Gauthier, M. Grivich, D. Petrusca, A. Sher, A. Litke, and E. Chichilnisky. The structure of multi-neuron firing patterns in primate retina. *The Journal of neuroscience*, 26(32):8254–8266, 2006.

[84] E. Simoncelli and B. Olshausen. Natural image statistics and neural representation. *Annual review of neuroscience*, 24(1):1193–1216, 2001.

[85] C. Sims. Implications of rational inattention. *Journal of Monetary Economics*, 50(3):665–690, 2003.

[86] C. Sims. The cost of misremembering: Inferring the loss function in visual working memory. *Journal of Vision*, 15(3):1–27, 2015.

[87] A. Skodras, C. Christopoulos, and T. Ebrahimi. The JPEG 2000 still image compression standard. *Signal Processing Magazine, IEEE*, 18(5):36–58, 2001.

[88] N. Slonim. *The information bottleneck: Theory and applications.* PhD thesis, Citeseer, 2002.

[89] J. Smith. *On the information capacity of peak and average power constrained Gaussian channels.* PhD thesis, University of California, 1969.

[90] J. G. Smith, *The information capacity of amplitude- and variance-constrained scalar Gaussian channels*, Information and Control **18** (1971), 203–219. MR0278817

[91] J. Sohl-Dickstein, P. Battaglino, and M. DeWeese. New method for parameter estimation in probabilistic models: minimum probability flow. *Physical Review Letters*, 107(22):220601, 2011.

[92] B. Sturmfels. Can biology lead to new theorems? *Annual report of the Clay Mathematics Institute*, pages 13–26, 2005.

[93] M. Talagrand, *Spin glasses: a challenge for mathematicians*, Ergebnisse der Mathematik und ihrer Grenzgebiete. 3. Folge. A Series of Modern Surveys in Mathematics [Results in Mathematics and Related Areas. 3rd Series. A Series of Modern Surveys in Mathematics], vol. 46, Springer-Verlag, Berlin, 2003. Cavity and mean field models. MR1993891

[94] F. Tanaka and S. Edwards. Analytic theory of the ground state properties of a spin glass. I. Ising spin glass. *Journal of Physics F: Metal Physics*, 10:2769, 1980.

[95] Joshua B Tenenbaum, Thomas L Griffiths, and Charles Kemp. Theory-based bayesian models of inductive learning and reasoning. *Trends in cognitive sciences*, 10(7):309–318, 2006.

[96] L. Theis, S. Gerwinn, F. Sinz, and M. Bethge, *In all likelihood, deep belief is not enough*, J. Mach. Learn. Res. **12** (2011), 3071–3096. MR2877594

[97] N. Tishby, F. Pereira, and W. Bialek. The information bottleneck method. *arXiv preprint physics/0004057*, 2000.

[98] A. M. Turing, *On computable numbers, with an application to the entscheidungsproblem*, Proc. London Math. Soc. **S2-42**, no. 1, 230, DOI 10.1112/plms/s2-42.1.230. MR1577030

[99] J. van Hateren. A theory of maximizing sensory information. *Biological Cybernetics*, 68(1):23–29, 1992.

[100] J. van Hateren and A. van der Schaaf. Independent component filters of natural images compared with simple cells in primary visual cortex. *Proceedings of the Royal Society of London. Series B: Biological Sciences*, 265(1394):359–366, 1998.

[101] J. von Neumann, *The computer and the brain*, Silliman Memorial Lectures, Vol. 36, Yale University Press, New Haven, Conn, 1958. MR0130066

[102] J. von Neumann, *First draft of a report on the EDVAC*, IEEE Ann. Hist. Comput. **15** (1993), no. 4, 27–75, DOI 10.1109/85.238389. Edited and with an introduction by Michael D. Godfrey. MR1241101

[103] G. Wallace. The JPEG still picture compression standard. *Communications of the ACM*, 34(4):30–44, 1991.

[104] Z. Wang and A. Bovik. Mean squared error: love it or leave it? a new look at signal fidelity measures. *IEEE Signal Processing Magazine*, 26(1):98–117, 2009.

[105] Z. Wang, A. Bovik, H. Sheikh, and E. Simoncelli. Image quality assessment: From error visibility to structural similarity. *IEEE Trans. Image Processing*, 13(4):600–612, 2004.

[106] B. Widrow and M. Hoff. Adaptive Switching Circuits. In *1960 IRE WESCON Convention Record, Part 4*, pages 96–104, New York, 1960. IRE.

[107] R. Yeung. *Information theory and network coding.* Springer Science & Business Media, 2008.

Redwood Center for Theoretical Neuroscience, Berkeley, CA, USA
E-mail address: `chillar@msri.org`

Massachusetts Institute of Technology, Cambridge, MA, USA
E-mail address: `semarzen@mit.edu`

Contemporary Mathematics
Volume **685**, 2017
http://dx.doi.org/10.1090/conm/685/13753

Proving Tucker's Lemma with a volume argument

Beauttie Kuture, Oscar Leong, Christopher Loa, Mutiara Sondjaja,
and Francis Edward Su

Abstract. Sperner's Lemma is a statement about labeled triangulations of a simplex. McLennan and Tourky (2007) provided a novel proof of Sperner's Lemma by examining volumes of simplices in a triangulation under time-linear simplex-linear deformation. We adapt a similar argument to prove Tucker's Lemma on a triangulated cross-polytope P. The McLennan-Tourky technique does not directly apply because this deformation may distort the volume of P. We remedy this by inscribing P in its dual polytope, triangulating it, and considering how the volumes of deformed simplices behave.

1. Introduction

Sperner's Lemma is a combinatorial result that is equivalent to the Brouwer fixed point theorem and has many useful applications in mathematics and economics, including Nash's proof of his famous equilibrium theorem [**7**]. Consider a d-dimensional simplex S with a *triangulation*, i.e., a subdivision into smaller simplices that meet face-to-face or not at all. Sperner's Lemma states that if such a triangulation has a *Sperner labeling* (each vertex is labeled by one of the vertices of S that span the minimal face that the vertex is on), then there exists an odd number of *fully-labeled* simplices (ones whose vertices have distinct labels).

Recently, McLennan and Tourky [**5**] gave a novel proof of Sperner's Lemma based on the following facts:

(1) A triangulation remains a triangulation under a small perturbation that keeps each vertex in its minimal face.
(2) The (signed) volume of a simplex spanned by $v_0, v_1, ..., v_d$ is $1/d!$ times the determinant of a matrix whose columns are $v_i - v_0$ for $i = 1, ..., d$.
(3) A polynomial function of t is constant if it is constant on a nonempty open t-interval.

Given a Sperner-labeled triangulation of a simplex, they construct a continuous deformation that moves each vertex of the triangulation linearly in time to the vertex of S that it is named after. By Fact (1), for small t the sum of deformed volumes of all simplices is equal to the total volume of S, so the sum is constant

2010 *Mathematics Subject Classification.* Primary 05-XX, 55-XX.

This work was conducted during the 2015 Mathematical Sciences Research Institute Undergraduate Program (MSRI-UP), supported by grants from the National Science Foundation (DMS 1156499) and the National Security Agency (H98230-15-1-0039).

for small t. By Fact (2), this volume sum is also a polynomial in t because the volume of each deformed simplex is the determinant of a matrix whose entries are linear in t. By Fact (3) the volume sum remains constant for all t, hence at $t = 1$ it is also non-zero and represents the sum of volumes of all deformed simplices. Thus some deformed simplex has non-zero volume, which means it was originally a fully-labeled simplex (because non-fully-labeled simplices become degenerate under the deformation). In fact, there must be one more fully-labeled simplex of positive volume than of negative volume, which implies that the number of fully-labeled simplices is odd.

While there are many proofs of Sperner's Lemma (e.g., see [**10**]), the novelty of the McLennan-Tourky approach is that it uses a geometric volume argument rather than an inductive combinatorial argument to find the desired fully-labeled simplex. In this paper, we show that this approach can be used to prove another classical lemma from topological combinatorics.

Like Sperner's Lemma, Tucker's Lemma [**3, 11**] is a combinatorial analog of a topological theorem, in this case, the Borsuk-Ulam theorem. It also has beautiful connections to other well-known results in mathematics and economics, such as the ham sandwich theorem, necklace-splitting problems [**6**], and fair division problems [**9**].

Let e_i denote the standard unit basis vector in $\mathbb{R}^d$ (with 1 in the i-th coordinate and 0 in every other coordinate) and define $e_{-i} = -e_i$. Let P denote the d-dimensional *cross-polytope*, the convex hull of the $2n$ vectors $\{e_i, e_{-i}\}_{i=1}^d$.

THEOREM 1 (Tucker's Lemma [**11**]). *Let T_P be a triangulation of the d-dimensional cross-polytope P whose restriction to the boundary ∂P is antipodally symmetric: if σ is a simplex in ∂P, then $-\sigma$ also is a simplex. Suppose that T_P has a* Tucker labeling*: each vertex v of T_P is assigned a label $\ell(v)$ in $\{\pm 1, \pm 2, \dots \pm d\}$ such that antipodal vertices in ∂P have labels summing to zero. Then there is a* complementary edge*: a pair of adjacent vertices in T_P have labels summing to zero.*

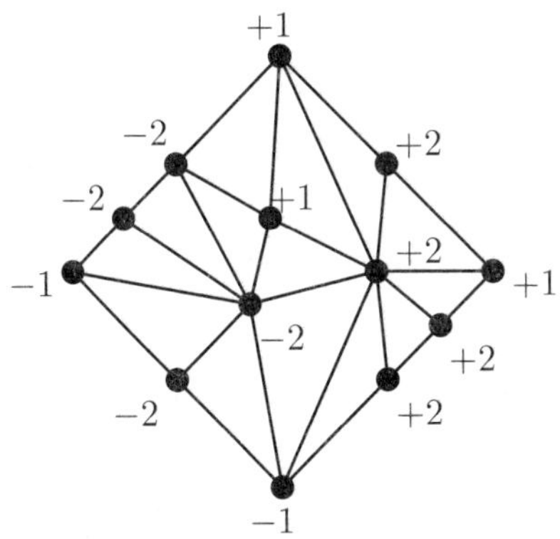

FIGURE 1. A Tucker labeling of the cross-polytope P must have a complementary edge.

Although the cross-polytope P in Tucker's Lemma is convenient for our purposes, the result holds if P is any centrally symmetric object topologically equivalent to a ball. Tucker [**11**] established his lemma for a triangulated 2-dimensional ball, and Lefschetz [**3**] gave a general proof for a d-dimensional ball. Baker [**1**] developed a cubical version. Freund-Todd [**2**] gave the first constructive proof of Tucker's Lemma by using a path-following argument on a cross-polytope. Prescott-Su [**8**] obtained a different constructive proof on a triangulated ball that proves also the more general Fan's Lemma.

At first glance, it may seem that the McLennan-Tourky approach can be directly applied to prove Tucker's Lemma—just deform the Tucker-labelled triangulation by moving every vertex to the extreme point with the same label. However, such a deformation, even for small time intervals, is not guaranteed to cover P, so the sum of the volumes of the deformed simplices will not necessarily remain constant. This is the new challenge that didn't exist with Sperner's Lemma.

We avoid this problem by placing the cross-polytope P inside a larger polytope C and applying the McLennan-Tourky technique to C. The deformation we construct will continue to cover C for an open t-interval, so the total volume of the simplices remains constant and equal to the total volume of C even after the deformed simplices no longer form a triangulation. We observe what happens inside P. A volume argument will show that if the boundary of P has no complementary edges, then the interior of P must be covered by deformed simplices of P at time $t = 1$. The only way P can be covered by simplices is if one of them had two labels that summed to zero. This would yield the desired complementary edge. The goal of this paper is to demonstrate how to make this intuition precise.

2. Volume Sums

Given a triangulation T of a d-dimensional polytope K and a time interval $[a, b]$, $a \neq b$, suppose $\Delta_T : [a, b] \times K \to K$ is a deformation that is:

(1) *time-linear*: $\Delta_T(t, x)$ is linear in t, with $\Delta_T(0, x) = x$, and
(2) *simplex-linear*: $\Delta_T(t, x)$ is linear in x when restricted to a simplex in T.

We use $\Delta_T(t, \sigma)$ to denote the image of a simplex σ at a particular time t. For a deformation Δ_T, we can define the volume sum

$$S_T(t) = \sum_{\sigma \in T} \operatorname{vol}(\Delta_T(t, \sigma))$$

which is the sum of the d-dimensional volumes of deformed simplices as a function of time. Throughout the paper the term *volume* will mean *signed volume* that takes orientation into account, so that at time $t = 0$ all simplex volumes are non-negative. For notational simplicity we write the sum over all simplices of T, though the only non-zero terms in the sum will come from d-dimensional simplices of T. This volume sum can be defined similarly for any subtriangulation of T.

McLennan and Tourky's proof of Sperner's Lemma rests on what we will call the McLennan-Tourky observation about such deformations.

THEOREM 2 (McLennan-Tourky observation). *Given a triangulation T of a polytope K and a time-linear simplex-linear deformation $\Delta_T : [a, b] \times K \to K$, if the deformation keeps each vertex of T in its minimal face in K, then the volume sum $S_T(t)$ is constant and equal to* $\operatorname{vol}(K)$ *for all $t \in [a, b]$.*

The observation follows immediately from Facts (1), (2), and (3) above and is proved in [**5**].

3. Tucker's Lemma

As in the hypothesis of Tucker's Lemma, let P be a d-dimensional cross-polytope and ∂P its boundary, endowed with:

(1) a triangulation T_P that is symmetric on ∂P, and
(2) a labeling $\ell : V(T) \to \{\pm 1, ..., \pm d\}$ that is anti-symmetric on ∂P: antipodal vertex labels sum to zero, i.e., $\ell(-v) = -\ell(v)$ for each vertex $v \in \partial P \cap V(T)$.

The d-dimensional cross-polytope P has $2d$ extreme points that come in complementary pairs: e_{+i}, e_{-i} for $i = 1, ..., d$. It also has 2^d facets, each spanned by d extreme points formed by choosing exactly one of each complementary pair.

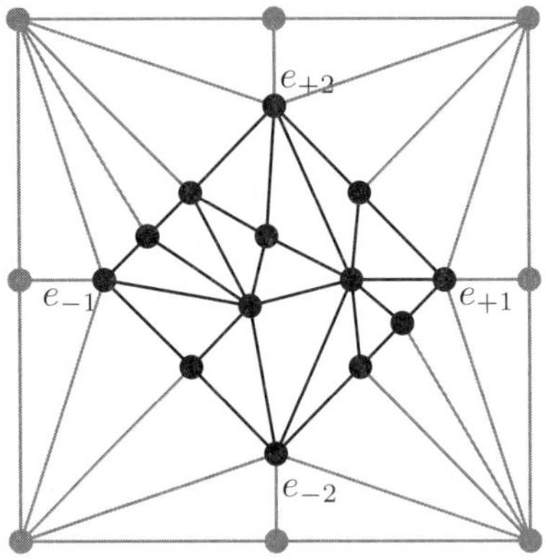

FIGURE 2. A triangulation T_P of a diamond (2-dimensional cross-polytope) P extended to a triangulation T of a square C.

Embed P in the *interior* of some polytope, which we may as well take to be the dual polytope of P, a cube C large enough to enclose P. Figure 2 shows the situation for dimension 2, and Figure 3 shows the situation for higher dimensions. Extend the triangulation T_P of P to a triangulation T of C in any convenient way.

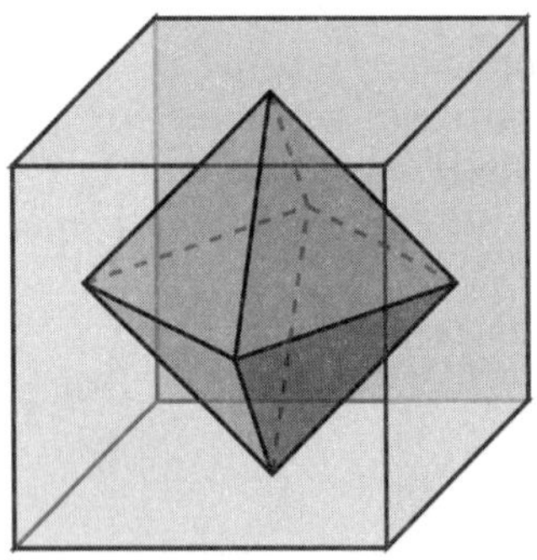

FIGURE 3. Embedding a cross-polytope P in a cube C.

Consider a deformation $\Delta_T : [0,1] \times C \to C$ defined on vertices of the triangulation T as follows:

$$\Delta_T(t,v) = \begin{cases} (1-t)v + te_{\ell(v)} & \text{if } v \in V(T_P) \\ v & \text{if } v \notin V(T_P) \end{cases}$$

and define $\Delta_T(t,x)$ for any $x \in C$ by linear extension. Then Δ_T is both time-linear and simplex-linear, and it fixes the boundary of C. At $t=0$, it is easy to see that Δ_T is the identity on all of C. If there are no complementary edges in T_P, then as t runs from 0 to 1, the simplices of T_P collapse onto whole facets of ∂P, since vertices in T_P map to the extreme points of P. The simplices outside P may deform but the boundary of C is fixed. So the McLennan-Tourky observation applies; the volume sum $S_T(t)$ remains constant over $[0,1]$ and equal to $\mathrm{vol}(C)$.

Now we consider a slightly different labeled triangulation of P, which we denote by T_P^*: it has exactly the same labeled triangulation structure on ∂P as T does, but only one vertex e_0 in the interior of P, placed at the origin and labeled 0. All maximal simplices of T_P^* are cones from e_0 to the simplices in ∂P. See Figure 4.

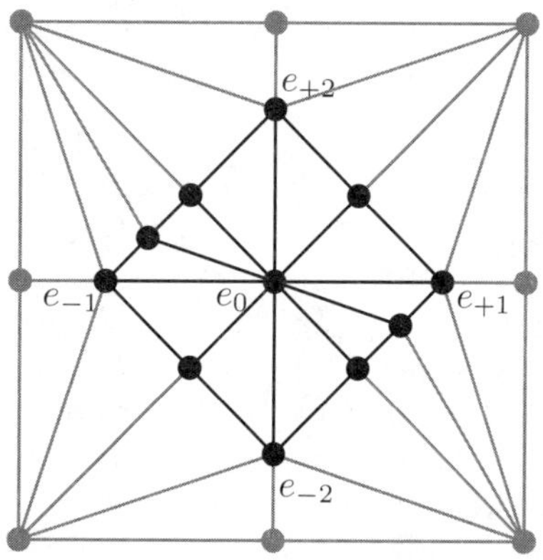

FIGURE 4. A triangulation T_P^* of P, extended to a triangulation T^* of C.

Extend T_P^* to a triangulation T^* of C in the exact same way as we extended T_P to T so that on $E = C \setminus \mathrm{int}(P)$, the triangulation T^* is identical to T. For convenience later, we denote the restrictions of T and T^* to E by T_E and T_E^*, respectively, and note that $T_E = T_E^*$. The deformation $\Delta_{T^*} : [0,1] \times C \to C$ can be defined as before:

$$\Delta_{T^*}(t,v) = \begin{cases} (1-t)v + te_{\ell(v)} & \text{if } v \in V(T_P^*) \\ v & \text{if } v \notin V(T_P^*) \end{cases}$$

on vertices of T^*, then we can extend the definition linearly over simplices. Then Δ_{T^*} is both time-linear and simplex-linear and fixes the boundary of C. At $t=0$, Δ_{T^*} is the identity on C. And as t runs from 0 to 1, the deformation Δ_{T^*} will keep e_0 fixed and move vertices in ∂P to one of the extreme points of P. The simplices outside P deform but the boundary of C is fixed. Thus the McLennan-Tourky observation applies and the volume sum $S_{T^*}(t)$ remains constant over $[0,1]$ and equal to $\mathrm{vol}(C)$.

Now compare these two deformations Δ_T and Δ_{T^*}. In both cases, their volume sums over C are equal, having remained constant at vol(C) as time runs from 0 to 1. So $S_T(t) = S_{T^*}(t) = \text{vol}(C)$. Breaking up the volume sums into sums over triangulations in E and P we have:

$$S_{T_P}(t) + S_{T_E}(t) = S_{T_P^*}(t) + S_{T_E^*}(t).$$

In both cases the volume sums over E may not remain constant, but because T and T^* are deforming in exactly the same way, i.e., $\Delta_T = \Delta_{T^*}$ on E, we see that their volume sums stay equal: $S_{T_E}(t) = S_{T_E^*}(t)$. Hence, the volumes sums over P must remain equal, and in particular, at time $t = 1$:

$$S_{T_P}(1) = S_{T_P^*}(1).$$

The following proposition shows the conclusion of Tucker's Lemma holds if we show $S_{T_P}(1) \neq 0$.

PROPOSITION 3. *If $S_{T_P}(1) \neq 0$, then T_P contains a complementary edge.*

PROOF. If the volume sum $S_{T_P}(1) = \sum_{\sigma \in P} \text{vol}(\Delta_{T_P}(1, \sigma))$ is non-zero, then at least one of the terms is non-zero: there must exist a simplex σ whose deformed volume at time 1 is non-zero. This can only happen if the vertices of σ have $d+1$ distinct labels. These are chosen from d complementary pairs of labels, so σ must have a complementary edge. □

So it will suffice to show $S_{T_P^*}(1) \neq 0$. We consider T_P^* instead of T_P for reasons that will be clear shortly.

We will appeal to the use of the *degree* of a simplicial map of spheres. It is a standard fact that every simplicial map f of a d-dimensional sphere with triangulation K to a d-dimensional sphere with triangulation L has a well-defined integer associated with it, called the *degree* of f. We review that briefly here. For any d-dimensional simplex σ in L, let $p(\sigma)$ be the number of pre-images of σ under f that map to σ with positive orientation (the outside of the sphere maps to the outside of the sphere), and let $n(\sigma)$ be the number of pre-images of σ under f that map to σ with negative orientation (the outside of the sphere maps to the inside of the sphere). Then the quantity $p(\sigma) - n(\sigma)$ is called the *degree* of f because it is independent of σ. This can be seen by noting that if σ and τ are adjacent d-simplices in L sharing a common $(d-1)$-face γ, then their pre-images of σ and τ in K are matched by paths of d-dimensional or $(d-1)$-dimensional simplices in K that are pre-images of γ. Pre-images of σ that are matched to each other will map to σ with opposite orientation so they contribute nothing to the sum $p(\sigma) - n(\sigma)$. Similarly, pre-images of τ that are matched to each other will map to τ with opposite orientation and contribute nothing to the sum $p(\tau) - n(\tau)$. What remains are pre-images of σ that are matched to pre-images of τ with the same orientation. So $p(\sigma) - n(\sigma) = p(\tau) - n(\tau)$ and the degree of f is well-defined.

PROPOSITION 4. *If T_P has no complementary edge in ∂P, then $S_{T_P^*}(1) \neq 0$.*

PROOF. Let T_P° denote the triangulation of P formed by taking facets of P and coning to e_0 at the origin. If there are no complementary edges in $T_{\partial P}$, the deformation Δ_{T^*} at time $t = 1$ will induce a simplicial map $T_P^* \to T_P^\circ$ whose restriction to ∂P is a simplicial map of spheres $f : T_{\partial P}^* \to T_{\partial P}^\circ$.

So f has a well-defined degree $\deg(f)$, which is the multiplicity of the preimages of any simplex σ, counted with orientation signs. (Our figures show examples for

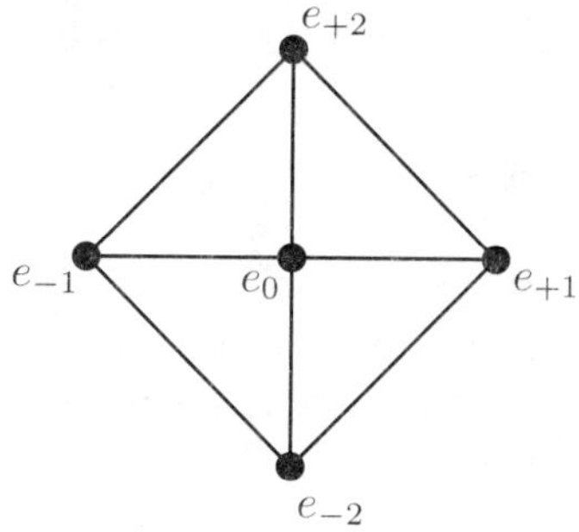

FIGURE 5. The triangulation T_P° of P.

a 2-dimensional cross-polytope P, and here the degree of f on the 1-dimensional boundary of P just becomes the *winding number* of $T_{\partial P}$ around $T_{\partial P}^\circ$.)

Any simplex in T_P° is a cone of e_0 over a simplex σ in $T_{\partial P}^\circ$ and that cone appears in $S_{T_P^*}(1)$ with multiplicity $\deg(f)$. Hence:

$$S_{T_P^*}(1) = \deg(f) \cdot \text{vol}(P).$$

The Tucker labeling condition causes $f : T_{\partial P}^* \to T_{\partial P}^\circ$ to be antipode-preserving, and it is a well-established fact the degree of such maps is odd (e.g., see Lefschetz [**3**] or Matoušek [**4**]). But since an odd integer and vol(P) are both nonzero, this means that $S_{T_P^*}(1)$ is nonzero, as desired. □

We remark that the purpose of introducing the triangulation T_P^* was to make the connection apparent between $\deg(f)$ and the volume sum $S_{T_P^*}(1)$, since the connection of $\deg(f)$ to the volume sum $S_{T_P}(1)$ is not so clear.

Propositions 3 and 4 together with the observation $S_{T_P}(1) = S_{T_P^*}(1)$ provide a proof Tucker's Lemma in any dimension, as desired.

Acknowledgement

The authors thank Duane Cooper for his encouragement.

References

[1] James K. Baker, *A combinatorial proof of Tucker's lemma for the n-cube*, J. Combinatorial Theory **8** (1970), 279–290. MR0253909

[2] Robert M. Freund and Michael J. Todd, *A constructive proof of Tucker's combinatorial lemma*, J. Combin. Theory Ser. A **30** (1981), no. 3, 321–325, DOI 10.1016/0097-3165(81)90027-3. MR618536

[3] Solomon Lefschetz, *Introduction to Topology*, Princeton Mathematical Series, vol. 11, Princeton University Press, Princeton, N. J., 1949. MR0031708

[4] Jiří Matoušek, *Using the Borsuk-Ulam theorem*, Universitext, Springer-Verlag, Berlin, 2003. Lectures on topological methods in combinatorics and geometry; Written in cooperation with Anders Björner and Günter M. Ziegler. MR1988723

[5] Andrew McLennan and Rabee Tourky, *Using volume to prove Sperner's lemma*, Econom. Theory **35** (2008), no. 3, 593–597, DOI 10.1007/s00199-007-0257-0. MR2398817

[6] Frédéric Meunier, *Discrete splittings of the necklace*, Math. Oper. Res. **33** (2008), no. 3, 678–688, DOI 10.1287/moor.1080.0311. MR2442647

[7] John Nash, *Non-cooperative games*, Ann. of Math. (2) **54** (1951), 286–295. MR0043432

[8] Timothy Prescott and Francis Edward Su, *A constructive proof of Ky Fan's generalization of Tucker's lemma*, J. Combin. Theory Ser. A **111** (2005), no. 2, 257–265, DOI 10.1016/j.jcta.2004.12.005. MR2156212

[9] Forest W. Simmons and Francis Edward Su, *Consensus-halving via theorems of Borsuk-Ulam and Tucker*, Math. Social Sci. **45** (2003), no. 1, 15–25, DOI 10.1016/S0165-4896(02)00087-2. MR1963117

[10] Francis Edward Su, *Rental harmony: Sperner's lemma in fair division*, Amer. Math. Monthly **106** (1999), no. 10, 930–942, DOI 10.2307/2589747. MR1732499

[11] A. W. Tucker, *Some topological properties of disk and sphere*, Proc. First Canadian Math. Congress, Montreal, 1945, University of Toronto Press, Toronto, 1946, pp. 285–309. MR0020254

Department of Mathematics, Pomona College, Claremont, California 91711
E-mail address: bak02013@mymail.pomona.edu

Department of Computational and Applied Mathematics, Rice University, Houston, Texas 77005
E-mail address: ofl2@rice.edu

Department of Mathematics, University of Illinois, Champaign, Illinois 61801
E-mail address: loa2@illinois.edu

Department of Mathematics, New York University, New York, New York 10012
E-mail address: sondjaja@cims.nyu.edu

Department of Mathematics, Harvey Mudd College, Claremont, California 91711
E-mail address: su@math.hmc.edu

Contemporary Mathematics
Volume **685**, 2017
http://dx.doi.org/10.1090/conm/685/13713

Factorization invariants in numerical monoids

Christopher O'Neill and Roberto Pelayo

ABSTRACT. Nonunique factorization in commutative monoids is often studied using factorization invariants, which assign to each monoid element a quantity determined by the factorization structure. For numerical monoids (co-finite, additive submonoids of $\mathbb{N}$), several factorization invariants have received much attention in the recent literature. In this survey article, we give an overview of the length set, elasticity, delta set, ω-primality, and catenary degree invariants in the setting of numerical monoids. For each invariant, we present current major results in the literature and identify the primary open questions that remain.

1. Motivating numerical monoids

Factorization theory has its genesis in studying the decomposition of natural numbers into their prime (irreducible) divisors. In this multiplicative monoid, the Fundamental Theorem of Arithmetic guarantees that such a decomposition is unique. Many other monoids, however, have elements that fail to admit a unique factorization into irreducibles; in such cases, algebraists are frequently interested in measuring, in various ways, how different the plural factorizations of an element can be.

One does not have to look far for examples. In fact, certain submonoids of the natural numbers have a rich non-unique factorization theory. *Numerical monoids* (co-finite, additive submonoids of the natural numbers) have garnered much recent interest, as their factorization theories are complicated enough to be interesting while controlled enough to be tractable. The goal of this paper is to introduce various perspectives and tools for studying factorizations in numerical monoids, and to present open problems that are of primary interest in this field.

Several natural examples of numerical monoids are found in the literature. Historically, Frobenius' coin-exchange problem asked what monetary values one could make with relatively prime coin denominations. A more endearing manifestation of numerical monoids is the McNugget monoid, which enumerates the number of chicken McNuggets one could buy by using the standard package sizes of 6, 9, and 20. As with the coin-problem and the McNugget monoid, all numerical monoids can be described using a finite set of generators. We provide the formal definition below. In what follows, we let $\mathbb{N}$ denote the set of non-negative integers.

2010 *Mathematics Subject Classification.* Primary 05C70.

DEFINITION 1.1. Let $n_1, n_2, \ldots, n_k$ be a collection of relatively prime positive integers. The *numerical monoid* S generated by these integers is the set of all non-negative, integral linear combinations of these numbers. In other words,

$$S = \langle n_1, n_2, \ldots, n_k \rangle = \{a_1n_1 + a_2n_2 + \cdots + a_kn_k : a_i \in \mathbb{N}\}.$$

REMARK 1.2. This set is an additive monoid with identity 0, and the condition that $\gcd(n_1, n_2, \ldots, n_k) = 1$ guarantees that there exist only finitely many natural numbers that are not elements of S. In fact, every co-finite additive submonoid of $\mathbb{N}$ can be generated by a finite set of relatively prime natural numbers; thus, numerical monoids are precisely the co-finite, additive submonoids of $\mathbb{N}$.

While generating sets of a fixed numerical monoid are non-unique, there always exists a unique minimal generating set (with respect to set-theoretic containment). As these minimal generators cannot be represented in terms of the other generators, they are precisely the *irreducible* elements of the numerical monoid. For the remainder of the paper, we will assume that $\{n_1, n_2, \ldots, n_k\}$ constitutes the unique minimal generating set for a numerical monoid S and that $n_1 < n_2 < \ldots < n_k$.

Both Frobenius' coin problem and the McNugget problem ask which natural numbers arise as non-negative, integral linear combinations of their respective generators. As such, understanding the largest natural number that is not an element of S (named for the aforementioned coin problem) is often useful.

DEFINITION 1.3. For a numerical monoid S, the values in $\mathbb{N} \setminus S$ are called the *gaps* of S, and $F(S) = \max(\mathbb{N} \setminus S)$ is called the *Frobenius number* of S.

EXAMPLE 1.4. For two relatively prime positive integers n_1, n_2, the numerical monoid $S = \langle n_1, n_2 \rangle$ has $F(S) = n_1n_2 - (n_1 + n_2)$ as its Frobenius number [**50**]. Furthermore, all but finitely many elements of S can be written in multiple ways as non-negative integral combinations of n_1 and n_2. The simplest example is the element $n_1n_2 \in S$, which can be seen as n_1 copies of n_2 or as n_2 copies of n_1. These two representations of n_1n_2 as an element of S are called *factorizations*; see Section 2 for precise definitions and notation.

The McNugget monoid $M = \langle 6, 9, 20 \rangle$ has Frobenius number 43; thus, for every $n > 43$, one can purchase exactly n McNuggets using the standard package sizes of 6, 9, and 20. As with the numerical monoid $\langle n_1, n_2 \rangle$, almost every element in M has multiple distinct factorizations. For example, 60 can be expressed as three copies of 20, ten copies of 6, or four copies of 6 plus four copies of 9. Each such expression corresponds to a way to purchase exactly 60 McNuggets.

The existence of a Frobenius number for a numerical monoid S guarantees that past some value, every natural number lies in S. Therefore, as a set, a numerical monoid S is a scattering of "small" elements (those less than $F(S)$) and an infinite ray after $F(S)$. What is less understood is the factorization theory of these elements, which is hinted at in Example 1.4. In general, the factorization theory of numerical monoids with 2 minimal generators is well-understood, while those numerical monoids with 3 or more irreducible elements have a much more complicated factorization structure with numerous open questions.

Factorization structures are often classified and quantified using a variety of important *factorization invariants*, which generally measure how far from unique a monoid element's factorizations are. The coming sections introduce several standard invariants, together with results and open questions in the context of numerical monoids. Section 2 describes the strongest (but most cumbersome) invariant, the set of factorizations. Then, Section 3 describes the length set and elasticity invariants, which measure, in different ways, how many irreducible elements appear in the factorizations of an element. Section 4 gives several important results for delta sets, which measure the gaps in lengths that are possible between distinct factorizations of a single element. While length sets, elasticity, and delta sets record only information about the length of factorizations of an element, the catenary degree, which is described in Section 5, measures how many irreducible elements different factorizations of the same element do not share in common. Lastly, as non-unique factorizations coincide with the presence of irreducible elements that are not prime, Section 6 describes ω-primality, which measures how "far from prime" a monoid element is.

2. Sets of factorizations

DEFINITION 2.1. Fix a numerical monoid $S = \langle n_1, \ldots, n_k \rangle$ and an element $n \in S$. A *factorization* of n is an expression $n = u_1 + \cdots + u_r$ of n as a sum of irreducible elements $u_1, \ldots, u_r$ of S. Collecting like terms, we often write factorizations of n in the form $n = a_1 n_1 + \cdots + a_k n_k$ (see Remark 1.2). Write

$$\mathsf{Z}_S(n) = \{(a_1, \ldots, a_k) : n = a_1 n_1 + \cdots + a_k n_k\} \subset \mathbb{N}^k$$

for the *set of factorizations* of $n \in S$. When there is no ambiguity, we often omit the subscript and simply write $\mathsf{Z}(n)$.

REMARK 2.2. The notation for factorizations in Definition 2.1 is motivated in part by connections to discrete geometry. In particular, each factorization of n corresponds to an integer point in $\mathbb{N}^k$, and the set $\mathsf{Z}_S(n)$ forms a codimension-1 linear cross-section of $\mathbb{N}^k$ that coincides with the set of integer points of a convex integral polytope. This viewpoint allows the use of techniques from lattice point enumeration and other tools from combinatorial commutative algebra when examining the factorization structure of numerical monoids [**8**, **36**].

The set of factorizations (Definition 2.3) is an example of a perfect factorization invariant in that it encapsulates enough information to uniquely determine the underlying monoid (Theorem 2.4). However, such complete information comes at a cost: extracting information from it (or even simply writing it down) is a nontrivial task. Invariants derived from the set of factorizations, such as the length set and elasticity invariants (Section 3), are more manageable and easier to work with, but necessitate a loss of information. Below, $2^{\mathbb{N}^k}$ denotes the power set of $\mathbb{N}^k$.

DEFINITION 2.3. Given a numerical monoid $S = \langle n_1, \ldots, n_k \rangle$, let

$$\mathcal{Z}(S) = \{\mathsf{Z}_S(n) : n \in S\} \subset 2^{\mathbb{N}^k}$$

denote the *set of factorizations* of S.

Theorem 2.4 follows from the fact that a minimal presentation of a numerical monoid S (as well as its defining congruence) can be recovered from $\mathcal{Z}(S)$. Example 2.5 exhibits one possible method of doing so; see [**45**, **46**] for more background on monoid congruences and minimal presentations. In fact, for a fixed $n \gg 0$, the set $\mathsf{Z}_S(n)$ alone contains enough information to recover the monoid structure of S. As such, it is perhaps not surprising that $\mathcal{Z}(S)$ is often hard to describe in complete detail.

THEOREM 2.4. *For two numerical monoids S and S', the following are equivalent.*
(a) $S = S'$,
(b) $S \cong S'$, *and*
(c) $\mathcal{Z}(S) = \mathcal{Z}(S')$.

PROOF. [**45**, Theorem 1.1]. □

EXAMPLE 2.5. Fix a numerical monoid S, and suppose that the two-element sets $\{(3,0,0),(0,2,0)\}$ and $\{(10,0,0),(0,0,3)\}$ are each a subset of some element of $\mathcal{Z}(S)$. Since each is a subset of $\mathbb{N}^3$, S must have three minimal generators, say $S = \langle n_1, n_2, n_3\rangle$. Moreover, since $(3,0,0)$ and $(0,2,0)$ are factorizations of the same element, we must have $3n_1 = 2n_2$, and similarly that $10n_1 = 3n_3$. Since $\gcd(n_1, n_2, n_3) = 1$, we conclude that $S = \langle 6, 9, 20\rangle$, the McNugget monoid from Example 1.4.

Larger elements in a numerical monoid S are more likely to have larger factorization sets. More precisely, if S has k minimal generators, the size of $\mathsf{Z}(n)$ grows on the order of n^{k-1}. Theorem 2.6, which specializes the asymptotic result in [**33**] to numerical monoids, can be found in [**38**].

THEOREM 2.6 ([**38**, Theorem 3.9]). *Given $S = \langle n_1, \ldots, n_k\rangle$ numerical, there exist periodic functions $a_0, \ldots, a_{k-2} : \mathbb{N} \to \mathbb{Q}$ whose periods divide* $\operatorname{lcm}(n_1, \ldots, n_k)$ *such that*

$$|\mathsf{Z}_S(n)| = \frac{1}{(k-1)!n_1 \cdots n_k} n^{k-1} + a_{k-2}(n)n^{k-2} + \cdots + a_1(n)n + a_0(n)$$

for all $n \in S$.

REMARK 2.7. Much of the recent literature on numerical monoids has focused on algorithms for explicit computation. It is worth noting that although Theorem 2.6 implies $|\mathsf{Z}(n)|$ is polynomial in n, is it NP-hard to compute this set in general, as doing so requires solving a system of Diophantine equations. See [**26**] for a survey of recent computational advances in this area.

Several of the factorization invariants introduced in the coming sections, including those derived directly from the set of factorizations, benefit from faster computation methods. In particular, several of these invariants admit algorithms that do not require computing large sets of factorizations; see, for instance, Remarks 4.20 and 6.5. Additional computational traction can also be obtained by specializing to certain families of monoids, such as those with a fixed number of generators or generating sets of a certain form; see, for instance, Theorems 3.12, 4.3 and 5.6.

Another computation technique in computing invariant values is to first compute the Apéry set of S (a set determined by the complement of S; see [**46**]). Although computing the Apéry set from a list of generators is also NP-hard in general, many invariants can then be computed directly from it. For instance, computing the Frobenius number of a numerical monoid is NP-hard in general [**42**], but if the Apéry set is known, its computation is trivial [**48**]. In this way, the complexity of computing an invariant (or several invariants) can be "transferred" to the computation of the Apéry set [**26, 35, 43**]. Additionally, for some special families of numerical monoids, the Apery set is known to admit a closed form; see [**44, 47**].

3. Length sets and elasticity

While the set of factorizations for a numerical monoid is a perfect invariant, it is often cumbersome to compute and encode. The next two sections focus on invariants obtained by passing from factorizations to their lengths. In this section, we introduce length sets and elasticity as useful invariants for numerical monoids. We begin with the definition of a length set.

DEFINITION 3.1. Fix a numerical monoid $S = \langle n_1, \ldots, n_k \rangle$, and fix $n \in S$. Given a factorization $\mathbf{a} = (a_1, a_2, \ldots, a_k) \in \mathsf{Z}_S(n)$, we denote by $|\mathbf{a}|$ the number of irreducibles in the factorization $\mathbf{a}$; that is, $|\mathbf{a}| = a_1 + \cdots + a_k$. The set of factorization lengths of n, denoted $\mathsf{L}_S(n) = \{|\mathbf{a}| : \mathbf{a} \in \mathsf{Z}_S(n)\}$, is called the *length set* of n. When there is no ambiguity, we often omit the subscript and simply write $\mathsf{L}(n)$.

EXAMPLE 3.2. The length set of an element is easily computed from its set of factorizations. In the numerical monoid $S = \langle 7, 10, 13 \rangle$, the element 20 has factorization set $\mathsf{Z}(20) = \{(1,0,1),(0,2,0)\}$ and length set $\mathsf{L}(20) = \{2\}$. This length set, however, is shared with other elements (e.g., $\mathsf{L}(14) = \{2\}$). This example highlights one manner in which information is frequently lost when passing from $\mathsf{Z}(n)$ to $\mathsf{L}(n)$.

Although distinct elements in a numerical monoid can have identical length sets, one may ask whether distinct numerical monoids can have different sets of length sets. In what follows, $2^{\mathbb{N}}$ denotes the power set of $\mathbb{N}$.

DEFINITION 3.3. For a numerical monoid S, its *set of length sets* is given by

$$\mathcal{L}(S) = \{\mathsf{L}(n) : n \in S\} \subset 2^{\mathbb{N}}.$$

Investigations into whether sets of length sets determine finitely generated, cancellative, commutative monoids up to isomorphism are widespread in factorization theory. The most famous conjecture is stated below for certain block monoids (that is, monoids of zero-sum sequences of abelian group elements under concatenation). Since a detailed introduction to block monoids would take us too far afield, we refer the interested reader to [**30**] for background on block monoids and [**31, 32**] for recent progress on Conjecture 3.4. In what follows, let $\mathcal{B}(G)$ denote the block monoid over a group G; $\mathcal{L}(\mathcal{B}(G))$ is defined analogously for this class of monoids. Note that for two abelian groups G and G', $\mathcal{B}(G) \cong \mathcal{B}(G')$ if and only if $G \cong G'$.

CONJECTURE 3.4. *Given two finite abelian groups G and G' with $|G|, |G'| > 3$, we have $\mathcal{L}(\mathcal{B}(G)) = \mathcal{L}(\mathcal{B}(G'))$ implies $\mathcal{B}(G) \cong \mathcal{B}(G')$.*

Theorem 3.5 answers an analogous question for numerical monoids in the negative.

THEOREM 3.5. [**2**, Corollary 3.4] *There exist numerical monoids $S \neq S'$ with $\mathcal{L}(S) = \mathcal{L}(S')$. In particular, sets of length sets do not characterize numerical monoids.*

REMARK 3.6. The distinct numerical monoids with equal sets of length sets described in [**2**] are called *arithmetical numerical monoids*, which are numerical monoids generated by an arithmetic sequence $a, a+d, \ldots, a+kd$ with $\gcd(a,d) = 1$ and $1 \leq k < a$. Theorem 3.5 is proven by computing the set of length sets of arithmetical numerical monoids in terms of the parameters a, d and k and showing that different values of these parameters yield the same set.

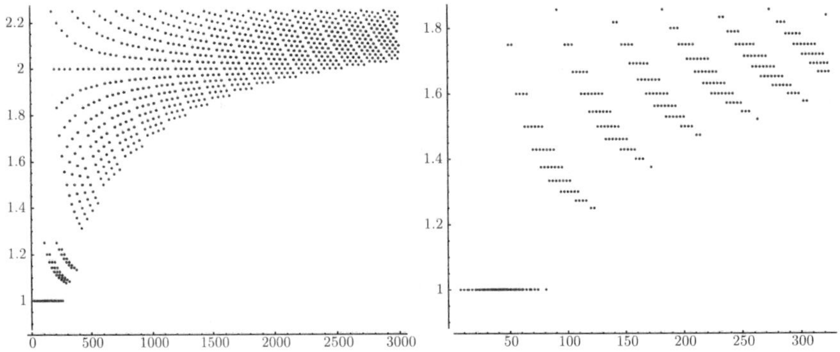

FIGURE 1. A plot showing the elasticities for the numerical monoid $\langle 20, 21, 45\rangle$ (left) and the arithmetical numerical monoid $\langle 7, 10, 13, 16\rangle$ (right), produced using SAGE [**49**]. Here, the point at (n, q) indicates that $\rho(n) = q$.

The description of the set of length sets for arithmetical numerical monoids upon which Theorem 3.5 is based is rare and, in general, it is difficult to obtain such a description for other families of numerical monoids. Part of the difficulty comes in computing every factorization length that an element in a numerical monoid can obtain. Many times, the most interesting factorization lengths for an element are the largest and smallest ones. The quotient of these two values gives a weaker factorization invariant called *elasticity*, which is defined below.

DEFINITION 3.7. Let S be a numerical monoid, $n \in S$ nonzero, and $\mathsf{L}(n)$ its length set. Let $\mathsf{M}(n) = \max \mathsf{L}(n)$ and $\mathsf{m}(n) = \min \mathsf{L}(n)$. The *elasticity of* n is given by

$$\rho(n) = \mathsf{M}(n)/\mathsf{m}(n).$$

The *set of elasticities of* S is given by $R(S) = \{\rho(n) : n \in S, n \neq 0\}$. The *elasticity of* S is given by $\rho(S) = \sup R(S)$.

In what follows, we see that $\rho(S) < \infty$ for numerical monoids and that, in fact, $\rho(S) = \max R(S)$. Moreover, elasticity is one of the few factorization invariants known to admit a closed form for any numerical monoid S.

THEOREM 3.8 ([**19**, Theorem 2.1 & Corollary 2.3]). *Let $S = \langle n_1, n_2, \ldots, n_k\rangle$ be minimally generated by $n_1, \ldots, n_k$ with n_1 and n_k the smallest and largest generators, respectively. Then, $\rho(S) = \rho(n_1 n_k) = n_k/n_1$ and the set of elasticities $R(S)$ has n_k/n_1 as its unique accumulation point.*

EXAMPLE 3.9. Figure 1 depicts the elasticity of elements in the numerical monoids $S = \langle 20, 21, 45\rangle$ and $S' = \langle 7, 10, 13, 16\rangle$. The maximal values attained are $9/4$ and $16/7$, respectively, as ensured by Theorem 3.8. Moreover, the monoid elements with maximal elasticity are precisely the multiples of the least common multiple of the first and last generators.

More can be said about the structure of the set $R(S)$ upon examining the asymptotic behavior of the maximum- and mininum-length factorization functions $\max \mathsf{L}_S(n)$ and $\min \mathsf{L}_S(n)$ for large monoid elements $n \in S$. See [**6**, Section 4] for more detail.

THEOREM 3.10. [**6**, Corollary 4.5] *Fix a numerical monoid* $S = \langle n_1, n_2, \ldots, n_k \rangle$.

(a) *For* $n \geq n_{k-1}n_k$, *we have*

$$\rho(n + n_1 n_k) = \frac{M(n) + n_k}{m(n) + n_1}.$$

(b) *The set* $R(S)$ *is the union of a finite set and a collection of* $n_1 n_k$ *monotone increasing sequences, each converging to* n_k/n_1.

REMARK 3.11. Both plots in Figure 1 demonstrate the claims in Theorem 3.10. Roughly speaking, each plot can be viewed as a union of "wedges" of points, where each wedge contains one point from each of the monotone increasing sequences described in Theorem 3.10(b). Moreover, the right-hand plot depicts the elasticies of an arithmetical numerical monoid, where elasticity values are often repeated in quick succession. It is this repetition of elasticitiy values that gives rise to Theorem 3.12.

Certainly, two numerical monoids with identical sets of length sets have equal sets of elasticities. Conversely, there exist pairs of numerical monoids with equal elasticity sets and unequal sets of length sets. For example, for $S = \langle 6, 10, 13, 14 \rangle$ and $S' = \langle 6, 11, 13, 14 \rangle$, a computation shows that $R(S) = R(S')$, while $\{4, 6\} \in \mathcal{L}(S)$ but $\{4, 6\} \notin \mathcal{L}(S')$; see [**6**, Example 3.11]. Thus, in general, sets of elasticities do not determine sets of length sets. Theorem 3.12 demonstrates that, within the class of arithmetical numerical monoids, equality of sets of length sets is equivalent to equality of sets of elasticities.

THEOREM 3.12. [**6**, Theorem 1.2] *For two distinct arithmetical numerical monoids* $S = \langle a, a + d, \ldots, a + kd \rangle$ *and* $S' = \langle a', a' + d', \ldots, a' + k'd' \rangle$, *the following are equivalent:*

(a) $R(S) = R(S')$

(b) $\mathcal{L}(S) = \mathcal{L}(S')$

Theorem 3.10 demonstrates that computing the set of elasticities is fairly tractable, while the set of length sets is, in general, much more difficult to describe. Therefore, finding classes of numerical monoids (beyond the arithmetical case) for which the set of elasticities is as strong an invariant as the set of length sets is of much interest.

PROBLEM 3.13. *Characterize which distinct numerical monoids* S *and* S' *satisfy* $R(S) = R(S')$ *and* $\mathcal{L}(S) \neq \mathcal{L}(S')$.

4. The delta set

In this section, we consider the *delta set* invariant (Definition 4.1). The delta set of a monoid element n, like its elasticity (Definition 3.7), is derived from the set of factorization lengths $\mathsf{L}(n)$.

DEFINITION 4.1. Fix a numerical monoid $S = \langle n_1, \ldots, n_k \rangle$, and fix $n \in S$. Writing $\mathsf{L}(n) = \{\ell_1 < \cdots < \ell_r\}$, the *delta set* of n is the set $\Delta_S(n) = \{\ell_i - \ell_{i-1} : 2 \leq i \leq r\}$ of successive differences of factorization lengths of n, and $\Delta(S) = \bigcup_{n \in S} \Delta_S(n)$. When there is no ambiguity, we often omit the subscript and simply write $\Delta(n)$.

EXAMPLE 4.2. Consider the numerical monoid $S = \langle n_1, n_2 \rangle$ with two relatively prime generators. If $n \in S$, then there exist $a, b \in N$ such that $n = an_1 + bn_2$. Notice that for any $k \in \mathbb{Z}$, $n = (a + kn_2)n_1 + (b - kn_1)n_2$ is a factorization so long as $a + kn_2$ and $b - kn_1$ are non-negative. In fact, it can be shown that all factorizations of n in S are of this form, corresponding to successive values of k. Since the length of such a factorization is $a + b + k(n_2 - n_1)$, all successive length differences in $\mathsf{L}(n)$ yield $n_2 - n_1$. Thus, for any $n \in S$, either $\Delta(n)$ is empty (when its length set is a singleton), or $\Delta(n) = \{n_2 - n_1\}$. In particular,

$$\Delta(S) = \{n_2 - n_1\}.$$

The above example is a special case of arithmetical numerical monoids, whose minimal generators form an arithmetic sequence (Remark 3.6). Such monoids have singleton delta sets consisting of the step size between these generators, a fact which plays a crutial role in characterizing their length sets (Theorems 3.5).

THEOREM 4.3. [**2**, Corollary 2.3] *Fix positive integers a, d, and k with* $\gcd(a, d) = 1$ *and* $1 \leq k < a$. *For* $S = \langle a, a + d, a + 2d, \ldots, a + kd \rangle$ *and* $n \in S$, *either* $\Delta(n)$ *is empty, or* $\Delta(n) = \{d\}$. *In particular,* $\Delta(S) = \{d\}$.

Even slight generalizations of arithmetical numerical monoids have elusive delta sets; this leads us to the following open problem.

PROBLEM 4.4. *Let a, h, d, and k be positive integers such that* $\gcd(a, d) = 1$ *and* $k < a$, *and consider* $S = \langle a, ah + d, ah + 2d, \ldots, ah + kd \rangle$. *Compute* $\Delta(S)$.

REMARK 4.5. The generators of the numerical monoid in Problem 4.4 form what is known as a *generalized arithmetic sequence*. Numerical monoids generated by generalized arithmetic sequences and arithmetic numerical monoids are closely connected: when $h = 1$, a generalized arithmetic sequence reduces to an arithmetic sequence; also, the last k generators of a generalized arithmetic sequence form an arithmetic sequence. These relationships, together with a precise membership criterion [**37**, Theorem 3.1], make working with this family of numerical monoids more tractable.

REMARK 4.6. Aside from the arithmetical setting, few natural families of numerical monoids are known to admit closed forms for their delta sets. In general, given a numerical monoid S, it can be hard to prove that $d \notin \Delta(S)$ for a given integer d, especially if $\min \Delta(S) < d < \max \Delta(S)$, since it must be shown that d does not lie in the delta set of even a single element of S. Example 4.7 demonstrates the sublety of this question.

EXAMPLE 4.7. The delta set of elements of the numerical monoid $S = \langle 17, 33, 53, 71 \rangle$ are depicted in Figure 2. Notice that $\Delta(S) = \{2, 4, 6\}$, but 6 only appears in $\Delta(266)$, $\Delta(283)$, and $\Delta(300)$. Indeed, for $n > 300$, $\Delta(n) \subset \{2, 4\}$; see Theorem 4.16.

REMARK 4.8. In contrast to numerical monoids, the delta set of any block monoid $M = \mathcal{B}(G)$ over a finite abelian group G is known to be an interval, and recent work has focused on determining its maximal element [**30**, Section 6.7]. In particular, any positive integer $d \leq \max \Delta(M)$ must also lie in $\Delta(M)$. Roughly speaking, block monoids have abundant irreducible elements, so if $d \in \Delta(a)$ for some $a \in M$, one can produce an element in M with $d - 1$ in its delta set by simply

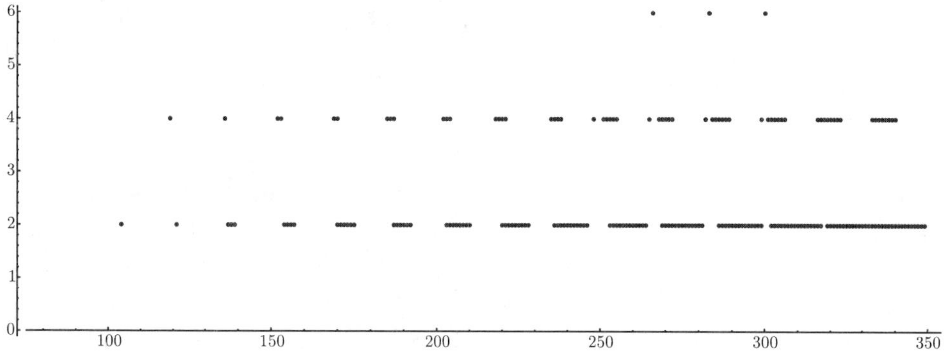

FIGURE 2. A plot showing the delta sets of elements in the numerical monoid $\langle 17, 33, 53, 71\rangle$ from Example 4.7, produced using SAGE [**49**]. Here, each point (n, d) indicates that $d \in \Delta(n)$.

adding the appropriate irreducible elements to a. This process effectively "cuts down" a length gap d by carefully filling in a particular factorization length.

The difficulty in controlling the phenomenon discussed in Remark 4.8 when considering numerical monoids motivates Problem 4.9, which is known as the *delta set realization problem for numerical monoids.*

PROBLEM 4.9. *Let $D \subset \mathbb{Z}_{>0}$ be a finite set of positive integers. Determine if there exists a numerical monoid S such that $\Delta(S) = D$.*

Currently, the only known restrictions on $\Delta(S)$ for a numerical monoid S are given in Theorem 4.10, each of which also holds for a much larger class of monoids.

THEOREM 4.10 ([**30**, Proposition 1.4.4]). *If M is an atomic, cancellative, commutative monoid with $\Delta(M) \neq \emptyset$, then, $\min(\Delta(M)) = \gcd(\Delta(M))$. If, additionally, M is finitely generated, then $\Delta(M)$ is finite.*

REMARK 4.11. Progress on Problem 4.9 has come from finding specific subsets that can be realized as delta sets. For example, every set of the form $\{d, 2d, \ldots, td\}$ is known to occur as the delta set of a numerical monoid [**11**]. Additionally, a family of numerical monoids whose delta sets are $\{d, td\}$ for any $t, d > 1$ is given [**22**].

Recent investigation into Problem 4.9 has greatly benefited from evidence produced by computer algebra systems. That said, until very recently, there was no known algorithm to compute the delta set of an entire numerical monoid S. Although the delta set $\Delta(n)$ of an element $n \in S$ can be computed by computing its set of factorizations $\mathsf{Z}(n)$, it is less straight forward to compute the entire set $\Delta(S)$, as this requires taking the union of the delta sets of all (infintely many) elements of S.

Before discussing how this issue can be resolved (see Corollary 4.17), we state Theorem 4.14, which demonstrates that the maximal element $\max \Delta(S)$ in the delta set of a numerical monoid S can be obtain by computing the delta sets of a special (computable) class of monoid elements, called *Betti elements* (Definition 4.12).

DEFINITION 4.12. Let $S = \langle n_1, n_2, \ldots, n_k\rangle$ be a numerical monoid and $n \in S$. Consider the graph G_n with vertex set given by its set of factorizations $\mathsf{Z}(n)$ and an edge connecting $\mathbf{a}, \mathbf{b} \in \mathsf{Z}(n)$ if $\mathbf{a}$ and $\mathbf{b}$ have disjoint support as vectors. That is, $\mathbf{a} = (a_1, \ldots, a_k)$ and $\mathbf{b} = (b_1, \ldots, b_k)$ are adjacent in G_n if for all i, a_i and b_i are never both non-zero. An element $n \in S$ is called a *Betti element* if G_n is disconnected. The *set of Betti elements of* S is denoted by $\text{Betti}(S)$.

REMARK 4.13. Every numerical monoid contains only finitely many Betti elements. When $S = \langle n_1, n_2 \rangle$ is a numerical monoid with two irreducible elements, $n_1 n_2$ is the unique Betti element. In the case when $S = \langle n_1, n_2, n_3 \rangle$ has three irreducible elements, there are at most three Betti elements, each of which is an integer multiple of one of the generators of S [**46**, Section 9.3].

THEOREM 4.14 ([**16**, Theorem 2.5]). *Let S be a numerical monoid. The largest element of $\Delta(S)$ is achieved at a Betti element of S. In other words,*

$$\max \Delta(S) = \max\{d : d \in \Delta(b) \textit{ for } b \in \text{Betti}(S)\}.$$

Betti elements are discussed in more depth in Section 5, as they are useful in finding maximal and minimal catenary degrees achieved in a numerical monoid (see Definitions 5.2 and 5.4).

EXAMPLE 4.15. As stated in Remark 4.13, for the numerical monoid $S = \langle n_1, n_2 \rangle$ with two irreducible elements, $n_1 n_2$ is the unique Betti element. In fact, it has precisely two factorizations, namely $(n_2, 0)$ and $(0, n_1)$. Since their difference in length equals $n_2 - n_1$, we have $\max \Delta(S) = n_2 - n_1$. Of course, by Theorem 4.3, $\Delta(S) = \{n_2 - n_1\}$.

Theorem 4.16 states that delta sets for numerical monoids are eventually periodic with period dividing $\text{lcm}(n_1, n_k)$. The value of N_S given in [**25**] is explicitly computed, albeit cumbersome to state. Most impressive about Theorem 4.16 is that it provides a concrete upper bound for the beginning of the periodic behavior, and thus a finite set of monoid elements the union of whose delta sets equals the delta set of the entire numerical monoid. As a consequence, Theorem 4.16 also yields the first known algorithm to compute the delta set of any numerical monoid; see Corollary 4.17.

THEOREM 4.16 ([**25**, Theorem 1]). *For $S = \langle n_1, \ldots, n_k \rangle$ numerical, there exists a number N_S such that for any $n \in S$ with $n \geq N_S$, $\Delta(n) = \Delta(n + \text{lcm}(n_1, n_k))$.*

COROLLARY 4.17. *Given a numerical monoid $S = \langle n_1, n_2, \ldots, n_k \rangle$, we have*

$$\Delta(S) = \bigcup_{n \in S \cap [0, N]} \Delta(n),$$

where $N = N_S + \text{lcm}(n_1, n_k)$.

EXAMPLE 4.18. Let $S = \langle 17, 33, 53, 71 \rangle$ as in Example 4.7. Readily visible in Figure 2 is the eventual periodicity of $\Delta : S \to 2^{\mathbb{N}}$ described by Theorem 4.16. In particular, Theorem 4.16 ensures that $\Delta(n) = \Delta(n + \text{lcm}(17, 71))$ for all $n \geq N_S = 6461$. Computing $\Delta(n)$ for smaller values of n demonstrates that, in fact, $\Delta(n) = \Delta(n + 17)$ holds for all $n \geq 319$.

REMARK 4.19. While Corollary 4.17 provides the delta set of the entire numerical monoid by taking the union of only finite many element-wise delta sets, the number of elements n for which $\Delta(n)$ needs to be computed becomes prohibitively large, especially when larger generators are used. Aside from recent improvements in the special case when S has 3 minimal generators [**27**], few refinements are known for the set of elements whose delta sets must be computed. Alternatively, we can utilize a dynamic algorithm for computing length sets to dynamically compute $\Delta(n)$ and $\Delta(S)$; see Remark 4.20.

REMARK 4.20. In the numerical monoid $S = \langle 6, 9, 20\rangle$, it is possible to compute $\mathsf{L}(60)$ directly from $\mathsf{L}(40)$, $\mathsf{L}(51)$, and $\mathsf{L}(54)$, without computing $\mathsf{Z}(60)$. The key observation is that one can produce a factorization of 60 from a factorization of 40, 51 or 54 by adding a single irreducible element, and this opereration always increases factorization length by exactly one. Consider the following computation:

$$\begin{aligned}\mathsf{L}(60) &= (\mathsf{L}(40)+1) \cup (\mathsf{L}(51)+1) \cup (\mathsf{L}(54)+1)\\ &= \{3\} \cup \{7,8,9\} \cup \{7,8,9,10\}\\ &= \{3,7,8,9,10\}.\end{aligned}$$

This provides an inductive (dynamic) method to compute all the length sets necessary to employ Corollary 4.17 without having to compute any factorization sets, and the resulting algorithm is significantly faster than one dependent on factorizations. Indeed, the set $\mathsf{Z}(n)$ of factorizations of n grow large very quickly [**33**], whereas $|\mathsf{L}(n)|$ grows linearly in n [**6**, Theorem 4.3]. For a more detailed analysis, see [**7**, Section 3].

5. The catenary degree

Many of the factorization invariants discussed above (e.g., length sets, delta sets, and elasticity) measure only information about the lengths of factorizations of an element. While helpful in measuring the non-uniqueness of factorizations of an element, these invariants lose certain information when passing from the set of factorizations to the set of lengths (Theorem 3.5). The *catenary degree*, which we describe in this section, is derived directly from factorizations and is not uniquely determined by sets of factorization lengths.

DEFINITION 5.1. Fix a numerical monoid $S = \langle n_1, \ldots, n_k\rangle$, and fix an element $n \in S$. For $\mathbf{a}, \mathbf{b} \in \mathsf{Z}(n)$, the *greatest common divisor of* $\mathbf{a}$ *and* $\mathbf{b}$ is given by

$$\gcd(\mathbf{a}, \mathbf{b}) = (\min(a_1, b_1), \ldots, \min(a_r, b_r)) \in \mathbb{N}^r,$$

and the *distance between* $\mathbf{a}$ *and* $\mathbf{b}$ (or the *weight of* $(\mathbf{a}, \mathbf{b})$) is given by

$$d(\mathbf{a}, \mathbf{b}) = \max(|\mathbf{a} - \gcd(\mathbf{a}, \mathbf{b})|, |\mathbf{b} - \gcd(\mathbf{a}, \mathbf{b})|).$$

The function $d : \mathsf{Z}(n) \times \mathsf{Z}(n) \to \mathbb{N}$ measures how many irreducible elements two factorizations do not share in common. As implied by the notation, $d(-,-)$ is a metric on the set $\mathsf{Z}(n)$ [**30**, Proposition 1.2.5.3]. In fact, it is common to represent $\mathsf{Z}(n)$ as the vertices of a complete graph, where the edge between factorizations $\mathbf{a}$ and $\mathbf{b}$ is decorated with $d(\mathbf{a}, \mathbf{b})$.

Below, we use the metric d to define the *catenary degree*, which measures the "overall" distance between distinct factorizations of a given monoid element.

DEFINITION 5.2. Given $\mathbf{a}, \mathbf{b} \in \mathsf{Z}(n)$ and $N \geq 1$, an *N-chain from* $\mathbf{a}$ *to* $\mathbf{b}$ is a sequence $\mathbf{a}_1, \ldots, \mathbf{a}_r \in \mathsf{Z}(n)$ of factorizations of n such that (i) $\mathbf{a}_1 = \mathbf{a}$, (ii) $\mathbf{a}_r = \mathbf{b}$, and (iii) $d(\mathbf{a}_{i-1}, \mathbf{a}_i) \leq N$ for all $i \leq r$. The *catenary degree of* n, denoted $\mathsf{c}(n)$, is the smallest $N \in \mathbb{N}$ such that there exists an N-chain between any two factorizations of n.

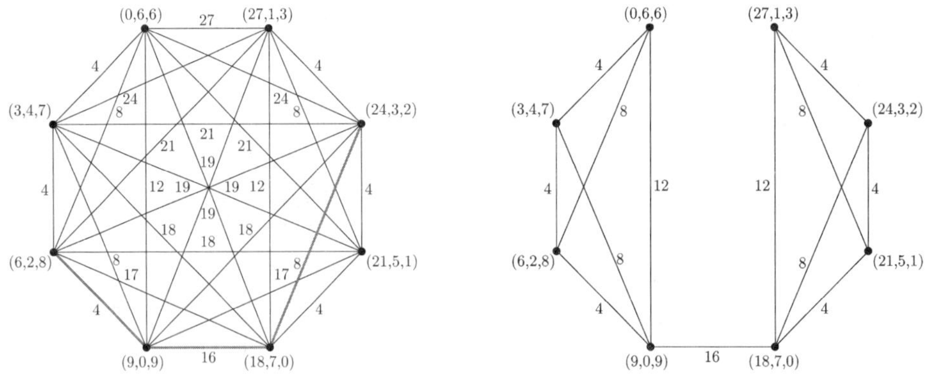

FIGURE 3. Computing the catenary degree of $450 \in S = \langle 11, 36, 39 \rangle$, as in Example 5.3. Each vertex is labeled with an element of $\mathsf{Z}(450)$, and each edge is labeled with the distance between the factorizations at either end. The diagram on the left depicts all edges, and the diagram on the right includes only those edges labeled at most $\mathsf{c}(450) = 16$. Both graphics were created using the computer algebra system SAGE [**49**].

With respect to the decorated complete graph representing $\mathsf{Z}(n)$, the catenary degree $\mathsf{c}(n)$ can be computed by repeatedly removing edges of maximal distance until the graph becomes disconnected. The distance decorating the last removed edge is equal to $\mathsf{c}(n)$. Example 5.3 describes such a computation.

EXAMPLE 5.3. Consider the numerical monoid $S = \langle 11, 36, 39 \rangle$. The left-hand picture in Figure 3 depicts the factorizations of $450 \in S$ along with all pairwise distances. There exists a 16-chain between any two factorizations of 450; one such 16-chain between $(6, 2, 8)$ and $(24, 3, 2)$ is depicted with bold red edges. Since every 16-chain between these factorizations contains the edge labeled 16, we have $\mathsf{c}(450) = 16$.

This can also be computed in a different way. In the right-hand picture of Figure 3, only distances of at most 16 are depicted, and the resulting graph is connected. Removing the edge labeled 16 yields a disconnected graph, so we again conclude $\mathsf{c}(450) = 16$.

While the catenary degree of an individual element contains information about its different factorizations, the range of catenary degrees achieved by all elements in a monoid has also been studied. Definition 5.4 gives two ways to measure how the catenary degree varies throughout a numerical monoid.

DEFINITION 5.4. The *set of catenary degrees of* S is the set $\mathsf{C}(S) = \{\mathsf{c}(m) : m \in S\}$, and the *catenary degree of* S is the maximal catenary degree: $\mathsf{c}(S) = \max \mathsf{C}(S)$.

By far, the more well-understood of these two invariants is $\mathsf{c}(S)$, which has been computed for several classes of numerical monoids. Theorem 5.5 gives a summary of some known results (additionally, see Theorem 5.8).

THEOREM 5.5 ([**34, 37**]). *Fix a numerical monoid* S.

(a) If $S = \langle n_1, n_2 \rangle$, *then* $\mathsf{c}(S) = n_2$.

(b) *If* $S = \langle a, ah+d, \ldots, ah+kd \rangle$ *for positive integers* a, h, k, d *with* $\gcd(a,d) = 1$ *and* $k < a$, *then* $\mathsf{c}(S) = \lceil \frac{a}{k} \rceil h + d$.

(c) *If* $S = \langle (\prod_{j=1}^{i} b_j)(\prod_{j=i+1}^{k} a_j) : 0 \le i \le k \rangle$, *where* $2 \le a_i < b_i$ *and* $\gcd(a_i, b_j) = 1$ *for all* $i \ne j \le k$, *then* $\mathsf{c}(S) = \max\{b_1, \ldots, b_k\}$.

While a closed form for $\mathsf{c}(n)$ for every element n is a numerical monoid is rare, such a formula has been found for numerical monoids generated by an arithmetic sequence.

THEOREM 5.6 ([**13**, Theorem 3.1]). *Suppose* $S = \langle a, a+d, \ldots, a+kd \rangle$ *for positive integers* a, k, d *with* $\gcd(a,d) = 1$ *and* $k < a$. *Then*

$$\mathsf{c}(n) = \begin{cases} 0 & |\mathsf{Z}(n)| = 1 \\ 2 & |\mathsf{L}(n)| = 1 \text{ and } |\mathsf{Z}(n)| > 1 \\ \mathsf{c}(S) & |\mathsf{L}(n)| > 1 \end{cases}$$

for each nonzero $n \in S$, *where* $\mathsf{c}(S) = \lceil \frac{a}{k} \rceil + d$. *In particular,* $\mathsf{C}(S) = \{0, 2, \mathsf{c}(S)\}$.

While it remains a lofty goal to give a closed form for the catenary degree of individual elements in the general setting, the set of catenary degrees may prove to be more approachable. Problem 5.7 is, in many ways, an analog of Problem 4.9 for the set of catenary degrees.

PROBLEM 5.7. *Determine which finite sets equal* $\mathsf{C}(S)$ *for some numerical monoid* S.

Many of the arguments that compute the maximal catenary degree $\mathsf{c}(S)$ for a finitely generated monoid S focus on the Betti elements of S (Definition 4.12). The factorizations of Betti elements contain enough information about $\mathsf{Z}(S)$ to give sharp bounds on the catenary degrees occuring in $\mathsf{C}(S)$ (Theorem 5.8), and can be readily computed using the `GAP` package `numericalsgps` [**23**].

THEOREM 5.8 ([**17**, **41**]). *For any finitely generated monoid* S, *we have*

$$\mathsf{c}(S) = \max \mathsf{C}(S) = \max\{\mathsf{c}(b) : b \in \mathrm{Betti}(S)\} \text{ and}$$
$$\min \mathsf{C}(S) = \min\{\mathsf{c}(b) : b \in \mathrm{Betti}(S)\}.$$

EXAMPLE 5.9. Let $S = \langle 11, 25, 29 \rangle$. The catenary degrees of S are plotted in Figure 4. The only Betti elements of S are 58, 150, and 154, which have catenary degrees 4, 12, and 14, respectively. The Betti elements can be seen in the plot as the first occurence of each of these three values. Notice that $\mathsf{c}(175) = 11$ is distinct from each of these values. By Theorem 5.8, every element of S with at least two distinct factorizations has catenary degree at least 4 and at most 14.

As with the delta set, understanding the long-term behavior of the catenary degree (when viewed as a function $\mathsf{c} : S \to \mathbb{N}$) is of particular interest. Theorem 5.10 shows that the catenary degree in numerical monoids is also eventually periodic, a phenomenon that can be seen in Figure 4.

THEOREM 5.10 ([**12**, Theorem 3.1]). *Fix a numerical monoid* $S = \langle n_1, \ldots, n_k \rangle$. *The catenary degree function* $\mathsf{c} : S \to \mathbb{N}$ *is eventually periodic, and moreover, its period divides* $\mathrm{lcm}(n_1, \ldots, n_k)$.

REMARK 5.11. Among the eventual periodicity results in this paper (Theorems 3.10, 4.16, 5.10, and 6.9), Theorem 5.10 provides the most room for improvement. There is currently no known bound on the start of this periodic behavior,

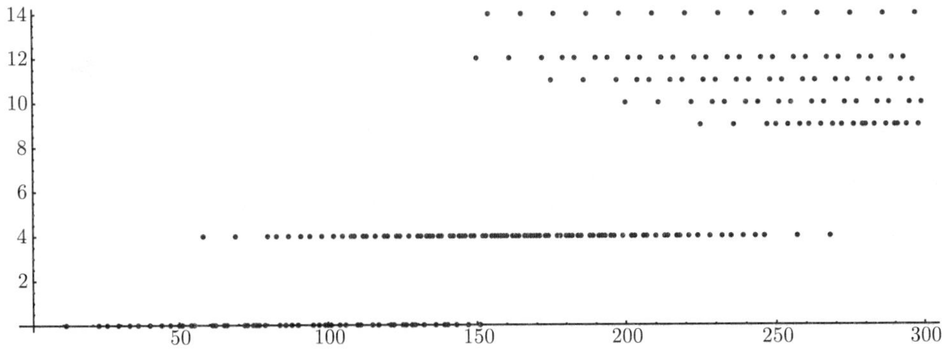

FIGURE 4. A SAGE plot [**49**] showing the catenary degrees for the numerical monoid $S = \langle 11, 25, 29\rangle$ discussed in Example 5.9.

as its proof in [**12**] relies on the fact that a nonincreasing sequence of positive integers is eventually constant. Additionally, the period is only known to divide the least common multiple of the monoid generators, but in practice this bound is far from sharp. It remains an interesting problem to refine understanding of these properties.

Examination of the delta set and the ω-primality invariant (intoduced in Section 6) both benefit from the use of dynamic algorithms; see Remarks 4.20 and 6.5. An answer to Problem 5.12 is likely within reach, and would aide in refining Theorem 5.10.

PROBLEM 5.12. *Find a dynamic algorithm to compute the catenary degree of numerical monoid elements.*

6. ω-primality

In cancellative, commutative monoids, non-unique factorizations arise from irreducible elements that are not prime. The rich factorization theory in numerical monoids stems in part from the fact that no element is prime. In this section, we develop the notion of ω-primality, which measures how "far from prime" an element is in a natural way (see [**3, 40**] for a more thorough introduction). We begin with a definition.

DEFINITION 6.1. Let $S = \langle n_1, n_2, \ldots, n_k\rangle$ be a numerical monoid. For $n \in S$, define the *ω-primality of n* to be $\omega(n) = m$ if m is the smallest positive integer with the property that whenever $\mathbf{a} \in \mathbb{N}^k$ satisfies $\sum_{i=1}^k a_i n_i - n \in S$ with $|\mathbf{a}| > m$, there exists a $\mathbf{b} \in \mathbb{N}^k$ with $b_i \le a_i$ for each i such that $\sum_{i=1}^k b_i n_i - n \in S$ and $|\mathbf{b}| \le m$.

In a multiplicative monoid M, an element $n \in M$ is prime if whenever $n \mid ab$ for $a, b \in M$, then either $n \mid a$ or $n \mid b$. Definition 6.1 is a generalization of primeness, written additively in the context of a numerical monoid. In fact, in a general cancellative, commutative monoid, $\omega(n) = 1$ if and only if n is prime.

In practice, the ω-primality of an element n of a numerical monoid $S = \langle n_1, \ldots, n_k\rangle$ is computed by finding those $\mathbf{a} \in \mathbb{N}^k$ that are maximal among those with the property that $\sum_{i=1}^{k} a_i n_i - n$ lies in S (Definition 6.2). Proposition 6.3 states their precise relationship to the ω-value of n.

DEFINITION 6.2. We say that $\mathbf{a} \in \mathbb{N}^k$ is a *bullet* for n if (i) $(\sum_{i=1}^{n} a_i n_i) - n \in S$ and (ii) $(\sum_{i=1}^{k} a_i n_i) - n_j - n \notin S$ whenever $a_j > 0$. A bullet $\mathbf{a}$ for n is *maximal* if $|\vec{a}| = \omega(n)$. The *set of bullets* for an element n is denoted by $\text{bul}(n)$.

In general, an element n has several bullets with different lengths. As stated in Proposition 6.3, the largest such length coincides with the $\omega(n)$, justifying the above definition for a *maximal bullet.*

PROPOSITION 6.3 ([**40**, Proposition 2.10]). *Given any numerical monoid S,*

$$\omega(n) = \max\{|\mathbf{a}| : \mathbf{a} \in \text{bul}(n)\}$$

for each element $n \in S$.

EXAMPLE 6.4. Let $S = \langle 6, 9, 20\rangle$ denote the numerical monoid from Example 1.4. Consider the following sets:

$$\begin{aligned}
\text{bul}(40) &= \{(0,0,2),(4,4,0),(7,2,0),(10,0,0),(1,6,0),(0,8,0)\},\\
\text{bul}(51) &= \{(0,7,0),(10,0,0),(4,3,0),(1,5,0),(0,0,3),(7,1,0)\},\\
\text{bul}(54) &= \{(9,0,0),(6,2,0),(0,6,0),(3,4,0),(0,0,3)\},\\
\text{bul}(60) &= \{(4,4,0),(7,2,0),(10,0,0),(1,6,0),(0,8,0),(0,0,3)\}.
\end{aligned}$$

We see that $(10,0,0) \in \text{bul}(40)$ since $10 \cdot 6 - 40 = 20 \in S$, but $9 \cdot 6 - 40 = 14 \notin S$. Similarly, $(7,2,0) \in \text{bul}(60)$ since $7 \cdot 6 + 2 \cdot 9 - 60 = 0 \in S$, but omitting a single 6 or 9 produces a difference lying outside of S. Once the above sets have been determined, Proposition 6.3 implies that $\omega(54) = 9$ and $\omega(40) = \omega(51) = \omega(60) = 10$.

REMARK 6.5. Several recent works examine methods of computing ω-primality in numerical monoids [**5**, **9**, **10**, **24**]. At the time of writing, the fastest known algorithm (which is implemented in the `GAP` package `numericalsgps` [**23**]) utilizes dynamic programming to compute bullets [**7**], similar to the dynamic algorithm used to compute the delta set (Remark 4.20). Example 6.6 demonstrates the idea behind this algorithm.

EXAMPLE 6.6. Resuming notation from Example 6.4, for each bullet $\mathbf{a} \in \text{bul}(54)$, either $\mathbf{a} \in \text{bul}(60)$ or $\mathbf{a} + \mathbf{e}_1 \in \text{bul}(60)$. Notice that it is impossible for both of these to lie in $\text{bul}(60)$ by Definition 6.2(ii). Similarly, bullets in $\text{bul}(51)$ and $\text{bul}(40)$ give rise to bullets in $\text{bul}(60)$. Moreover, each bullet for 60 is the "image" of a bullet for 54, 51 or 40 in this way. Much like the recurrence relation for sets of factorizations (Remark 4.20), the set of bullets of an element $n \in S$ is determined by the "images" of the bullets of $n - 6$, $n - 9$ and $n - 20$; see [**7**] for more detail.

In general, the existence of non-unique factorizations coincides with the existence of non-prime irreducible elements. As such, the ω-primality invariant was originally constructed to measure how far irreducible elements were from being prime (see, for example, [**3**, **4**]). Although some progress has been made in the setting of numerical monoids [**5**, **18**, **21**], Problem 6.7 remains largely open.

PROBLEM 6.7. *Determine the ω-values of numerical monoid generators.*

REMARK 6.8. Both the catenary degree (Definition 5.2) and a variation called the tame degree share a surprising connection to ω-primality. Given $n \in S$, the *tame degree* $\mathsf{t}(n)$ equals the minimum value m such that every factorization of n

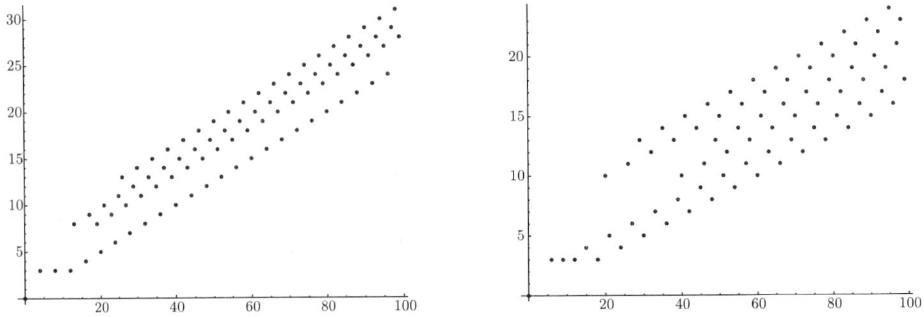

FIGURE 5. A plot showing the ω-function for the numerical monoids $\langle 4, 13, 19\rangle$ (left) and $\langle 6, 9, 20\rangle$ (right), produced using SAGE [**49**].

with at least one zero entry is distance at most m (in the sense of Definition 5.1) from a fully supported factorization of n, or $\mathsf{t}(n) = 0$ if n has no fully supported factorizations [**15, 17**]. As with the catenary degree, the tame degree of S is defined as $\mathsf{t}(S) = \sup\{\mathsf{t}(n) : n \in S\}$ and is known to be finite. It is not hard to show that $\mathsf{c}(S) \le \mathsf{t}(S)$, but what is perhaps more surprising is that $\mathsf{c}(S) \le \omega(S) \le \mathsf{t}(S)$, where $\omega(S) = \max\{\omega(n_1), \ldots, \omega(n_k)\}$ is the maximum ω-value obtained at an irreducible [**10**]. This connection has been explored in some of the recent literature, in part as a method of bounding $\omega(S)$ [**10, 28, 29**].

Recent investigations focusing on the ω-values of all non-unit elements in monoids have uncovered interesting asymptotic behavior. In the setting of numerical monoids, ω-primality grows in a periodic, linear fashion when viewed as a function $\omega : S \to \mathbb{N}$.

THEOREM 6.9 ([**7, 24, 39**]). *Let $S = \langle n_1, n_2, \ldots, n_k\rangle$ be a numerical monoid. The ω-primality function is eventually quasilinear. More specifically, for $N_0 = \frac{F(S)+n_2}{n_2/n_1 - 1}$, we have $\omega(n) = \frac{1}{n_1}n + a_0(n)$ for every $n > N_0$, where $a_0 : \mathbb{N} \to \mathbb{Q}$ is n_1-periodic.*

Theorem 6.9 implies that the graph of the function $\omega : S \to \mathbb{N}$ eventually has the form of n_1 lines with common slope $1/n_1$. Figure 5 demonstrates this phenomenon.

REMARK 6.10. We resume notation from Theorem 6.9. In contrast to Theorem 4.16, the period of the function a_0 is known to be exactly n_1 [**7**]. Additionally, the value of N_0, though not tight, is in general a relatively good bound on the start of the quasilinear behavior of ω_S [**7**, Remark 6.8]. Each of these improvements on the original statement of Theorem 6.9 in [**24, 39**] are heavily motivated by data produced by the dynamic algorithm described in Remark 6.5.

References

[1] F. Aguiló-Gost and P. A. García-Sánchez, *Factoring in embedding dimension three numerical semigroups*, Electron. J. Combin. **17** (2010), no. 1, Research Paper 138, 21. MR2729387

[2] Jeff Amos, S. T. Chapman, Natalie Hine, and João Paixão, *Sets of lengths do not characterize numerical monoids*, Integers **7** (2007), A50, 8. MR2373112

[3] David F. Anderson and Scott T. Chapman, *How far is an element from being prime?*, J. Algebra Appl. **9** (2010), no. 5, 779–789, DOI 10.1142/S0219498810004221. MR2726554

[4] David F. Anderson and Scott T. Chapman, *On bounding measures of primeness in integral domains*, Internat. J. Algebra Comput. **22** (2012), no. 5, 1250040, 15, DOI 10.1142/S0218196712500403. MR2949206

[5] David F. Anderson, Scott T. Chapman, Nathan Kaplan, and Desmond Torkornoo, *An algorithm to compute ω-primality in a numerical monoid*, Semigroup Forum **82** (2011), no. 1, 96–108, DOI 10.1007/s00233-010-9259-5. MR2753835

[6] T. Barron, C. O'Neill, and R. Pelayo, *On the set of elasticities in numerical monoids*, to appear, Semigroup Forum. Available at arXiv: math.CO/1409.3425.

[7] T. Barron, C. O'Neill, and R. Pelayo, *On dymamic algoriths for factorization invariants in numerical monoids*, preprint, available at arXiv: math.AC/1507.07435.

[8] Matthias Beck and Sinai Robins, *Computing the continuous discretely*, Undergraduate Texts in Mathematics, Springer, New York, 2007. Integer-point enumeration in polyhedra. MR2271992

[9] Víctor Blanco, *A mathematical programming approach to the computation of the omega invariant of a numerical semigroup*, European J. Oper. Res. **215** (2011), no. 3, 539–550, DOI 10.1016/j.ejor.2011.07.004. MR2832848

[10] V. Blanco, P. A. García-Sánchez, and A. Geroldinger, *Semigroup-theoretical characterizations of arithmetical invariants with applications to numerical monoids and Krull monoids*, Illinois J. Math. **55** (2011), no. 4, 1385–1414 (2013). MR3082874

[11] Craig Bowles, Scott T. Chapman, Nathan Kaplan, and Daniel Reiser, *On delta sets of numerical monoids*, J. Algebra Appl. **5** (2006), no. 5, 695–718, DOI 10.1142/S0219498806001958. MR2269412

[12] Scott T. Chapman, Marly Corrales, Andrew Miller, Chris Miller, and Dhir Patel, *The catenary and tame degrees on a numerical monoid are eventually periodic*, J. Aust. Math. Soc. **97** (2014), no. 3, 289–300, DOI 10.1017/S1446788714000330. MR3270769

[13] S. T. Chapman, M. Corrales, A. Miller, C. Miller, and D. Patel, *The catenary degrees of elements in numerical monoids generated by arithmetic sequences*, forthcoming.

[14] Scott T. Chapman, M. Fontana, A. Geroldinger, and B. Olberding (editors), *Multiplicative Ideal Theory and Factorization Theory*, Springer, to appear.

[15] S. T. Chapman, P. A. García-Sánchez, and D. Llena, *The catenary and tame degree of numerical monoids*, Forum Math. **21** (2009), no. 1, 117–129, DOI 10.1515/FORUM.2009.006. MR2494887

[16] S. T. Chapman, P. A. García-Sánchez, D. Llena, A. Malyshev, and D. Steinberg, *On the delta set and the Betti elements of a BF-monoid* (English, with English and Arabic summaries), Arab. J. Math. (Springer) **1** (2012), no. 1, 53–61, DOI 10.1007/s40065-012-0019-0. MR3040913

[17] S. T. Chapman, P. A. García-Sánchez, D. Llena, V. Ponomarenko, and J. C. Rosales, *The catenary and tame degree in finitely generated commutative cancellative monoids*, Manuscripta Math. **120** (2006), no. 3, 253–264, DOI 10.1007/s00229-006-0008-8. MR2243561

[18] S. T. Chapman, P. A. García-Sánchez, Z. Tripp, and C. Viola, *Measuring primality in numerical semigroups with embedding dimension three*, J. Algebra Appl. **15** (2016), no. 1, 1650007, 16, DOI 10.1142/S0219498816500079. MR3393936

[19] Scott T. Chapman, Matthew T. Holden, and Terri A. Moore, *Full elasticity in atomic monoids and integral domains*, Rocky Mountain J. Math. **36** (2006), no. 5, 1437–1455, DOI 10.1216/rmjm/1181069375. MR2285293

[20] Scott T. Chapman, Rolf Hoyer, and Nathan Kaplan, *Delta sets of numerical monoids are eventually periodic*, Aequationes Math. **77** (2009), no. 3, 273–279, DOI 10.1007/s00010-008-2948-4. MR2520501

[21] Scott T. Chapman, Walter Puckett, and Katy Shour, *On the omega values of generators of embedding dimension-three numerical monoids generated by an interval*, Involve **7** (2014), no. 5, 657–667, DOI 10.2140/involve.2014.7.657. MR3245842

[22] S. Colton and N. Kaplan, *The realization problem for delta sets of numerical semigroups*, Available at arXiv: math.AC/1503.08496.

[23] M. Delgado, P. García-Sánchez, J. Morais, *NumericalSgps, A package for numerical semigroups*, Version 0.980 dev (2013), (GAP package), `http://www.fc.up.pt/cmup/mdelgado/numericalsgps/`.

[24] J. I. García-García, M. A. Moreno-Frías, and A. Vigneron-Tenorio, *Computation of the ω-primality and asymptotic ω-primality with applications to numerical semigroups*, Israel J. Math. **206** (2015), no. 1, 395–411, DOI 10.1007/s11856-014-1144-6. MR3319645

[25] J. I. García-García, M. A. Moreno-Frías, and A. Vigneron-Tenorio, *Computation of delta sets of numerical monoids*, Monatsh. Math. **178** (2015), no. 3, 457–472, DOI 10.1007/s00605-015-0785-9. MR3411252

[26] P. García-Sánchez, *An overview of the computational aspects of nonunique factorization invariants*, preprint. Available at arXiv: math.AC/1504.07424

[27] P. García-Sánchez, D. Llena, and A. Moscariello, *Delta sets for numerical semigroups with embedding dimension three*, preprint. Available at arXiv: math.AC/1504.02116

[28] P. A. García Sánchez, I. Ojeda, and A. Sánchez-R.-Navarro, *Factorization invariants in half-factorial affine semigroups*, Internat. J. Algebra Comput. **23** (2013), no. 1, 111–122, DOI 10.1142/S0218196713500033. MR3040805

[29] Pedro A. García-Sánchez and Caterina Viola, *When the catenary degree agrees with the tame degree in numerical semigroups of embedding dimension three*, Involve **8** (2015), no. 4, 677–694, DOI 10.2140/involve.2015.8.677. MR3366018

[30] Alfred Geroldinger and Franz Halter-Koch, *Non-unique factorizations*, Pure and Applied Mathematics (Boca Raton), vol. 278, Chapman & Hall/CRC, Boca Raton, FL, 2006. Algebraic, combinatorial and analytic theory. MR2194494

[31] A. Geroldinger and W. A. Schmid, *A characterization of class groups via sets of lengths*, Available at arXiv: math.AC/1503.04679.

[32] A. Geroldinger and Q. Zhong, *A characterization of class groups via sets of lengths II*, Available at arXiv: math.AC/1506.05223.

[33] Franz Halter-Koch, *On the asymptotic behaviour of the number of distinct factorizations into irreducibles*, Ark. Mat. **31** (1993), no. 2, 297–305, DOI 10.1007/BF02559488. MR1263556

[34] C. Kiers, C. O'Neill and V. Ponomarenko, *Numerical semigroups on compound sequences*, to appear, Communications in Algebra. Available at arXiv: math.AC/1503.05993

[35] Guadalupe Márquez-Campos, Ignacio Ojeda, and José M. Tornero, *On the computation of the Apéry set of numerical monoids and affine semigroups*, Semigroup Forum **91** (2015), no. 1, 139–158, DOI 10.1007/s00233-014-9631-y. MR3369384

[36] Ezra Miller and Bernd Sturmfels, *Combinatorial commutative algebra*, Graduate Texts in Mathematics, vol. 227, Springer-Verlag, New York, 2005. MR2110098

[37] Mehdi Omidali, *The catenary and tame degree of numerical monoids generated by generalized arithmetic sequences*, Forum Math. **24** (2012), no. 3, 627–640, DOI 10.1515/form.2011.078. MR2926638

[38] C. O'Neill, *On factorization invariants and Hilbert functions*, preprint. Available at arXiv: math.AC/1503.08351

[39] Christopher O'Neill and Roberto Pelayo, *On the linearity of ω-primality in numerical monoids*, J. Pure Appl. Algebra **218** (2014), no. 9, 1620–1627, DOI 10.1016/j.jpaa.2014.01.002. MR3188860

[40] Christopher O'Neill and Roberto Pelayo, *How do you measure primality?*, Amer. Math. Monthly **122** (2015), no. 2, 121–137, DOI 10.4169/amer.math.monthly.122.02.121. MR3324682

[41] C. O'Neill, V. Ponomarenko, R. Tate, and G. Webb, *On the set of catenary degrees of finitely generated cancellative commutative monoids*, preprint. Available at arXiv: math.AC/1506.07587.

[42] J. L. Ramírez Alfonsín, *The Diophantine Frobenius problem*, Oxford Lecture Series in Mathematics and its Applications, vol. 30, Oxford University Press, Oxford, 2005. MR2260521

[43] Jorge L. Ramírez Alfonsín and Øystein J. Rødseth, *Numerical semigroups: Apéry sets and Hilbert series*, Semigroup Forum **79** (2009), no. 2, 323–340, DOI 10.1007/s00233-009-9133-5. MR2538729

[44] J. C. Rosales, *Numerical semigroups with Apéry sets of unique expression*, J. Algebra **226** (2000), no. 1, 479–487, DOI 10.1006/jabr.1999.8202. MR1749900

[45] J. C. Rosales and P. A. García-Sánchez, *Finitely generated commutative monoids*, Nova Science Publishers, Inc., Commack, NY, 1999. MR1694173

[46] J. C. Rosales and P. A. García-Sánchez, *Numerical semigroups*, Developments in Mathematics, vol. 20, Springer, New York, 2009. MR2549780

[47] J. C. Rosales, P. A. García-Sánchez, J. I. García-García, and M. B. Branco, *Arf numerical semigroups*, J. Algebra **276** (2004), no. 1, 3–12, DOI 10.1016/j.jalgebra.2004.03.007. MR2054385

[48] Ernst S. Selmer, *On the linear Diophantine problem of Frobenius*, J. Reine Angew. Math. **293/294** (1977), 1–17. MR0441855

[49] Sage: Open Source Mathematics Software, available at `www.sagemath.org`.

[50] J. Sylvester, *Question 7382*, Mathematical Questions from the Educational Times **41** (1884) 21.

Mathematics Department, Texas A&M University, College Station, Texas 77840
E-mail address: `coneill@math.tamu.edu`

Mathematics Department, University of Hawaiʻi at Hilo, Hilo, Hawaii 96720
E-mail address: `robertop@hawaii.edu`

Contemporary Mathematics
Volume **685**, 2017
http://dx.doi.org/10.1090/conm/685/13714

A survey of discrete methods in (algebraic) statistics for networks

Sonja Petrović

ABSTRACT. Sampling algorithms, hypergraph degree sequences, and polytopes play a crucial role in statistical analysis of network data. This article offers a brief overview of open problems in this area of discrete mathematics from the point of view of a particular family of statistical models for networks called exponential random graph models. The problems and underlying constructions are also related to well-known concepts in commutative algebra and graph-theoretic concepts in computer science. We outline a few lines of recent work that highlight the natural connection between these fields and unify them into some open problems. While these problems are often relevant in discrete mathematics in their own right, the emphasis here is on statistical relevance with the hope that these lines of research do not remain disjoint. Suggested specific open problems and general research questions should advance algebraic statistics theory as well as applied statistical tools for rigorous statistical analysis of networks.

1. Introduction

The development of a rich literature on networks in the past decade has left ample opportunities for complementary mathematically rigorous results that should serve as the foundation for statistical modeling and fast computation of network features. In this review, we focus on exponential families for random graphs, models over equivalence classes of graphs summarized by a selected set of graph or network summary statistics. Here, the word 'model' is used in the statistical sense, and the terms 'graph' and 'network' are used interchangeably. Introductory paragraphs of Sections 3 and 4 offer an answer to the following question: Why consider statistics when studying random graphs and networks? A summarized answer can be phrased as follows: *how* a network is generated is crucial to properly calculate statistical network properties and specify the distributions being sampled. This includes properties that depend on degrees of nodes in the network, higher-order degree correlations, or other, possibly global, summary statistics; cf. an excellent example and the concluding paragraph in [**Jac08**, Section 4.2.1].

The reader may wonder why the focus on exponential families for random graphs, or ERGMs. While being able to write a random graph model as an exponential family is not impressive in and of itself, understanding the geometry, algebra,

2010 *Mathematics Subject Classification.* Primary 05C80, 60J10, 90B15, 91D30.
Key words and phrases. Random graphs, network models, alternating cone, balanced graphs, balanced hypergraphs, vertex bi-coloring, Markov bases, algebraic statistics, exponential families.

and the discrete structures supporting the model offer various statistical insights that are the theme of this chapter. A recent computer science article [**BDE**] shows that ERGMs are 'hard', that is, their normalizing constants are incomputable in general. While this important result about inapproximability in polynomial time formalizes the inherent complexity of this family of models, as a statistical model family they are still very broad and rich and have desirable properties. As such, they in the very least offer the theoretical foundation for studying random graphs and networks, and a bedrock for algorithmic exploratory analysis of sampling distributions of various graph summary statistics.

The statistics literature on understanding and developing network models is ever-growing; a partial list of references is offered below in context. Within this realm, though, the basic problem of establishing rigorous procedures for statistical inference is still a challenge, due to the varying complexity of the models and types of properties of data they capture, as well as networks being a novel data type in terms of traditional statistics.

The motivating problem for this discussion is thus a basic one of *statistical inference*, tightly related to several fundamental tasks required for statistical analysis of (any kind of) data: parameter estimation[1], sampling from the distributions in the model, testing model/data fit, and model selection. The broad goal of statistical inference is to decide, with a high degree of confidence, whether an observed data sample $x = x_1, \dots, x_N$ can be regarded as a draw from a distribution $p_{\theta_0} \in \mathcal{M}$ coming from a candidate statistical model $\mathcal{M}$, specified by the unknown parameter θ_0. Of course, this entails two crucial steps: 1) estimation: use the observed data x to produce an optimal estimate $\hat{\theta} = \hat{\theta}(x)$ of θ_0; and 2) goodness-of-fit testing: assess whether $p_{\hat{\theta}}$ can be considered a satisfactory generative model for the observed data x. Surprisingly, these fundamental tasks pose a family of open problems, in particular for discrete sparse small-sample data such as networks [**KK15**, **HGH08**, **HL81**, **Hab81**, **Agr92**, **YLZ**, **CKHG15**, **CDS11**]. These problems and their natural connection to discrete structures are the focus of this short overview.

In the remainder of the chapter, we will use the β-model for random graph, defined in Example 2.1, as a running example to illustrate the structure and implications of various statistical modeling questions that discrete mathematics tools can answer.

2. Preliminaries

We begin with some technical preliminaries on linear ERGMs and statistical considerations about these models. A *statistical model* $\mathcal{M}$ is a family of probability distributions indexed by a set of parameters $\Theta \subset \mathbb{R}^n$. In exponential random graph models, or ERGMs for short, one first selects the network characteristic T of interest in a particular problem. This selection is done so that T represents some interpretable and meaningful summary statistic of a network. The resulting model is then the collection of probability measures

$$\mathcal{M} = \{p_\theta : \theta \in \Theta\},$$

[1]Sometimes called 'fitting' in computer science or statistical physics literature, but it is different from model fit testing.

indexed by points in $\Theta \subseteq \mathbb{R}^n$ such that, for any $\theta \in \Theta$, the probability of observing a given network[2] $G = g$ takes on the exponential form

$$p_\theta(g) = \exp\{\langle T(g), \theta\rangle - \psi(\theta)\}, \tag{2.1}$$

where $\psi(\theta) := \sum_g \exp\{\langle T(g), \theta\rangle\}$ is a normalizing function called the log-partition function, and $T(g)$ is the vector of minimal sufficient statistics for the model. The exponential form of the probabilities above is a central theme in statistical theory, and statistical models of this form, known as exponential families, are known to exhibit optimal statistical performance and appear with a growing appeal in the machine learning community as well. Consistent with the running examples, we are considering *linear* exponential families for now, that is, $\langle T(g), \theta\rangle$ is a linear map of the state space. General exponential families can use non-linear maps, of course. For a more detailed statistical treatment of this family of models the reader is referred to classical references [**Bro86**], [**BN14**] and [**LC83, LC98**], as well as a more recent book [**BD15**]. ERGMs are, effectively, models over *equivalence classes of networks*, where two networks g_1 and g_2 are regarded as probabilistically equivalent whenever $T(g_1) = T(g_2)$. Conditioning on the values of T permits the reduction of the data through sufficient statistics and thus eliminates nuisance parameters [**Agr92**], which can be very helpful in applications.

EXAMPLE 2.1 (The β-model for graphs). The β-model for graphs is a well-known statistical model for random undirected graphs. In its original form, the β-model considers simple graphs, that is, does not allow loops or multiple edges; but the general version allows edges to appear with bounded multiplicity. It essentially assigns parameters $\beta_1, \ldots, \beta_n$ to the n vertices in the network that measure their 'friendliness', or propensity to attract edges. The edges are then assumed to be independent and appear with probability proportional to the product of the parameters of its vertices; in symbols:

$$Prob(G = g) \propto \prod_{\{i,j\} \in E(g)} \beta_i\beta_j = \prod_{i=1}^{n} \beta_i^{\deg(i)},$$

where '$\propto$' refers to the fact that the resulting quantity should be normalized by the log-partition function. In this model, the sufficient statistics vector $T(g)$ is the degree sequence of the graph g. Indeed, the model is usually given directly in the exponential family form (2.1) with $T(g) = (d_1, \ldots, d_n)$ being the degree sequence of g:

$$p_\beta(g) = \exp\{\sum_{i=1}^{n} d_i\beta_i - \psi(\beta)\}. \tag{2.2}$$

For a brief history of the model, see for example the introduction and references given in [**RPF13**], which deals with the general version of the β-model and the special case of the simple graph version studied in [**CDS11**].

The β-model (2.2) is indeed one of the simplest interpretable undirected random graph models of relevance in statistical applications. In fact, the study of the degree sequences and, in particular, of the degree distributions of real networks is a classical topic in network analysis, which has received extensive treatment in the statistics

[2] As is standard in statistics, G represents a random variable and the lowercase g its realization.

literature (starting with, e.g., [**HL81**], [**FW81**], [**FMW85**]), the physics literature (e.g. [**NSW01**], [**New03**], [**PN04**], [**WAD09**]) as well as in the social network literature (e.g., [**RPKL07**], [**Goo07**], [**HM07**], and references therein). See also the monograph by [**GZFA09**] and the books by [**Kol09**] and [**New10**]. Its known properties can be used as a blueprint for consideration of more complex models. Specifically, it is a very nice and well-understood ERGM from the following points of view that simultaneously serve as an outline of the remainder of this chapter:

(1) Exact inference for model fitting: exact testing is used for testing goodness of fit of a model when applicability of standard statistical asymptotic methods is unclear. Typically, this is the case for small sample sizes N, and also largely remains a problem for network data in particular. These topics are the content of Section 3.
   (a) For the β-model, exact testing depends on sampling graphs with a prescribed degree sequence, for which there are several algorithms throughout the statistics, computer science and graph theory literatures. The general problem is wide open for other ERGMs.
   (b) To construct an appropriate Markov chain for exact testing, sampling graphs with fixed properties should be done within the context of a statistical model.
   (c) As we will see, *all* linear ERGMs reduce to hypergraph degree sequence problems.

(2) Parameter estimation and noisy data: Viewing random graph models through the lens of exponential families offers a way to capture the issue of existence of maximum likelihood estimators (MLE), a fundamental problem that is largely unexplored. The problem and its statistical implications are explained in Section 4.
   (a) For the β-model, the geometry of MLE existence is captured by the convex hull of all degree sequences. The extreme points of that polytope correspond to threshold graphs and facet-defining inequalities have been characterized. In general, the geometry of MLE existence is captured by the model polytope, whose lattice points are realizable sufficient statistics vectors. These polytopes are not known for other ERGMs.
   (b) In data privacy considerations, or in dealing with noisy data, observed sufficient statistics may contain errors and thus in particular may not be realizable by any graph. For the β -odel, it is known when an integer sequence is graphical, i.e., when there exists a graph that realizes the sequence as its degree sequence. The general problem of characterizing realizable sufficient statistic vectors is wide open yet crucial for establishing reliability of inference.
   (c) Efficient facet description of the model polytope for other examples of ERGMs would be crucial to develop appropriate tools for dealing with noise in the data.

(3) The model has desirable asymptotic properties [**CDS11**], closely related to graphons [**LS06**]. As the asymptotics are different than the asymptotics mentioned in item (1) above, we will not discuss these issues in this Chapter, but instead point the interested reader to [**WO**] and for a quick list of references to [**RPF13**, Introduction].

Here are examples of some other models that build on the β-model for graphs and for which many of these questions remain open. In the interest of space and readability, equations repeating the structure (2.1) for each case are omitted but can easily be found in the references.

EXAMPLE 2.2. The *joint degree matrix model* is the ERGM of the form (2.1) where the sufficient statistic $T(g)$ is the joint degree matrix, or JDM for short, of g. The JDM counts the number of edges between nodes of given degrees, for all degree pairs. The model, as an ERGM, was introduced to the statistics literature in recent work [**SR14b**], motivated by previous lines of research outside of mainstream statistics, such as [**EMT15**], which offers a fast mixing algorithm and a fantastic introduction and overview of the joint degree matrix as a network statistic, and [**SP12**].

As pointed out in [**EMT15**], fixing the value of the JDM of a network is stronger than just fixing the degree sequence, though it uniquely defines the degree sequence.

EXAMPLE 2.3. The *β-model for hypergraphs* was introduced in [**SSR$^+$14**]. It looks identical to the β-model for graphs, except the edges in the network g can be of size larger than 2. In particular, there are three variants: uniform, layered uniform, and general. In the uniform variant, hyperedges of fixed size k occur independently with probability proportional to $\beta_{i_1} \cdots \beta_{i_k}$ for all k-tuples $i_1, \dots, i_k$ of vertices on the graph. The sufficient statistic in the this case is the hypergraph k-degree sequence, that is, the number of edges of size k to which each vertex belongs.

In the layered variant, the set of possible hyperedges is extended to include edges of all sizes up to and including some fixed size k. The sufficient statistics are thus layered hypergraph degree sequences: the number of hyperedges *of each size*, $2, \dots, k$, to which a vertex belongs. In the general variant, all edge sizes are allowed, and the model is more complicated; for details see [**SSR$^+$14**].

EXAMPLE 2.4. The *p_1 model for random graphs* assigns probabilities to *directed* edges in a random graph according to the propensity of the nodes to receive, send, or reciprocate edges. The sufficient statistics vector $T(g)$ of the model consists of: the in-degrees of all nodes, out-degrees of all nodes, and the number of reciprocated edges, where a reciprocated edge is of the form $i \leftrightarrow j$. The model has a history in statistic and applications, see for example [**HL81**], [**FW81**] and [**PRF10**]. The last reference interprets the model through the lens of algebraic statistics and studies its linear ERGM structure in terms of algebra and geometry; cf. [**RPF13**].

A fairly recent extended review article on statistical network models [**GZFA09**] offers many other examples of models, beyond linear ERGMs, whose geometric and combinatorial properties are yet to be explored.

3. Sampling algorithms in exact testing

Testing the fit of the model means deciding whether the model provides a plausible probabilistic representation of the available data. Unfortunately, most tests for goodness of fit are based on large sample approximations that are not applicable to data such as sparse contingency tables used in, for example, cross-classification of categorical variables, or data such as random graphs that can be naturally thought of as contingency tables through incidence matrices. Fundamental examples of inadequacy of the use of asymptotic approximations were already pointed out over two decades ago in statistics literature [**Hab88**], [**Agr92**]: when sample size N is small, or for example if some data table cell entries are much smaller than others, *exact testing* should be performed instead. Even so, a large part of the literature on, say, network computation and modeling does not address the lack of model fitting and testing methodology[3] beyond heuristic algorithms [**Han03, CKHG15, HGH08**]. This is largely due to the inherent model complexity or degeneracy and the lack of tools that can handle network models and sparse small-sample data.

The role and relevance of discrete mathematics in exact testing is as follows. In an exact test for a model with sufficient statistics vector T, the observed data x with $T(x) = t_{obs}$ is compared to a reference set

$$\mathcal{F}_{t_{obs}} := \{y : T(y) = t_{obs}\}$$

of data with the same value of the sufficient statistic. This reference set is also called the *fiber* in algebraic statistics literature. It is a fiber of the algebraic map that computes the sufficient statistics; in the β-model case the map is linear, since node degrees can be computed as row/column sums of the incidence matrix. Note that, by definition of sufficiency, the probability of any data point in the fiber is determined completely by the value t_{obs}, hence the fiber is a good reference set to consider for testing. Using a suitable choice of a goodness-of-fit statistic that measures, for example, the distance of the data x from the expected value under the model, the simulated data points in $\mathcal{F}_{t_{obs}}$ are compared to the observed data. If a relatively large number of those are closer to the expected value than the observed data, than one declares poor model fit. These are, roughly, the essential elements of an exact test, whose name refers to the exact conditional distribution of data points in the fiber $\mathcal{F}_{t_{obs}}$, where we are conditioning on the observed value t_{obs} of the sufficient statistics.

To visualize the elements of an exact conditional test, consider the following example. Suppose the observed graph is g_1 in Figure 1; see [**OHT13**, Figure 13] and [**GPS**, Section 4.1]. Given the degree sequence of g_1, the expected value under the model is visualized as a weighted graph whose edge weights are the expected values of the random variables representing the edges. This is computed using the MLE of the parameters; see Equation (4.1). One may view these expected values as computing the relative frequency of the edges in the sample space of all graphs on 8 nodes, weighted by the probabilities of each of those graphs under the model. Intuitively, the exact test is meant to assesses the relative closeness of the observed graph to the expected graph, and thus it answers the question: is g_1 more like the expected graph than the graphs g_2, g_3, g_4 from Figure 2 and the other 587 graphs in the fiber?

There are two important points to be made before we discuss the key problem. First, choosing a good goodness-of-fit statistic is in general an open problem, and is difficult in particular in random graph models for which edges are not independent random variables. For a discussion of the special case of 2-way contingency table models, see [**Agr92**, §3.1]. Second, a good example of a goodness-of-fit statistics

[3] The reader is urged to distinguish between this statistical questions of model fitting ('is the model *correct*?') and parameter estimation ('assuming the model is correct, find the parameter value(s) that best fit the data'). The latter is often alluded to within the context of 'model validation' in the computer science literature.

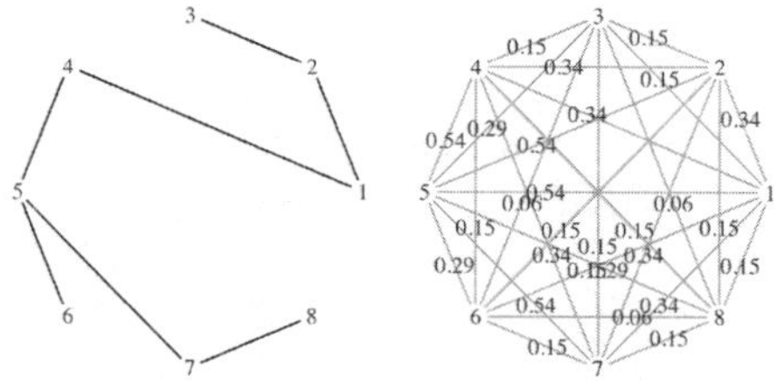

FIGURE 1. Observed graph g_1 (left) with degree sequence (2,1,1,2,3,1,2,1), and the expected probability-weighted graph under the β-model.

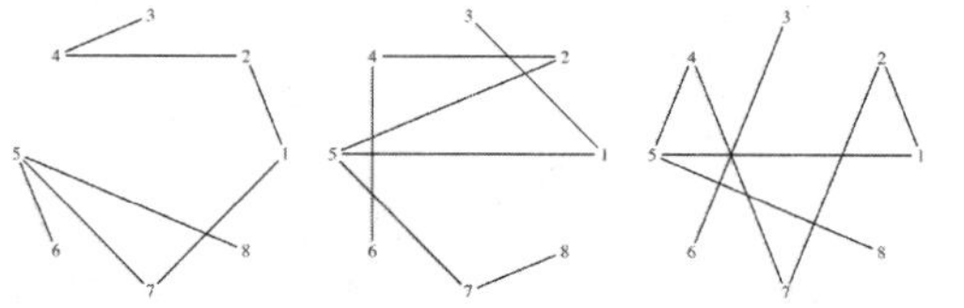

FIGURE 2. Three out of the 590 other graphs in the fiber $\mathcal{F}_{(2,1,1,2,3,1,2,1)}$: g_2 (left), g_3 (center), g_4.

to use for random graphs with independent edges is the chi-square statistic, which simply computes the square of the weighted Euclidean distance in $\mathbb{R}^n$ between the observed graph and the expected graph (or MLE). It is defined as:

$$\chi^2(g) := \sum_{i,j} \frac{(\hat{g}_{ij} - g_{ij})^2}{\hat{g}_{ij}},$$

where g_{ij} is 1 if the edge $\{i, j\}$ is present in g and 0 otherwise, and $\hat{g}_{ij} \in (0,1)$ is the expected value of the random variable representing the edge $\{i, j\}$ under the MLE, as illustrated in Figure 1. The values of the chi-square statistic for the four graphs above are:

$$\chi^2(g_1) = 19.49, \quad \chi^2(g_2) = 17.42, \quad \chi^2(g_3) = 21.56, \quad \chi^2(g_4) = 26.53.$$

To determine whether the value $\chi^2(g_1)$ is too large, one should enumerate the fiber and compute the value of the statistic for all 591 graphs in it.

As it is computationally prohibitive to enumerate the fiber except in very small cases, one tries to instead sample from the fiber and estimate the number of data points that are more extreme than the observed, with respect to the chosen goodness-of-fit statistic. Thus the **key problem** in the exact testing procedure is that of exploring the fiber through sampling from the conditional distribution given t_{obs}.

PROBLEM 3.1 (General Problem 1). *Fix a random graph model $\mathcal{M}$ with sufficient statistics T. Develop an efficient algorithm that samples from the space of graphs with arbitrary fixed value of $T(g) = t_{obs}$.*

A fiber sampling algorithm can be a direct sampler or it can be based on a Markov chain and used through the Metropolis algorithm, which is a standard algorithm and can be found, for example, in [**RC99**]. The algebraic statistics literature focuses on the latter. The reason for this is the natural connection to algebraic geometry. Namely, the theoretical feasibility of constructing Markov chains to solve

General Problem 1 for arbitrary *linear* ERGMs, as well as finiteness of the complexity of steps needed to perform the Markov chain random walk on fibers of *every* log-linear model, form a cornerstone of traditional algebraic statistics through a fundamental theorem from [**DS98**]; see also [**DSS09**, Theorem 1.3.6]. Its applied solution and relevance in exact testing for any linear ERGM hinges upon the development of efficient sampling algorithms for graphs and hypergraphs, as outlined here.

The steps or moves needed to perform the Markov chain above for a a given model are called *Markov moves*. A *Markov basis* for a model is a collection of moves guaranteed to connect each fiber of that model. A move is an element of the *Graver basis* of the model if it contains no proper sub-moves; see Figure 3 for a running example.

REMARK 3.2. In order to be of statistical relevance, and make any inference from the observed data, Problem 3.1 should be solved within the context of a statistical model. In other words, when a network feature, or a set thereof, is fixed in sampling, it is understood that the feature can serve as a sufficient statistic of some model; the statisticians then ask to identify the model and study its various properties in order to determine feasibility and reliability of inference and model fitting.

EXAMPLE 3.3. For the β-model for random graphs from Example 2.1, the reference set $\mathcal{F}_{t_{obs}}$ is the set of graphs whose degree sequence d equals the degree sequence $t_{obs} := d(g)$ of the observed graph g. It is easy to see that in the β-model the conditional probability distribution of graphs given a fixed observed degree sequence is uniform; in general, of course, this distribution depends on the model. Therefore, in the β-model, as well as others we discuss in this chapter as the reader may easily verify, graphs should be sampled uniformly, because conditioning on the sufficient statistics results in the uniform distribution on the fiber.

Of course, Problem 3.1 has been solved for the β-model for graphs. Indeed, various algorithms for constructing random graphs with a fixed degree sequence - in other words, sampling the fibers of the β-model - have been proposed over the years, with or without the context of the model, and the reader is probably familiar with the graph theory literature on this topic, most relevant part of which is cited throughout this Chapter. On the statistics side, [**BD10**] derive a sequential importance sampling algorithm for this model. A different random generation algorithm for simple connected graphs can be found in [**VL05**].

As mentioned above, one could think of solving the sampling problem in different ways: by an algorithm that constructs, uniformly at random, a graph that realizes the given degree sequence (where care should be taken that such an algorithm can, in fact, discover the entire fiber!); or by finding a set of moves, sometimes also called 'edge swaps' in graph theory, that can serve as a basis for a Markov chain Monte Carlo (MCMC) sampling algorithm on the fiber. Here the word 'basis' is used in a technical sense but can also be understood intuitively: the moves should guarantee to discover the entire fiber and define a symmetric, aperiodic chain on *any* given fiber of the model. The Markov chain approach offers a nice alternative and an interesting connection to other areas of mathematics. As such, it is largely the focus of the remainder of this section.

3.1. Ingredients for a Markov chain fiber sampler. In order to use the MCMC method to sample the fiber of an observed graph, three important questions should be answered. First, a *Markov basis* should be specified: a set of moves that the algorithm uses to go from any graph to any other graph in any fiber. Second, the *stationary distribution* of the proposed Markov chain should be the correct conditional distribution; recall that in the β-model example, it should be uniform. Third, *mixing time* considerations cannot be avoided in this approach - one can easily propose seemingly straightforward Markov chains that have extremely unreasonable behavior in that they will take a very long time to converge or to explore the fiber.

One way to answer the first two questions simultaneously is through the use of algebraic statistics. The third question on mixing time is discussed subsequently.

There is an algebraic geometry or, if you will, commutative algebra equivalent to a Markov basis for a linear ERGM stemming from the fundamental theorem from [**DS98**]. Namely, every such statistical model lies in a natural algebraic variety: A *variety* is the set of all solutions to a system of polynomial equations, and those encoding parametric statistical models can equivalently be described parametrically – there is an algebra-geometry duality here; see the excellent introductory explanation in [**SKKT00**, §2.4 and 2.5]. Focusing on the algebra, the parametrization is encoded via a homomorphism between two polynomial rings, say $\phi : \mathbb{C}[E] \to \mathbb{C}[\theta_1, \ldots, \theta_n]$, where E is some set of variables, and θ_i are parameters. The indeterminates in the two polynomial rings carry a special meaning in algebraic statistics: E represent random variables of interest, while $\theta = (\theta_1, \ldots, \theta_n)$ is the vector of unknown model parameters. For example, in the case of ERGMs whose edges are independent random variables, $E = E(G)$ are edges of the random graph G. The kernel of the map ϕ is an ideal in the polynomial ring $\mathbb{C}[E]$, and this ideal is the defining ideal of the algebraic variety whose real positive part contains the statistical model. Readers interested in further algebraic details of this correspondence may consult [**DSS09**, page 25].

EXAMPLE 3.4. In the specific case of the β-model for graphs with no statistical sampling constraints, the graph G is the complete graph because all edges are allowed to appear in the random graph realization g. The homomorphism ϕ maps each edge e_{ij} to the product of the parameters $\beta_i\beta_j$. For example, consider the random graph β-model on $n = 8$ nodes. The coordinate map corresponding to the parameterization of the model is

$$\begin{aligned} \phi_{K_8} : \mathbb{C}[e_{12}, e_{13}, \ldots, e_{78}] &\to \mathbb{C}[\beta_1, \ldots, \beta_8] \\ e_{ij} &\mapsto \beta_i\beta_j. \end{aligned}$$

Note that here we have forgotten the exponential and the normalizing constant because they do not affect the algebraic structure of the parametrization, so by abuse of notation we drop the 'exp' from the model specification. An example of an equation that vanishes on the model is $e_{17}e_{24}e_{34}e_{58} - e_{14}e_{23}e_{45}e_{78}$. The algebraic relation holds because $\phi_{K_8}(e_{17}e_{24}e_{34}e_{58}) = \phi_{K_8}(e_{14}e_{23}e_{45}e_{78})$ since, of course, the two graphs on edge sets $\{17, 24, 34, 58\}$ and $\{14, 23, 45, 78\}$ have the same degree sequence, which is the sufficient statistic for the model; refer to Figure 3. Note that the equation vanishing on the model is equivalent to saying that equation is in the kernel of the coordinate map: $e_{17}e_{24}e_{34}e_{58} - e_{14}e_{23}e_{45}e_{78} \in \ker \phi_{K_8}$.

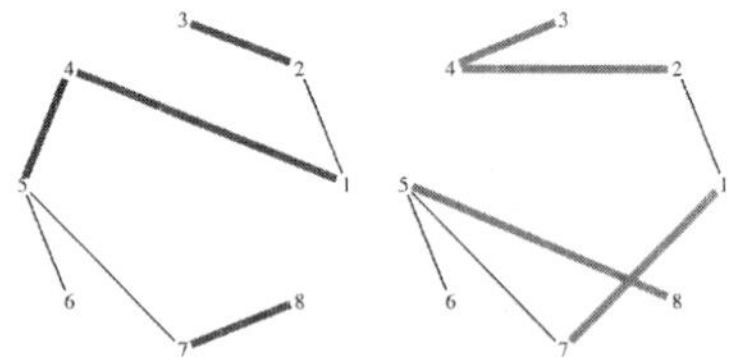

FIGURE 3. A Markov move under the β-model for the graph $g = g_1$ from Figure 1: the graph g_1 with edges h_2 to be removed in blue (left), and the resulting graph g_2 from Figure 2 after h_2 is replaced by edges in h_1 which is shown in red. Note that this move $h_1 - h_2$ is not a minimal one, cf. Figure 4. Instead, it is an example of a Graver move. If structural zeros were present in the model, moves like this one may be needed for connecting the fiber.

To summarize: probability distributions in the model correspond to real positive points on an algebraic variety, where the parametric description of the model gives rise to the parametrization of the variety. This variety is defined implicitly by the equations that vanish on the model. Coordinates of the polynomial ring that is the domain of ϕ correspond to the joint probabilities of random variables in the model. Equations that vanish on the variety represent algebraic relations among the joint probabilities. To visualize what these relations mean, consider $\prod_{\{i,j\}\in E^+} e_{ij} - \prod_{\{i,j\}\in E^-} e_{ij}$, for two sets of edges E^+ and E^-. This is an equation that vanishes for all points on the model (i.e., a relation on the edges) if and only if the graphs whose edges sets are E^+ and E^- have the same probability under the model. In turn this happens if and only if the two subgraphs have the same values of the sufficient statistics vector. In the linear ERGM case, the variety has a special structure called toric, due to linearity, and as a consequence the defining equations are all binomials.

The crucial observation is then that such binomials correspond to moves on the fibers of the model, and in fact *the moves comprise a Markov basis if and only if the binomials suffice to generate all equations that vanish on the variety* [**DS98**]. This fact is sometimes called the fundamental Theorem of Markov Bases. This correspondence should be interpreted in the following way: $h_1 - h_2$ is a defining equation of the variety corresponding to the model if and only if replacing the subgraph $h_2 \subset g$ by the subgraph $h_1 \subset g$ is a move on the fiber of the model, meaning that the move does not change the value of the sufficient statistics vector, as illustrated in Figure 3.

In fact, the correspondence applies to general log-linear models for discrete data, not only networks, and is one of the cornerstones of algebraic statistics. In the general case the moves $h_1 - h_2$ need not be squarefree and need not represent graphs; instead each h_1 and h_2 can be an arbitrary monomial in the indeterminates E and this machinery works for general contingency table models. Simple graphs are just 0/1 contingency tables, while graphs can in general be represented by non-negative tables, where each positive entry indicates presence of an edge, with multiplicity if the model allows for it. Given the very extensive literature on this topic, for example summarized in the recent book [**AHT12**] and references given therein, let us not spend more time studying details of this equivalence here. Instead, we pause

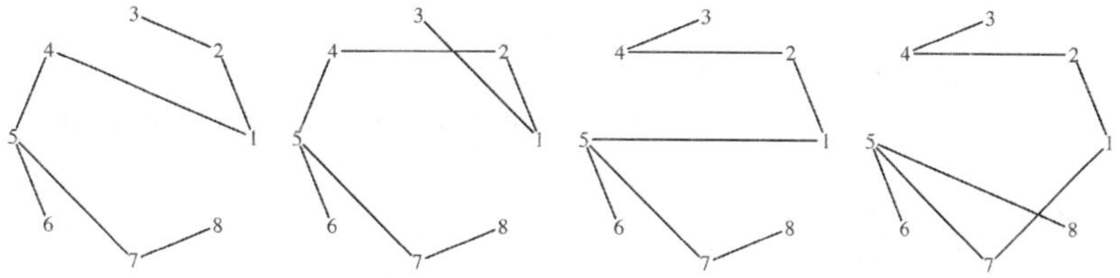

FIGURE 4. A walk using only quadratic moves from g_1 to g_2 on the fiber $\mathcal{F}_{(2,1,1,2,3,1,2,1)}$.

to note the important consequences: By a fundamental theorem in commutative algebra called the Hilbert basis theorem, a Markov basis exists and is finite for *every* linear ERGM. Furthermore, using the moves from a Markov basis given by the algebraic correspondence theorem to generate moves in the Metropolis algorithm *automatically* converges to the correct stationary distribution on the fiber. Note that, by definition, any set of moves containing a minimal Markov basis has the same property.

The β-model for graphs from Example 3.3 somehow seems to have a parallel history in various literature, as reflected upon in the following Remark.

REMARK 3.5. In commutative algebra, Markov bases for the β-model exist under the name of *generators of toric ideals of graphs.* This name comes from the fact that the moves from the β-model can be encoded by *closed alternating walks on graphs*: sets of edges partitioned into two sets, E^+ and E^-, such that they constitute a closed walk when traversed in alternating order $e_1, \dots, e_m$ with $e_i \in E^+$ for even i and $e_i \in E^-$ for odd i. For example, the move corresponding to removing blue and adding red edges in Figure 3 represents one such alternating walk, with E^+ and E^- corresponding to red and blue edges, respectively. The reader will note that this move can be obtained as a sequence of three moves represented by closed 4-cycles illustrated in Figure 4. Equivalently, the polynomial $e_{17}e_{24}e_{34}e_{58} - e_{14}e_{23}e_{45}e_{78}$ lies in the ideal generated by the three polynomials representing the 4-cycles: $e_{17}e_{24}e_{34}e_{58} - e_{14}e_{23}e_{45}e_{78} = e_{24}e_{34}(e_{17}e_{58} - e_{15}e_{78}) + e_{45}e_{78}(e_{13}e_{24} - e_{14}e_{23}) + e_{24}e_{78}(e_{15}e_{34} - e_{13}e_{45})$.

The commutative algebra community has known the structure of these generators, without calling them moves or relating them to MCMC methods, for at least 20 years, for example, in [**Vil95**, **Vil00**] and [**OH99**], as well as [**Stu96**, Chapter 9] and [**RTT12**].

Further, these results have been known in the algebraic statistics literature for some time as well. Notable examples include works such as [**SS06**, Theorem 4] on toric geometry of compatible full conditionals, [**Mor13**, Theorem 3.2] on relations among conditional probabilities, [**SW12**] that studies algebra and geometry of statistical ranking models and the ascending model in particular, and [**PRF10**] that study the structure of Markov and Graver bases for the directed model p_1 whose restriction (to the part of the random graph consisting only of reciprocated edges) is the β-model. More recently, [**OHT13**] offer the exact test implementation whose main ingredient is a random generation of Markov and Graver bases elements for the β-model.

Interestingly, bicolored or alternating closed walks on graphs make a later appearance in computer science literature as well. [**BPS09**] define the *alternating cone* of a graph to be the set of characteristic vectors of alternating walks on the graph, cf. [**BPS07**] and the subsequent book chapter [**Bha12**]. In algebraic terms, this is the cone of exponent vectors of the binomials in the toric ideal of that graph. Results from toric algebra of graphs and the definition of the alternating cone should easily imply that [**BPS09**, Conjecture 3.1] is actually true, and can furthermore be generalized for hypergraphs appropriately. [**Sri**] is a very nice set of notes on open problems related to threshold graphs; cf. Conjecture 4. To prove the conjecture, note that a 'balanced sum of cycles' is certainly a balanced graph, thus it is a lattice point in the cone of balanced subgraphs. Each of these lattice points is an integer exponent of some binomials in the toric ideal of the graph. It is well known that the Graver basis is an integral generating set of the lattice, in that every lattice point in the cone can be written as a integral sum of the Graver exponents. Finally, Graver bases of these cones for graphs are encoded by alternating walks by [**Vil95**], see also recent literature cited above. More generally, for hypergraphs, the walks are interpreted as bi-colored [**PS14**] and again form a Graver basis; see [**PTV**] for the general non-uniform case. By definition, these are exactly balanced subgraphs. Thus, the exponent vector of a *balanced sum of cycles is an integral sum of balanced subgraphs* represented by Graver elements. The present author is not aware of whether the conjecture has been proved in other literature; only the *fractional* sum part of the conjecture is proved in the works on alternating cones cited above.

Apart from the β-model for random graphs, sampling algorithms for the space of graphs with fixed properties have evolved disjointly from statistics. At this point, solutions to the following two subproblems of Problem 3.1 are of special importance:

PROBLEM 3.6 (An instance of General Problem 1: p_1 model moves). *Determine a set of moves sufficient to connect the space of graphs, consisting of both directed and bidirected (reciprocated) edges, with a given directed degree sequence and number of reciprocated edges.*

The problem above was solved in algebraic statistics [**PRF10**] and an implementation of a dynamic algorithm was offered in [**GPS**]; see the subsection on mixing times below for a related open problem.

PROBLEM 3.7 (An instance of General Problem 1: Hypergraph β-model moves). *Determine a set of moves sufficient to connect the space of hypergraphs, uniform, layered or general, with a given degree sequence.*

Of course, one can 'solve' these problems in various ways. From the point of view of (algebraic) statistics applied to networks, the most interesting types of solutions to Problems 3.7 and 3.1 in general will *at least address* general models with sampling constraints and forbidden edges, or derive bases of bounded complexity with good mixing time properties for a subclass of the models. The literature offers some examples of such solutions, but many problems remain open. These issues are discussed next.

Sampling constraints. If, in applications, the random sampler of the model fibers is allowed to step through the extended state space of non-simple graphs, where edges can have a multiplicity, then the problem is solved. For example, simple-edge-swap solution to Problem 3.7 follows from standard textbook results in algebraic geometry; namely, the generating set of the toric ideal for the model is known to consist of quadratic binomials, as the algebraic closure of the model is defined by the toric ideal of the second hypersimplex, known to be generated by quadrics. However, inherent sampling constraints can compound the issue of computing a set of moves guaranteed to connect each fiber, for example, the sampler may need to step through only simple graphs, or graphs with bounded multiplicity. In some statistical modeling considerations other constraints may come in the form of forbidden edges or forbidden neighbors; such constraints in statistics are called *structural zeros* of the model. Of course sampling and modeling constraints restrict the fiber, as expected, but also make the required moves more complicated: consider the situation when the edges used in the intermediate steps of the walk in Figure 4, such as the edge $\{1, 3\}$ for example, are structural zeros.

Interestingly, [**HTE**$^+$**09**] characterize when a sequence is realizable as a degree sequence of a simple graph such that a given set of edges from an arbitrary node is avoided. This allows the authors to construct a swap-free algorithm for sampling the restricted fiber of the β-model. If one is interested in sampling a restricted fiber for any general linear ERGM or any log-linear model, [**OHT13**] and [**AHT12**] show that one needs a much larger set of moves to guarantee connectivity, i.e. minimal Markov bases may not do the job. For example, the set known to algebraists to be inherently more complicated, called the Graver basis of the toric ideal [**DSS09**, §1.3], [**AHT12**, §4.6] will suffice to connect the fiber, however it is also so complex and large that it is not feasible to compute it for any reasonable example beyond toy-size networks; to give an idea of such size, for the p_1 model this basis can only be directly computed for networks on less than 7 nodes. What [**OHT13**] then do is provide an algorithm that constructs one step in the chain – a Markov move – in a dynamic fashion and show its performance on several small network datasets. Since the moves generated belong to the Graver basis, the method is provably applicable in the case of structural zeros and sampling constraints. This method does rely on the understanding of the algebraic machinery underlying the model, but appears to be inefficient in discovering the fiber and the resulting Markov chain has a high rejection rate. In contrast, simulations indicate that the chain from [**GPS**, §4.1] does not have that problem, but there is no formal proof of its good properties. This leads us to the next topic regarding sampling fibers.

Mixing time considerations. Considerations of mixing times/convergence are crucial in any MCMC sampling scheme [**RC09**] [**BGJM11**]. Notably, using minimal Markov bases may not produce good mixing as seen in some experiments and pointed out recently in [**Win**], which formalizes how mixing properties of Markov bases depend on the structure and connectedness in particular of the underlying *fiber graph*. Moreover, it explains that results on mixing times of chains built on minimal Markov bases are in general prohibitive to obtain, and suggests instead the construction of related expander graphs for the problem - a worthwhile pursuit. In earlier work, [**GPS**] construct a chain on the fiber graph for the p_1 model in in such a way that it is, in fact, a complete graph, with loops having seemingly low probability. In simulations, the chain behaves quite well and indicates fast mixing. Nevertheless, a formal proof of rapid mixing or any mixing properties are left as open questions, as the authors there did not study the structure of the transition matrix for that chain.

This is where recent advances in discrete literature, for some reason disjoint from statistics literature on networks, offer example results for the β-model that, in this author's opinion, should be connected to statistics - algebraic statistics in particular - and used as a guide for studying Markov bases of other linear and general ERGMs. Namely, [**KTV99**] conjectured that the Markov chain based on swaps mixes rapidly, and the proofs for arbitrary regular graphs and half-regular bipartite graphs were given in [**CDG07**] and [**MES13**], respectively, while irregular graphs are studied in [**Gre15**]. Recently, [**EMT15**] offer insight to the general case via the study of swap-based chains on the fibers of the JDM model from Example 2.2; they prove fast mixing on a subset of the JDM-model fibers and offer an excellent review of related literature, while irreducibility of the Markov chain on the JDM realizations was proved in [**CDEM15**]. To translate what this means for Markov bases, note that in algebraic statistics language, 'swaps' of two edges at a time are quadratic Markov moves, or quadratic generators of the toric ideal, since taking two edges at a time and replacing them by two other edges translates to a quadratic monomial in Example 3.4.

Crucially, however, the difference is not just linguistic: in general in algebraic statistics, Markov moves are sampled from a set of all possible moves that apply to the model, that have been computed a priori. Such sampling can clearly lead to more rejections. On the other hand, 'swaps' are constructed from observed edges, so rarely is a non-applicable swap proposed! In particular, the literature cited here provides a set of examples of *rapidly mixing sequences of fiber graphs*, and thus also hope that Markov-move-based constructions, adapted appropriately, can still lead to good chains for fiber exploration and goodness-of-fit tests; compare to examples reported in [**GPS**] or [**OHT13**] for an illustration.

In the remainder of this subsection, let us clarify how these chains should be constructed 'with care', and why an out-of-the-box Markov basis may not be good enough. For example, [**Win**] observes that enlarging minimal Markov bases to, say, Graver bases doesn't improve mixing time. Intuitively, if the chain can be constructed so that the fiber graph has a relatively small number of loops, then rapid mixing may be possible. Such a fiber graph would resemble the idea of using moves corresponding to graph-theoretic 'swaps': namely, not attempting to construct or apply Markov moves that work on just some fiber, but, instead, focus on the given observed graph and its particular fiber. We refer to such moves as *data-dependent*. Incidentally, the chain on the p_1 model used a large set of moves that is guaranteed to contain the Graver bases *in combination with* constructing data-dependent moves dynamically, involving both 'large' and 'small' steps on the fibers of the p_1 and β-models, shows excellent mixing properties in simulations on networks of various sizes. In fact, the Graver move from Figure 3 was constructed using the particular implementation from [**GPS**], without going through the intermediate steps in Figure 4. In the language of Example 3.5, what we construct is a random walk on the realizable lattice points of the alternating cone using a superset of the Graver basis for the toric ideal of the complete graph, or for an arbitrary graph whose missing edges represent structural zeros in the p_1 model. This leads us to the final remark about Markov bases construction, which is intimately related to mixing time, but deserves to be singled out.

Data-dependent samplers. That knowing an entire Markov basis for a model may still not be sufficient to run goodness-of-fit tests efficiently is a well-known fact in algebraic statistics. Namely, Markov bases are data-independent [**DFR$^+$08**, Problem 5.5]. To paraphrase [**AHT12**]: since a Markov basis is common for every fiber, the set of moves connecting the particular fiber of the observed data will usually be significantly smaller than the entire basis for the model. The subsequent paper [**Win**] offers another intuitive explanation of why data-dependent fiber samplers are necessary: "The conclusions we draw [...] are that an adaptation of the Markov basis has to take place depending on the right-hand side", meaning that distinct fibers require different adaptations, confirming previous observations. To handle precisely this issue, [**Dob12**] suggested generating only moves needed to complete one step of the random walk, and then covering the fiber by sets of local moves; this was the first scalable method for exact tests on tables beyond decomposable models [**DS04**] and applicable to log-linear models where sufficient statistics are table marginals. However, many models of interest are not captured by marginals [**SZP14, HL81, KPP$^+$, GZFA09**], leaving a gaping hole in the methodology for other categorical models, particularly for sparse data. This leads us to consider more general models for networks, with a long-term aim to design a data-dependent sampler beyond the one in [**GPS**] and [**OHT13**] for the p_1 and β-models.

3.2. General linear ERGMs. The situation may now seem hopeless for general non-degree-based models, because even if we establish fast mixing for Markov chains on β-model fibers, there are infinitely many other models that are left to be considered. However, not all is lost: it turns out that as long as the model is log-linear and the sufficient statistics are given by a linear map from the sample space, sampling hypergraphs with prescribed degree sequences, uniform or not, and with prescribed forbidden edges actually suffices for performing the exact test. This takes care of a number of interesting models! The answer lies in the structure all log-linear models including linear ERGMs: each of these models is encoded by a hypergraph, defined in [**PS14**], Section 3.1 of which provides further details and a simple example for the independence model.

DEFINITION 3.8. corresponds to the parameters of the statistical model, and whose edge set is determined by joint probabilities of all possible states of the random variables $Z_1, \dots, Z_m$. More precisely, $\{\theta_j\}_{j \in J}$ is an edge in the parameter hypergraph if the index set J describes one of the probabilities in the model, that is, there exist values $i_1, \dots, i_m$ such that $Prob(Z_1 = i_1, \dots, Z_m = i_m) = \prod_{j \in J} \theta_j$.

EXAMPLE 3.9. Let $\mathcal{M}_\beta$ be the β-model for random graphs on n vertices $v_1, \dots, v_n$. The parameter hypergraph $\mathcal{M}_\beta$ is the complete graph on vertices $\beta_1, \dots, \beta_n$. To see why, recall the probability formula of the edges in Example 2.1: the random variables $Z_1, \dots, Z_m$ represent the $m = \binom{n}{2}$ edges of the graph. The probability of the joint state where $Z_k = 1$ and $Z_l = 0$ for $l \neq k$, that is, the event of the occurrence of only the edge $Z_k := \{i, j\}$ in the graph, is proportional to $\beta_i \beta_j$. [4] Note that $\mathcal{H}_\mathcal{M}$ being complete does not mean that all edges $\{v_i v_j\}$ in the random graph are observed, but instead that all edges $\{\beta_i \beta_j\}$ of $\mathcal{M}_\beta$ correspond to random graph edges that are *allowed*. In particular this is why the graph on the right of

[4]Of course we take the minimal presentation here; knowing $Prob(Z_1 = 1, Z_2 = 0)$ and $Prob(Z_1 = 0, Z_2 = 1)$, we need not consider the joint probability $Prob(Z_1 = 1, Z_2 = 1)$.

Figure 1 is complete with non-zero probability weights on every edge- there are no structural zeros in that example.

Similarly, for the hypergraph β-model, the parameter hypergraph is the complete hypergraph, either uniform, layered, or general, respectively in each of the there cases of the model. Recently, the hypergraph $\mathcal{H}_{p_1}$ of the p_1 random graph model was described in [**GPS**] and consists of edges of sizes 1, 3 and 7, for probabilities of no connection between two nodes, or a directed edge, or a reciprocated edge, respectively; for example, a directed edge $i \to j$ occurs with probability $\lambda_{ij}\alpha_i\beta_j$ and is thus encoded by a 3-edge on $\mathcal{H}_{p_1}$.

REMARK 3.10. The parameter hypergraph runs the danger of being lost in translation, so to understand its construction, recall that the sufficient statistics of $\mathcal{M}$ are computed using a linear map. Let M be the matrix of that linear map. *The parameter hypergraph $\mathcal{H}_{\mathcal{M}}$ of the model is simply the hypergraph whose vertex-edge incidence matrix is this matrix M.*
For the β-model, the definition may appear trivialized, but further log-linear model examples show this isn't the case in general; see for example Figure 2 in [**GPS**] for two further examples of parameter hypergraphs.

What about the fibers of $\mathcal{M}$? Consider again the β-model for graphs: two random graphs g_1 and g_2 occur with the same probability and are in the same fiber of the model if and only if their images under the linear map M are the same. If g_1 and g_2 are encoded by a set of red and blue edges on $\mathcal{M}_\beta$, respectively, they have the same probability if and only if the same multiset of parameters β_i's are covered by the red and blue edges. In other words, the red edge set and the blue edge set are graphs with the same degree sequence. This degree sequence correspondence for $\mathcal{M}_\beta$ seems trivial; remarkably, it holds for all linear ERGMs.

DEFINITION 3.11. Let E be a multiset collection of edges in a hypergraph H. E is *balanced* with respect to a given bicoloring of H if for each vertex v covered by E, the number of red edges containing v equals the number of blue edges containing v. In other words, the 'blue degree' of E equals its 'red degree'.

The definition above appears as Definition 2.7 in [**PS14**] for the case of uniform hypergraphs, and Definition 5.7 in [**PTV**] for an arbitrary hypergraph. The motivation for this definition and interpretation of the binomials as bicolored edge sets was to study toric algebra of hypergraphs by generalizing the algebraic literature on graphs. The bicoloring construction idea simply generalizes Villarreal's [**Vil95**] 'closed alternating walks' on graphs, see [**RTT12**] for a further classification and characterization of the walks. A straightforward argument now shows the following.

THEOREM 3.12 (See Theorem 2.8 of [**PS14**] for the uniform case). *Let $\mathcal{H}_{\mathcal{M}}$ be the parameter hypergraph of the model $\mathcal{M}$. Then any balanced collection of edges $E \subset E(\mathcal{H})$ constitutes a move in the toric ideal associated to $\mathcal{H}$. In particular, the set of* balanced bicolored subgraphs *connect all fibers of the model $\mathcal{M}$.*

Note that a balanced bicolored hypergraph is simply a set of two hypergraphs, not necessarily simple, that have the same degree sequence! Examples and structure of balanced edge sets on a hypergraph, as well as extension to the non-uniform case, can be found in [**PTV**, §5.1 and 5.2].

Thus far, we have seen how all linear ERGMs that are encoded by 0/1 matrices M - 'legal' incidence matrices - give rise to parameter hypergraphs. But what about models whose probability form is more general? Surprisingly, by a recent result [**PTV**, Theorem 6.2], *all* toric ideals and therefore all linear ERGMs and log-linear models for discrete data are such that there exists a hypergraph that encodes their Graver bases and preserves the complexity. Therefore, we arrive at the following crucial fact:

The parameter hypergraph encodes sufficient statistics of any linear ERGM by a hypergraph degree sequence.

This further reinforces the present author's view that discrete mathematics literature on sampling graphs and hypergraphs with fixed properties should be imported to statistics and considered a crucial tool in linear ERGM model fitting in particular. The following General Problem summarizes the discussion of this section:

PROBLEM 3.13 (A restatement of General Problem 1 with constraints). *Fix a general linear ERGM with model hypergraph $\mathcal{H}$, so that the incidence matrix of $\mathcal{H}$ computes the sufficient statistics vector $T(g)$ for the random graph g. Develop an efficient algorithm that samples from the set of sub-hypergraphs of $\mathcal{H}$ with arbitrary fixed degree sequence.*

Of course, as discussed above, we know that '2-switches', or quadratic Markov moves, suffice to connect all realizations of hypergraph degree sequences because they correspond to the the defining ideal of the variety of Veronese-type corresponding to the r-th hypersimplex [**Stu96**, Chapter 14], but intermediate steps may go through non-simple hypergraphs, that is, those that have edges with multiplicities larger than 1. Further, this construction ignores possible forbidden edges. A related problem for the restricted class of complete k-uniform hypergraphs can be found in the problem collection [**Wes**, Problem 2]: For $k \geq 4$, determine whether there exists a function $f(k)$ and a set of operations, each on at most $f(k)$ vertices, that can be used to transform one k-realization of a k-graphic sequence into another. Even if this problem is solved, it would only impact the uniform case of the hypergraph β-model, for which $\mathcal{H}_\mathcal{M}$ is complete, that is, there are no forbidden edges. For applications, statistical sampling constraints and forbidden edges really impact the solution. For example, the famous 'bad news Theorem' for Markov bases in algebraic statistics from [**DLO06**] shows that Markov bases for already the simplest non-trivial model on 3 discrete random variables, called the no-three-way interaction model, are arbitrarily complicated. What this means is that there are fibers of observable data points that require moves of arbitrarily large degree if the number of states of at least two of the 3 variables are not bounded. For completeness, let me add also the complementing 'good news Theorem' for Markov bases famous in algebraic statistics: [**HS07**] show that the Markov bases for that same no-three-way interaction model have bounded complexity if the number of states of two of the 3 variables is bounded. Incidentally, the parameter hypergraph of the no-three-way model is 3-uniform and 3-partite, it is not complete: there are as many edges as in $K_{p,q}$ where p and q are sizes of two of its parts, and it is regular in each part. So the presence of forbidden edges in $\mathcal{H}_\mathcal{M}$ compound the issue of connecting the fiber in the worst possible way. However, positive results can be obtained for restricted classes of problems, references to which are scattered throughout this section.

4. Polytopes play an important role in ERGMs

The parameter estimation problem. In the estimation problem, given an observation of the joint states of the random variables, the goal is to estimate the unknown probability distribution p_{θ_0} from a model $\mathcal{M}$ that 'best explains' the data. A basic method commonly used is that of maximum likelihood estimation: a maximum likelihood estimate (MLE) of θ_0 is a parameter vector $\hat{\theta} \in \Theta$ that makes the given data most likely to have been observed. It is defined as:

$$\widehat{\theta} = \text{argmax}_{\theta \in \mathbb{R}^n} p_\theta(x). \tag{4.1}$$

Computing it amounts to maximizing the log-likelihood function, which maps the parameters indexing the probability distributions in a model to the likelihood of observing the given data. While there are several estimators that can be used, an MLE is a very straightforward one that enjoys nice statistical properties such as consistency and asymptotic efficiency, and computing it is a seemingly simple optimization problem. However, there are several obstacles. One if them is that the MLE of the natural parameters of the model may not exist, meaning that we are unable to make inferences about any or a subset of the parameters from the given data. As expected, the estimation problem has been studied in detail in the statistics literature for many classes of models. In the last decade, it has been shown that for discrete exponential families, the geometric structure of the model captures important information about parameter estimation including MLE existence; see [**RFZ09**], [**DFR$^+$08**], cf. [**Gey09**], [**Han03**]. Luckily, in case of exponential families,[5] if the MLE exists it is unique. The following polyhedral object captures the MLE information in exponential family models.

DEFINITION 4.1. Let $\mathcal{T} = \{T(g) \in \mathbb{R}^d\}$ be the range of sufficient statistics T, where g ranges over the set of all observable networks in the sample space. The *model polytope* associated with the family $\mathcal{M}$ is

$$\mathrm{P} = \text{convhull}(\mathcal{T}) \subset \mathbb{R}^d.$$

We may assume P to be full-dimensional.

EXAMPLE 4.2. The model polytope P for the β-model for random graphs on n vertices is the polytope of degree sequences of graphs on n vertices. [**RPF13**] use the explicit description of facets from [**MP95**] to give a characterization of when the sufficient statistic vector is on the boundary of the polytope, and study the statistical consequences.

The complexity of the MLE problem for the model $\mathcal{M}$ is captured by the geometry of the polytope P and by the combinatorics of its face lattice as follows: *the MLE for the observed data exists if and only if the observed sufficient statistic vector lies in the relative interior of the model polytope* P. For instance, [**RPF13**] studies MLE existence for the β-model for graphs and links to the graph theory literature on degree sequences. Nonetheless, partly because the link with polyhedral

[5] The general case is not so lucky at all: an MLE may not exist because the likelihood function is not globally bounded; further, if an MLE exists, it may not be unique. Also, the likelihood function may be highly multimodal; a simple example is provided already by a standard example of mixture of normal distributions. In that case, standard tools such as hill-climbing algorithms (e.g. Newton-Rhapson) may fail to find the maximum and converge to a local optimum only, and would not be aware of it. Finally, MLE may be located on the boundary of the model, which is poorly understood except in few cases in very recent literature [**KRS15**].

geometry has remained largely unexplored in both the mathematical and statistical literature, methodologies for estimation and model validation under a non-existent MLE with proven statistical performance have yet to be developed. A first step in this direction is to solve the following.

PROBLEM 4.3 (General Problem 2). *Fix a random graph model with sufficient statistics vector $T(g)$ and sample space $\mathcal{G}_n$. Often, $\mathcal{G}_n$ will be the set of simple graphs on n vertices. Determine the model polytope $P = conv\{T(g) : g \in \mathcal{G}_n\}$. What are the extreme points? What is the facial structure? What is the proportion of realizable lattice points on the boundary vs. in the interior?*

The following can be thought of a subproblem of 4.3, but due to its difficulty and relevance for developing algorithms to detect MLE non-existence, we single it out.

PROBLEM 4.4 (A sequel to General Problem 2). *Develop an efficient algorithm that can determine whether a given point lies on the relative interior of P; i.e., obtain a facet description of the model polytope that is 'efficient'.*

While the general theoretical link between statistical estimation and polyhedral geometry is in place for discrete exponential families as outlined in [**RFZ09**] and references therein, statistics now relies on discrete methods to provide the required tools. Let us gather here some of the relevant results. We emphasize again that for discrete exponential families if an MLE exists then it is unique. Estimation algorithms will behave well when off the boundary of the polytope. For example, [**SSR$^+$14**] define the hypergraph β-model and show that fixed point algorithms will converge to the MLE geometrically fast as long as the sufficient statistic is on the relative interior of the polytope. [**RPF13**] use the polytope of graphical degree sequences to offer a complementary study to [**CDS11**] of the estimation problem in the β-model. Another simple of interest is called the edge-triangle model. Its sufficient statistic vector consists of two numbers: the number of edges and the number of triangles in the graph. Its polytope and estimation issues are addressed in [**RFZ09**], the asymptotics of the MLE existence problem, polytope and its fan in [**YRF13**], while [**HCT15**] study its degeneracy from the point of view of statistical mechanics, dealing with the issue of non-realizable points that exist within the model polytope in such a way that they are meant to represent data averages but are by definition non-interpretable.

However, Problem 4.3 has not been solved for most models listed in this review. For example, the polytope of hypergraph degree sequences has been studied in the literature and some partial results are known: [**Liu13**] shows that the set of hypergraph degree sequences for uniform hypergraphs is non-convex, and thus Erdös-Gallai-type theorems do not hold. [**MS02**] also offers a very nice review of the problem, and shows that vertices of the polytope are known by Theorem 2.5. Namely, they are r-threshold sequences, where r is the uniformity of the hypergraph. Further, *some* of the facet inequalities are known.

Data privacy and noisy sufficient statistics. Statistical analyses of network data sometimes require projection of a noisy degree sequence onto the model polytope, in particular its realizable points. The noisy sequence need not be a realizable degree sequence vector at all. The example below is considered from the point of view of data privacy and confidentiality, but note that the problem is also relevant for noisy sampling, for example when there are measurement errors and the reported observed sufficient statistic does not represent a realizable sequence.

The basic concept of data privacy and confidentiality is to minimize the risk of releasing sensitive information about, say in our case, people in a social network, while maximizing statistical utility of the data. Having statistical utility means that the released information can be used for statistical analyses. A straightforward example of such information that one may want to release is the observed value of the sufficient statistic. However, in the age where information about any particular data set may be gathered from different sources, releasing a sufficient statistic, e.g., the degree sequence in the β-model case, may not be 'private enough'. For this reason, in the interest of preserving data privacy, noise is added to the observed sufficient statistic in order to have it releasable to the public while not jeopardizing sensitive network information. Of course this presents various problems for statistical inference; thus the noise is added in a principled way. There are several well-known privacy-preserving mechanisms. The statistical task is then to determine how to perform reliable inference using the noisy statistic; for an example and overview of the relevant privacy literature for ERGMs see [**SKK14**]. One of the steps in [**KS16**] that is required for computing a private parameter estimator of the released data, with good statistical properties, is to project the noisy sufficient statistic onto the lattice points in the model polytope. To this end, one needs to understand the following.

PROBLEM 4.5 (General Problem 3). *Fix a random graph model with sufficient statistics vector $T(g)$, sample space $\mathcal{G}_n$. Often, $\mathcal{G}_n$ will be the set of graphs on n vertices. Let $P = conv\{T(g) : g \in \mathcal{G}_n\}$ be the model polytope. Develop an (efficient) algorithm to compute the projection of a noisy sufficient statistic vector onto the realizable vectors in P.*

Obviously, determining if a lattice point is a realizable sufficient statistic for the model $\mathcal{M}$ is a problem on its own. In fact realizability results of Havel-Hakimi [**Hav55, Hak62**] type can be used to solve Problem 4.5:

EXAMPLE 4.6. For the β-model, [**KS16, KS12**] use the Havel-Hakimi decomposition to compute the realizable lattice point in the polytope of degree sequences that is closest to the given noisy sequence. In order to achieve a more efficient implementation, they in fact use the polytope of degree partitions [**BSS06**], that is, of sorted degree sequences. This polytope can be thought of as an asymmetrized version of the degree sequence polytope.

This directly leads to the analogous problems for the other models, which are singled out for convenience.

PROBLEM 4.7 (An instance of General Problem 3). *Generalize Havel-Hakimi to the p_1 model polytope.*

Remarkably, the directed part of this problem has been solved in [**EMT10**], where they also study the case of some forbidden edges. The authors there also prove a result useful to design an MCMC algorithm to find random realizations of prescribed directed degree sequences, which relates back to Problem 3.6 (solved in [**PRF10**]). However, this sampling algorithm offers a partial solution to the problem in the general case, though it should be easy to modify it to fit the statistics framework. Namely, in the general variant of the p_1 model, reciprocated edges are of the form $i \to j$ and $i \leftarrow j$ simultaneously combined into one bidirected edge $i \leftrightarrow j$. The sufficient statistics of the p_1 model include the number of reciprocated edges! This means that in sampling or finding random realizations, this number also matters: reciprocated edges are allowed to be 'pulled apart' while sampling, but the total number of them is to remain constant. Recall again that forbidden edges in this case correspond to structural zeros in the p_1 model.

PROBLEM 4.8 (An instance of General Problem 3). *Generalize Havel-Hakimi to the three variants of the β hypergraph model polytope: uniform, layered or general hypergraph degree sequences.*

The uniform instance of this problem is essentially [**BEF$^+$13**, Problem 1.1], who also emphasize that the only known characterization of k-graphic sequences is due to Dewdney in 1975, but it does not yield an efficient algorithm. Two of the variants of the β hypergraph model work with nonuniform hypergraphs.

5. Concluding remarks

There exist, of course, families of ERGMs beyond the ones discussed in this brief overview, including those based on global summary statistics not related to node degrees, for example the cores decomposition [**KPP$^+$**], or other types of models such as graphical models for networks [**SR14a**]. A uniform sampler for the fibers of the former model family has not been developed, while the algebra and geometry of the latter model family has not been explored. Clearly these are interesting problems in the statistical analysis of networks and, in particular, algebraic statistics. But the main points of this chapter are:

- Several basic examples of linear ERGMs already offer many interesting open problems in discrete mathematics, instances and special cases of which are already known and interesting in their own right from the graph-theoretic point of view.
- These models also provide a theoretical foundation for exploring distributions of various network summary statistics that could serve to develop a rigorous testing and model fitting framework for networks, nowadays generally lacking despite the exploding literature on network analysis.

The first type of problem discussed here is the sampling problem given a fixed set of characteristics of a network. Within the context of statistical modeling of networks, the value of sampling and therefore of exact testing can be summarized as follows. Comparing to the reference set $\mathcal{F}_t$ avoids the use of asymptotics, offers an alternative to model fit testing, and makes it unnecessary to use large-sample approximations to sampling distributions, in particular when their adequacy has not been determined. As demonstrated on several cases in the literature, tools from algebraic statistics offer a valid and critical approach for analyzing such models. However, within this realm of mathematical research, several problems remain, including problems related to scalability and applicability of the algebraic methods. The second type of problem discussed here is related to the model polytope defined to be the convex hull of all sufficient statistics for the model. The polytope captures the difficulty of parameter estimation and MLE existence and is a crucial step in statistical inference. Both problems are key steps in the general scheme of testing goodness of fit of the model and model selection, detailed statistical considerations of which are beyond the scope of the present chapter.

To close, it is worth pointing out that algebraic statistics is in fact a much broader field, not confined to the study of log-linear models for discrete data. Indeed, the field builds upon the rich and long history of the use of algebraic tools in statistics, starting with Fisher [**Fis25**] and algebra for confounding [**PW83**], but has expanded to encompass at least three distinct fields of mathematics: commutative and computational algebra, via solving systems of polynomial/rational equations; combinatorics of graphs, hypergraphs, and simplicial complexes; and geometry, algebraic, convex, and polyhedral. The field is roughly two decades old, yet it has many facets, and recent theoretical advances show that there is a significant impact on applicability and behavior of various statistical methods, including but not limited to the following problems: parameter estimation and reliability of inference for discrete exponential families [**RFZ09, PRF10**] and Gaussian models [**ZUR**]; general methods for experimental design [**PRW01**]; parameter identifiability studied in numerous references of [**APRS11**], [**FSD11**] and complexity of parameter estimation for Gaussian graphical models [**DFD12**], [**DGP12**], [**GS**]; Bayesian model selection and singular learning theory [**Wat10, Wat09**], sampling from conditional or marginal distributions on contingency tables with implications to cell bounds and data privacy [**Sla10, SZP14**], exact tests for marginal table models [**Dob03**], and model fitting via exact tests for degree-based network models [**OHT13, SSR$^+$14**], as well as offering new models and algorithms for networks not based on degrees [**KPP$^+$**]. This bibliography is far from complete, but should give enough information on some recent advances in the field to the interested reader.

Acknowledgements

Supported in part by AFOSR Grant #FA9550-14-1-0141.

The author is endlessly grateful to Stephen E. Fienberg for his guidance in the study of statistical network models, and Alessandro Rinaldo, Despina Stasi and Elizabeth Gross for continued enlightening collaborations and motivating discussions.

Many thanks to Tobias Windisch for a thorough reading and many important corrections of the initial version of this manuscript. In addition, two anonymous referees have helped improve the text tremendously. The author is also grateful to Zoltán Toroczkai and Éva Czabarka for introduction to the relevant graph theory literature, and Péter L. Erdös and István Miklós for clarifying questions and remarks.

References

[Agr92] Alan Agresti, *A survey of exact inference for contingency tables*, Statist. Sci. **7** (1992), no. 1, 131–177. With comments and a rejoinder by the author. MR1173420

[AHT12] Satoshi Aoki, Hisayuki Hara, and Akimichi Takemura, *Markov bases in algebraic statistics*, Springer Series in Statistics, Springer, New York, 2012. MR2961912

[APRS11] Elizabeth Allman, Sonja Petrović, John Rhodes, and Seth Sullivant, *Identifiability of two-tree mixtures under group-based models*, IEEE/ACM Transactions in Computational Biology and Bioinformatics **8** (2011), no. 3, 710–722.

[BD10] Joseph Blitzstein and Persi Diaconis, *A sequential importance sampling algorithm for generating random graphs with prescribed degrees*, Internet Math. **6** (2010), no. 4, 489–522, DOI 10.1080/15427951.2010.557277. MR2809836

[BD15] Peter J. Bickel and Kjell A. Doksum, *Mathematical statistics—basic ideas and selected topics. Vol. 1*, 2nd ed., Texts in Statistical Science Series, CRC Press, Boca Raton, FL, 2015. MR3445928

[BDE] Michael J. Bannister, William E. Devanny, and David Eppstein, *ERGMs are Hard*, Preprint, arXiv:1412.1787 [cs.DS].

[BEF+13] Sarah Behrens, Catherine Erbes, Michael Ferrara, Stephen G. Hartke, Benjamin Reiniger, Hannah Spinoza, and Charles Tomlinson, *New results on degree sequences of uniform hypergraphs*, Electron. J. Combin. **20** (2013), no. 4, Paper 14, 18. MR3139399

[BGJM11] *Handbook of Markov chain Monte Carlo*, Chapman & Hall/CRC Handbooks of Modern Statistical Methods, CRC Press, Boca Raton, FL, 2011. Edited by Steve Brooks, Andrew Gelman, Galin L. Jones and Xiao-Li Meng. MR2742422

[Bha12] *Graph-theoretic concepts in computer science*, Lecture Notes in Computer Science, vol. 7551, Springer, Heidelberg, 2012. Revised selected papers from the 38th International Workshop (WG 2012) held in Jerusalem, June 26–28, 2012; Edited by Martin Charles Golumbic, Michal Stern, Avivit Levy and Gila Morgenstern. MR3236881

[BN14] O. Barndorff-Nielsen, *Information and exponential families in statistical theory*, Wiley Series in Probability and Statistics, John Wiley & Sons, Ltd., Chichester, 2014. Reprint of the 1978 original [MR0489333]. MR3221776

[BPS07] Amitava Bhattacharya, Uri N. Peled, and Murali K. Srinivasan, *Cones of closed alternating walks and trails*, Linear Algebra Appl. **423** (2007), no. 2-3, 351–365, DOI 10.1016/j.laa.2007.01.013. MR2312413

[BPS09] Amitava Bhattacharya, Uri N. Peled, and Murali K. Srinivasan, *The cone of balanced subgraphs*, Linear Algebra Appl. **431** (2009), no. 1-2, 266–273, DOI 10.1016/j.laa.2009.02.029. MR2522574

[Bro86] Jack Carl Kiefer, *Collected papers. Supplementary volume*, Springer-Verlag, New York; published with the cooperation of Institute of Mathematical Statistics, Hayward, CA, 1986. Edited by Lawrence D. Brown, Ingram Olkin, Jerome Sacks and Henry P. Wynn. MR856410

[BSS06] Amitava Bhattacharya, S. Sivasubramanian, and Murali K. Srinivasan, *The polytope of degree partitions*, Electron. J. Combin. **13** (2006), no. 1, Research Paper 46, 18 pp. (electronic). MR2223521

[CDEM15] Éva Czabarka, Aaron Dutle, Péter L. Erdős, and István Miklós, *On realizations of a joint degree matrix*, Discrete Appl. Math. **181** (2015), 283–288, DOI 10.1016/j.dam.2014.10.012. MR3284534

[CDG07] Colin Cooper, Martin Dyer, and Catherine Greenhill, *Sampling regular graphs and a peer-to-peer network*, Combin. Probab. Comput. **16** (2007), no. 4, 557–593, DOI 10.1017/S0963548306007978. MR2334585

[CDS11] Sourav Chatterjee, Persi Diaconis, and Allan Sly, *Random graphs with a given degree sequence*, Ann. Appl. Probab. **21** (2011), no. 4, 1400–1435, DOI 10.1214/10-AAP728. MR2857452

[CKHG15] Nicole Bohme Carnegie, Pavel N. Krivitsky, David R. Hunter, and Steven M. Goodreau, *An approximation method for improving dynamic network model fitting*, J. Comput. Graph. Statist. **24** (2015), no. 2, 502–519, DOI 10.1080/10618600.2014.903087. MR3357392

[DFD12] Rina Foygel, Jan Draisma, and Mathias Drton, *Half-trek criterion for generic identifiability of linear structural equation models*, Ann. Statist. **40** (2012), no. 3, 1682–1713, DOI 10.1214/12-AOS1012. MR3015040

[DFR+08] Adrian Dobra, Stephen E. Fienberg, Alessandro Rinaldo, Aleksandra Slavkovic, and Yi Zhou, *Algebraic statistics and contingency table problems: log-linear models, likelihood estimation, and disclosure limitation*, Emerging applications of algebraic geometry, IMA Vol. Math. Appl., vol. 149, Springer, New York, 2009, pp. 63–88, DOI 10.1007/978-0-387-09686-5_3. MR2500464

[DGP12] Elizabeth Gross, Mathias Drton, and Sonja Petrović, *Maximum likelihood degree of variance component models*, Electron. J. Stat. **6** (2012), 993–1016, DOI 10.1214/12-EJS702. MR2988436

[DLO06] Jesús A. De Loera and Shmuel Onn, *Markov bases of three-way tables are arbitrarily complicated*, J. Symbolic Comput. **41** (2006), no. 2, 173–181, DOI 10.1016/j.jsc.2005.04.010. MR2197153

[Dob03] Adrian Dobra, *Markov bases for decomposable graphical models*, Bernoulli **9** (2003), no. 6, 1093–1108, DOI 10.3150/bj/1072215202. MR2046819

[Dob12] Adrian Dobra, *Dynamic Markov bases*, Journal of Computational and Graphical Statistics (2012), 496–517.

[DS98] Persi Diaconis and Bernd Sturmfels, *Algebraic algorithms for sampling from conditional distributions*, Ann. Statist. **26** (1998), no. 1, 363–397, DOI 10.1214/aos/1030563990. MR1608156

[DS04] Adrian Dobra and Seth Sullivant, *A divide-and-conquer algorithm for generating Markov bases of multi-way tables*, Comput. Statist. **19** (2004), no. 3, 347–366. MR2096204

[DSS09] Mathias Drton, Bernd Sturmfels, and Seth Sullivant, *Lectures on algebraic statistics*, Oberwolfach Seminars, vol. 39, Birkhäuser Verlag, Basel, 2009. MR2723140

[EMT10] Péter L. Erdös, Istvan Miklós, and Zoltán Toroczkai, *A simple Havel–Hakimi type algorithm to realize graphical degree sequences of directed graphs*, Electronic Journal of Combinatorics **17** (2010), no. 1.

[EMT15] Péter L. Erdős, István Miklós, and Zoltán Toroczkai, *A decomposition based proof for fast mixing of a Markov chain over balanced realizations of a joint degree matrix*, SIAM J. Discrete Math. **29** (2015), no. 1, 481–499, DOI 10.1137/130929874. MR3319846

[Fis25] Sir Ronald Aylmer Fisher, *Statistical methods for research workers*, Edinburgh: Oliver & Boyd, 1925.

[FMW85] Stephen E. Fienberg, Michael M. Meyer, and Stanley S. Wasserman, *Statistical analysis of multiple sociometric relations*, Journal of the American Statistical Association **80** (1985), 51–67.

[FSD11] Mathias Drton, Rina Foygel, and Seth Sullivant, *Global identifiability of linear structural equation models*, Ann. Statist. **39** (2011), no. 2, 865–886, DOI 10.1214/10-AOS859. MR2816341

[FW81] Paul W. Holland and Samuel Leinhardt, *An exponential family of probability distributions for directed graphs*, J. Amer. Statist. Assoc. **76** (1981), no. 373, 33–65. With comments by Ronald L. Breiger, Stephen E. Fienberg, Stanley Wasserman, Ove Frank and Shelby J. Haberman and a reply by the authors. MR608176

[Gey09] Charles J. Geyer, *Likelihood inference in exponential families and directions of recession*, Electron. J. Stat. **3** (2009), 259–289, DOI 10.1214/08-EJS349. MR2495839

[Goo07] Steven M. Goodreau, *Advances in exponential random graph (p^*) models applied to a large social network*, Social Networks **29** (2007), no. 2, 231–248.

[GPS] Elizabeth Gross, Sonja Petrović, and Despina Stasi, *Goodness-of-fit for log-linear network models: Dynamic Markov bases using hypergraphs*, Annals of the Institute of Statistical Mathematics. DOI 10/1007/s10463-016-0560-2

[Gre15] Catherine Greenhill, *The switch Markov chain for sampling irregular graphs*, 26th Annual ACM-SIAM Symposium on Discrete Algorithms (New York-Philadelphia), ACM, 2015, pp. 1564 – 1572.

[GS] Elizabeth Gross and Seth Sullivant, *The maximum likelihood threshold of a graph*, Preprint available at arXiv:1404.6989. To appear in Bernoulli.

[GZFA09] Anna Goldenberg, Alice X. Zheng, Stephen E. Fienberg, and Edoardo M. Airoldi, *A survey of statistical network models*, Foundations and Trends in Machine Learning **2** (2009), no. 2, 129–233.

[Hab81] Paul W. Holland and Samuel Leinhardt, *An exponential family of probability distributions for directed graphs*, J. Amer. Statist. Assoc. **76** (1981), no. 373, 33–65. With comments by Ronald L. Breiger, Stephen E. Fienberg, Stanley Wasserman, Ove Frank and Shelby J. Haberman and a reply by the authors. MR608176

[Hab88] Shelby J. Haberman, *A warning on the use of chi-squared statistics with frequency tables with small expected cell counts*, J. Amer. Statist. Assoc. **83** (1988), no. 402, 555–560. MR971386

[Hak62] S. L. Hakimi, *On realizability of a set of integers as degrees of the vertices of a linear graph. I*, J. Soc. Indust. Appl. Math. **10** (1962), 496–506. MR0148049

[Han03] Mark S. Handcock, *Assessing degeneracy in statistical models for social networks*, Working paper 39., Center for Statistics and the Social Sciences, University of Washington, Seattle, 2003.

[Hav55] Václav Havel, *Eine Bemerkung über die Existenz der endlichen Graphen* (Czech, with Russian and German summaries), Časopis Pěst. Mat. **80** (1955), 477–480. MR0089165

[HCT15] Szabolcs Horvát, Éva Czabarka, and Zoltán Toroczkai, *Reducing degeneracy in maximum entropy models of networks*, Physical Review Letters **114** (2015), no. 158701.

[HGH08] David R. Hunter, Steven M. Goodreau, and Mark S. Handcock, *Goodness of fit of social network models*, J. Amer. Statist. Assoc. **103** (2008), no. 481, 248–258, DOI 10.1198/016214507000000446. MR2394635

[HL81] Paul W. Holland and Samuel Leinhardt, *An exponential family of probability distributions for directed graphs*, J. Amer. Statist. Assoc. **76** (1981), no. 373, 33–65. With comments by Ronald L. Breiger, Stephen E. Fienberg, Stanley Wasserman, Ove Frank and Shelby J. Haberman and a reply by the authors. MR608176

[HM07] Mark S. Handcock and Martina Morris, *A simple model for complex networks with arbitrary degree distribution and clustering*, Statistical Network Analysis: Models, Issues and New Directions (Edo Airoldi, D. Blei, Stephen E. Fienberg, A. Goldenberg, E. Xing, and A. Zheng, eds.), Lecture Notes in Computer Science, vol. 4503, Springer, 2007, pp. 103–114.

[HS07] Serkan Hoşten and Seth Sullivant, *A finiteness theorem for Markov bases of hierarchical models*, J. Combin. Theory Ser. A **114** (2007), no. 2, 311–321, DOI 10.1016/j.jcta.2006.06.001. MR2293094

[HTE+09] Hyunju Kim, Zoltán Toroczkai, Péter L. Erdős, István Miklós, and László A. Székely, *Degree-based graph construction*, J. Phys. A **42** (2009), no. 39, 392001, 10, DOI 10.1088/1751-8113/42/39/392001. MR2539311

[Jac08] Matthew O. Jackson, *Social and economic networks*, Princeton University Press, Princeton, NJ, 2008. MR2435744

[KK15] Pavel N. Krivitsky and Eric D. Kolaczyk, *On the question of effective sample size in network modeling: an asymptotic inquiry*, Statist. Sci. **30** (2015), no. 2, 184–198, DOI 10.1214/14-STS502. MR3353102

[Kol09] Eric D. Kolaczyk, *Statistical analysis of network data*, Springer Series in Statistics, Springer, New York, 2009. Methods and models. MR2724362

[KPP+] Vishesh Karwa, Michael Pelsmajer, Sonja Petrović, Despina Stasi, and Dane Wilburne, *Statistical models for cores decomposition of an undirected random graph.* Preprint available at arXiv:1410.7357

[KRS15] Kaie Kubjas, Elina Robeva, and Bernd Sturmfels, *Fixed points EM algorithm and nonnegative rank boundaries*, Ann. Statist. **43** (2015), no. 1, 422–461, DOI 10.1214/14-AOS1282. MR3311865

[KS12] Vishesh Karwa and Aleksandra Slavković, *Differentially private graphical degree sequences and synthetic graphs*, Privacy in Statistical Databases. (2012), 273–285.

[KS16] Vishesh Karwa and Aleksandra Slavković, *Inference using noisy degrees: differentially private β-model and synthetic graphs*, Ann. Statist. **44** (2016), no. 1, 87–112, DOI 10.1214/15-AOS1358. MR3449763

[KTV99] Ravi Kannan, Prasad Tetali, and Santosh Vempala, *Simple Markov-chain algorithms for generating bipartite graphs and tournaments*, Random Structures Algorithms **14** (1999), no. 4, 293–308, DOI 10.1002/(SICI)1098-2418(199907)14:4⟨293::AID-RSA1⟩3.3.CO;2-7. MR1691976

[LC83] Erich Leo Lehmann and George Casella, *Theory of point estimation*, 1 ed., London: Chapman and Hall, 1983.

[LC98] E. L. Lehmann and George Casella, *Theory of point estimation*, 2nd ed., Springer Texts in Statistics, Springer-Verlag, New York, 1998. MR1639875

[Liu13] Ricky Ini Liu, *Nonconvexity of the set of hypergraph degree sequences*, Electron. J. Combin. **20** (2013), no. 1, Paper 21, 7. MR3035031

[LS06] László Lovász and Balázs Szegedy, *Limits of dense graph sequences*, J. Combin. Theory Ser. B **96** (2006), no. 6, 933–957, DOI 10.1016/j.jctb.2006.05.002. MR2274085

[MES13] István Miklós, Péter L. Erdős, and Lajos Soukup, *Towards random uniform sampling of bipartite graphs with given degree sequence*, Electron. J. Combin. **20** (2013), no. 1, Paper 16, 51. MR3035026

[Mor13] Jason Morton, *Relations among conditional probabilities*, J. Symbolic Comput. **50** (2013), 478–492, DOI 10.1016/j.jsc.2012.02.005. MR2996892

[MP95] N. V. R. Mahadev and U. N. Peled, *Threshold graphs and related topics*, Annals of Discrete Mathematics, vol. 56, North-Holland Publishing Co., Amsterdam, 1995. MR1417258

[MS02] N. L. Bhanu Murthy and Murali K. Srinivasan, *The polytope of degree sequences of hypergraphs*, Linear Algebra Appl. **350** (2002), 147–170, DOI 10.1016/S0024-3795(02)00272-0. MR1906751

[New03] M. E. J. Newman, *The structure and function of complex networks*, SIAM Rev. **45** (2003), no. 2, 167–256 (electronic), DOI 10.1137/S003614450342480. MR2010377

[New10] M. E. J. Newman, *Networks*, Oxford University Press, Oxford, 2010. An introduction. MR2676073

[NSW01] Mark E. J. Newman, S. H. Strogatz, and D. J. Watts, *Random graphs with arbitrary degree distributions and their applications*, Physical Review E **026118** (2001).

[OH99] Hidefumi Ohsugi and Takayuki Hibi, *Toric ideals generated by quadratic binomials*, J. Algebra **218** (1999), no. 2, 509–527, DOI 10.1006/jabr.1999.7918. MR1705794

[OHT13] Mitsunori Ogawa, Hisayuki Hara, and Akimichi Takemura, *Graver basis for an undirected graph and its application to testing the beta model of random graphs*, Ann. Inst. Statist. Math. **65** (2013), no. 1, 191–212, DOI 10.1007/s10463-012-0367-8. MR3011620

[PN04] Juyong Park and M. E. J. Newman, *Statistical mechanics of networks*, Phys. Rev. E (3) **70** (2004), no. 6, 066117, 13, DOI 10.1103/PhysRevE.70.066117. MR2133807

[PRF10] Sonja Petrović, Alessandro Rinaldo, and Stephen E. Fienberg, *Algebraic statistics for a directed random graph model with reciprocation*, Algebraic methods in statistics and probability II, Contemp. Math., vol. 516, Amer. Math. Soc., Providence, RI, 2010, pp. 261–283, DOI 10.1090/conm/516/10180. MR2730754

[PRW01] Giovanni Pistone, Eva Riccomagno, and Henry P. Wynn, *Computational commutative algebra in discrete statistics*, Algebraic methods in statistics and probability (Notre Dame, IN, 2000), Contemp. Math., vol. 287, Amer. Math. Soc., Providence, RI, 2001, pp. 267–282, DOI 10.1090/conm/287/04791. MR1873681

[PS14] Sonja Petrović and Despina Stasi, *Toric algebra of hypergraphs*, J. Algebraic Combin. **39** (2014), no. 1, 187–208, DOI 10.1007/s10801-013-0444-y. MR3144398

[PTV] Sonja Petrović, Apostolos Thoma, and Marius Vladoiu, *Bouquet algebra of toric ideals.*, Submitted. Preprint available at http://arxiv.org/abs/1507.02740.

[PW83] Giovanni Pistone and Henry P. Wynn, *Generalised confounding with Gröbner bases*, Biometrika **83** (1996), no. 3, 653–666, DOI 10.1093/biomet/83.3.653. MR1423881

[RC99] Christian P. Robert and George Casella, *Monte Carlo statistical methods*, Springer Texts in Statistics, Springer-Verlag, New York, 1999. MR1707311

[RC09] Christian P. Robert and George Casella, *Introducing Monte Carlo methods with R*, Springer-Verlag, 2009.

[RFZ09] Alessandro Rinaldo, Stephen E. Fienberg, and Yi Zhou, *On the geometry of discrete exponential families with application to exponential random graph models*, Electron. J. Stat. **3** (2009), 446–484, DOI 10.1214/08-EJS350. MR2507456

[RPF13] Sonja Petrović, Alessandro Rinaldo, and Stephen E. Fienberg, *Maximum likelihood estimation in the Beta model*, Annals of Statistics **41** (2013), no. 3, 1085–1110.

[RPKL07] Garry Robins, Pip Pattison, Yuval Kalish, and Dean Lusher, *An introduction to exponential random graph (p^*) models for social networks*, Social Networks **29** (2007), no. 2, 173–191.

[RTT12] Enrique Reyes, Christos Tatakis, and Apostolos Thoma, *Minimal generators of toric ideals of graphs*, Adv. in Appl. Math. **48** (2012), no. 1, 64–78, DOI 10.1016/j.aam.2011.06.003. MR2845507

[SKK14] Aleksandra Slavković, Vishesh Karwa, and Pavel N. Krivitsky, *Differentially private exponential random graphs*, Privacy in Statistical Databases, Lecture Notes in Computer Science, vol. 8744, Springer, 2014, pp. 143–155.

[SKKT00] Karen E. Smith, Lauri Kahanpää, Pekka Kekäläinen, and William Traves, *An invitation to algebraic geometry*, Universitext, Springer-Verlag, New York, 2000. MR1788561

[Sla10] Aleksandra Slavković, *Partial information releases for confidential contingency table entries: Present and future research efforts*, Journal of Privacy and Confidentiality **1** (2010), no. 2.

[SP12] Isabelle Stanton and Ali Pinar, *Constructing and sampling graphs with a prescribed joint degree distribution*, ACM J. Exp. Algorithmics **17** (2012), Article 3.5, 25, DOI 10.1145/2133803.2330086. MR2981864

[SR14a] Kayvan Sadeghi and Alessandro Rinaldo, *Hierarchical models for independence structures of networks*, To appear in the Journal of the American Statistical Association, 2014.

[SR14b] ———, *Statistical models for degree distributions of networks*, NIPS 2014 Workshop "From Graphs to Rich Data", available at arXiv:1411.3825, 2014.

[Sri] Murali K. Srinivasan, *Some problems motivated by the notion of threshold graphs*, Online, available at http://www.math.iitb.ac.in/ mks/papers/threshold.pdf.

[SS06] Aleksandra B. Slavkovic and Seth Sullivant, *The space of compatible full conditionals is a unimodular toric variety*, J. Symbolic Comput. **41** (2006), no. 2, 196–209, DOI 10.1016/j.jsc.2005.04.006. MR2197155

[SSR+14] Despina Stasi, Kayvan Sadeghi, Alessandro Rinaldo, Sonja Petrovic, and Stephen Fienberg, *β models for random hypergraphs with a given degree sequence*, Proceedings of COMPSTAT 2014—21st International Conference on Computational Statistics, Internat. Statist. Inst., The Hague, 2014, pp. 593–600. MR3372442

[Stu96] Bernd Sturmfels, *Gröbner bases and convex polytopes*, University Lecture Series, vol. 8, American Mathematical Society, Providence, RI, 1996. MR1363949

[SW12] Bernd Sturmfels and Volkmar Welker, *Commutative algebra of statistical ranking*, J. Algebra **361** (2012), 264–286, DOI 10.1016/j.jalgebra.2012.03.028. MR2921622

[SZP14] Aleksandra Slavković, Xiaotian Zhu, and Sonja Petrović, *Fibers of multi-way contingency tables given conditionals: relation to marginals, cell bounds and Markov bases*, Ann. Inst. Statist. Math. **67** (2015), no. 4, 621–648, DOI 10.1007/s10463-014-0471-z. MR3357932

[Vil95] Rafael H. Villarreal, *Rees algebras of edge ideals*, Comm. Algebra **23** (1995), no. 9, 3513–3524, DOI 10.1080/00927879508825412. MR1335312

[Vil00] Rafael H. Villarreal, *Monomial algebras*, CRC Press, 2000.

[VL05] Fabien Viger and Matthieu Latapy, *Efficient and simple generation of random simple connected graphs with prescribed degree sequence*, Computing and combinatorics, Lecture Notes in Comput. Sci., vol. 3595, Springer, Berlin, 2005, pp. 440–449, DOI 10.1007/11533719_45. MR2190867

[WAD09] Walter Willinger, David Alderson, and John C. Doyle, *Mathematics and the Internet: a source of enormous confusion and great potential*, Notices Amer. Math. Soc. **56** (2009), no. 5, 586–599. MR2509062

[Wat09] Sumio Watanabe, *Algebraic geometry and statistical learning theory*, Cambridge Monographs on Applied and Computational Mathematics, vol. 25, Cambridge University Press, Cambridge, 2009. MR2554932

[Wat10] Sumio Watanabe, *Asymptotic equivalence of Bayes cross validation and widely applicable information criterion in singular learning theory*, J. Mach. Learn. Res. **11** (2010), 3571–3594. MR2756194

[Wes] Douglas West, *Hypergraphic Sequences (2012) - (presented by Mike Ferrara - REGS 2012) `http://www.math.illinois.edu/~dwest/regs/hypergraphic.html`.*

[Win] Tobias Windisch, *Rapid mixing and Markov bases*, Preprint available at arXiv:1505.03018.

[WO] Patrick Wolfe and Sofia C. Olhede, *Nonparametric graphon estimation*, Preprint, available at arXiv:1309.5936.

[YLZ] Ting Yan, Chenlei Leng, and Ji Zhu, *Asymptotics in directed exponential random graph models with an increasing bi-degree sequence*, Ann. Statist. **44** (2016), no. 1, 31–57, DOI 10.1214/15-AOS1343. MR3449761

[YRF13] Mei Yin, *Critical phenomena in exponential random graphs*, J. Stat. Phys. **153** (2013), no. 6, 1008–1021, DOI 10.1007/s10955-013-0874-x. MR3131507

[ZUR] Piotr Zwiernik, Caroline Uhler, and Donald St. P. Richards, *Maximum likelihood estimation for linear Gaussian covariance models*, Submitted, preprint available at http://arxiv.org/abs/1408.5604. To appear in the Journal of the Royal Statistical Society Series B.

Department of Applied Mathematics, Illinois Institute of Technology, Chicago, Illinois 60616

E-mail address: `sonja.petrovic@iit.edu`

Selected Published Titles in This Series

685 **Heather A. Harrington, Mohamed Omar, and Matthew Wright, Editors,** Algebraic and Geometric Methods in Discrete Mathematics, 2017

682 **Gregory Arone, Brenda Johnson, Pascal Lambrechts, Brian A. Munson, and Ismar Volić, Editors,** Manifolds and K-Theory, 2017

681 **Shiferaw Berhanu, Nordine Mir, and Emil J. Straube, Editors,** Analysis and Geometry in Several Complex Variables, 2017

680 **Sergei Gukov, Mikhail Khovanov, and Johannes Walcher, Editors,** Physics and Mathematics of Link Homology, 2016

679 **Catherine Bénéteau, Alberto A. Condori, Constanze Liaw, William T. Ross, and Alan A. Sola, Editors,** Recent Progress on Operator Theory and Approximation in Spaces of Analytic Functions, 2016

678 **Joseph Auslander, Aimee Johnson, and Cesar E. Silva, Editors,** Ergodic Theory, Dynamical Systems, and the Continuing Influence of John C. Oxtoby, 2016

677 **Delaram Kahrobaei, Bren Cavallo, and David Garber, Editors,** Algebra and Computer Science, 2016

676 **Pierre Martinetti and Jean-Christophe Wallet, Editors,** Noncommutative Geometry and Optimal Transport, 2016

675 **Ana Claudia Nabarro, Juan J. Nuño-Ballesteros, Raúl Oset Sinha, and Maria Aparecida Soares Ruas, Editors,** Real and Complex Singularities, 2016

674 **Bogdan D. Suceavă, Alfonso Carriazo, Yun Myung Oh, and Joeri Van der Veken, Editors,** Recent Advances in the Geometry of Submanifolds, 2016

673 **Alex Martsinkovsky, Gordana Todorov, and Kiyoshi Igusa, Editors,** Recent Developments in Representation Theory, 2016

672 **Bernard Russo, Asuman Güven Aksoy, Ravshan Ashurov, and Shavkat Ayupov, Editors,** Topics in Functional Analysis and Algebra, 2016

671 **Robert S. Doran and Efton Park, Editors,** Operator Algebras and Their Applications, 2016

670 **Krishnendu Gongopadhyay and Rama Mishra, Editors,** Knot Theory and Its Applications, 2016

669 **Sergiĭ Kolyada, Martin Möller, Pieter Moree, and Thomas Ward, Editors,** Dynamics and Numbers, 2016

668 **Gregory Budzban, Harry Randolph Hughes, and Henri Schurz, Editors,** Probability on Algebraic and Geometric Structures, 2016

667 **Mark L. Agranovsky, Matania Ben-Artzi, Greg Galloway, Lavi Karp, Dmitry Khavinson, Simeon Reich, Gilbert Weinstein, and Lawrence Zalcman, Editors,** Complex Analysis and Dynamical Systems VI: Part 2: Complex Analysis, Quasiconformal Mappings, Complex Dynamics, 2016

666 **Vicenţiu D. Rădulescu, Adélia Sequeira, and Vsevolod A. Solonnikov, Editors,** Recent Advances in Partial Differential Equations and Applications, 2016

665 **Helge Glöckner, Alain Escassut, and Khodr Shamseddine, Editors,** Advances in Non-Archimedean Analysis, 2016

664 **Dihua Jiang, Freydoon Shahidi, and David Soudry, Editors,** Advances in the Theory of Automorphic Forms and Their L-functions, 2016

663 **David Kohel and Igor Shparlinski, Editors,** Frobenius Distributions: Lang-Trotter and Sato-Tate Conjectures, 2016

662 **Zair Ibragimov, Norman Levenberg, Sergey Pinchuk, and Azimbay Sadullaev, Editors,** Topics in Several Complex Variables, 2016

661 **Douglas P. Hardin, Doron S. Lubinsky, and Brian Z. Simanek, Editors,** Modern Trends in Constructive Function Theory, 2016

For a complete list of titles in this series, visit the
AMS Bookstore at **www.ams.org/bookstore/conmseries/**.